MATHEMATICAL THOUGHT AND ITS OBJECTS

In *Mathematical Thought and Its Objects*, Charles Parsons examines the notion of object, with the aim of navigating between nominalism, which denies that distinctively mathematical objects exist, and forms of Platonism that postulate a transcendent realm of such objects. He introduces the central mathematical notion of structure and defends a version of the structuralist view of mathematical objects, according to which their existence is relative to a structure and they have no more of a "nature" than that confers on them.

Parsons also analyzes the concept of intuition and presents a conception of it distantly inspired by that of Kant, which describes a basic kind of access to abstract objects and an element of a first conception of the infinite. An intuitive model witnesses the possibility of the structure of natural numbers. However, the full concept of number and knowledge of numbers involve more that is conceptual and rational. Parsons considers how one can talk about numbers, even though they are not objects of intuition. He explores the conceptual role of the principle of mathematical induction and the sense in which it determines the natural numbers uniquely.

Parsons ends with a discussion of reason and its role in mathematical knowledge, attempting to do justice to the complementary roles in mathematical knowledge of rational insight, intuition, and the integration of our theory as a whole.

Charles Parsons is Edgar Pierce Professor of Philosophy, Emeritus, at Harvard University. He is a former editor of the *Journal of Philosophy*. He is the author of *Mathematics in Philosophy* and co-editor of the posthumous works of Kurt Gödel.

Mathematical Thought and Its Objects

Charles Parsons
Harvard University

CAMBRIDGE
UNIVERSITY PRESS

32 Avenue of the Americas, New York NY 10013-2473, USA

Cambridge University Press is part of the University of Cambridge.

It furthers the University's mission by disseminating knowledge in the pursuit of education, learning and research at the highest international levels of excellence.

www.cambridge.org
Information on this title: www.cambridge.org/9780521452793

© Cambridge University Press 2008

This publication is in copyright. Subject to statutory exception and to the provisions of relevant collective licensing agreements, no reproduction of any part may take place without the written permission of Cambridge University Press.

First published 2008

A catalogue record for this publication is available from the British Library

Library of Congress Cataloguing in Publication data

Parsons, Charles, 1933–
Mathematical thought and its objects / Charles Parsons.
   p. cm.
Includes bibliographical references and index.
ISBN 978-0-521-45279-3 (hardback)
1. Mathematics – Philosophy.   2. Object (Philosophy)   3. Logic. . I. Title.
QA8.4.P366   2008
510.1–dc22   2007016310

ISBN 978-0-521-45279-3 Hardback
ISBN 978-0-521-11911-5 Paperback

Cambridge University Press has no responsibility for the persistence or accuracy of URLs for external or third-party internet websites referred to in this publication, and does not guarantee that any content on such websites is, or will remain, accurate or appropriate.

For Marjorie

Contents

Preface		*page* xi
Sources and Copyright Acknowledgments		xix
1	**Objects and logic**	**1**
	§1. Abstract objects	1
	§2. The concept of an object in general: Actuality	3
	§3. Intuitability	8
	§4. Logic and the notion of object	10
	§5. Is whatever is an object?	13
	§6. Being and existence	23
	§7. Abstract objects and their concrete representations	33
2	**Structuralism and nominalism**	**40**
	§8. The structuralist view of mathematical objects	40
	§9. The concept of structure	43
	§10. Dedekind on the natural numbers	45
	§11. Eliminative structuralism and logicism	50
	§12. Nominalism	56
	§13. Nominalism and second-order logic	61
	§14. Structuralism and application	73
3	**Modality and structuralism**	**80**
	§15. Mathematical modality	80
	§16. Modalism	92
	§17. Difficulties of modalism: Rejection of eliminative structuralism	96
	§18. A noneliminative structuralism	100

4 A problem about sets 117

§19. An objection — 117
§20. Ontological conceptions of set — 119
§21. The iterative conception of set — 122
§22. "Intuitive" arguments for axioms of set theory — 124
§23. The Replacement and Power Set axioms — 130

5 Intuition 138

§24. Intuition: Basic distinctions — 138
§25. Intuition and perception — 143
§26. Objections to the very idea of mathematical intuition — 148
§27. Toward a viable conception of intuition: Perception and the abstract — 152
§28. Hilbertian intuition — 159
§29. Intuitive knowledge: A step toward infinity — 171
§30. The objections revisited — 179

6 Numbers as objects 186

§31. What are the natural numbers? — 186
§32. Cardinality and the genesis of numbers as objects — 190
§33. Finite sets and sequences — 199
§34. Sets and sequences, intuition and number — 205
§35. Difficulties concerning intuition of finite sets — 211
§36. Well, then, what are the numbers? Structuralism put in its place — 218
§37. Intuition of numbers denied — 222
§38. Appendix 1: Theories of sets and sequences — 225
§39. Appendix 2: Relative substitutional semantics for the language of hereditarily finite sets — 231

7 Intuitive arithmetic and its limits 235

§40. Arithmetic as about strings: Finitism — 235
§41. The elementary axioms — 244
§42. Logic and intuition — 247
§43. Induction — 252
§44. Primitive recursion — 254
§45. The limits of intuitive knowledge — 260
§46. Appendix — 262

8	**Mathematical induction**	**264**
	§47. Induction and the concept of natural number	264
	§48. The problem of the uniqueness of the number structure: Nonstandard models	272
	§49. Uniqueness and communication	279
	§50. Induction and impredicativity	293
	§51. Predicativity and inductive definitions	307
9	**Reason**	**316**
	§52. Reason and "rational intuition"	316
	§53. Rational intuition and perception	325
	§54. Arithmetic	328
	§55. Set theory	338
Bibliography		343
Index		365

Preface

The present work is largely concerned with a limited number of themes in the philosophy of mathematics. The first is the notion of *object* as it is deployed in mathematics. I begin in Chapter 1 with a general discussion of the notion of object, not on the whole focused on mathematics. One of the motives of this discussion is to defuse too-high expectations of what the existence of objects of some mathematical type such as numbers would entail. We proceed to discuss issues surrounding the structuralist view of mathematical objects, which has had a lot of currency in the last forty years or so but has much earlier roots. The general idea of this view is that mathematical objects do not have a richer "nature" than is given by the basic relations of some structure in which they reside. The problem of giving a viable formulation developing this idea is not trivial and raises a lot of issues. That is the concern of Chapters 2 and 3. Chapter 2 is mainly devoted to pursuing a program that uses the structuralist idea to eliminate explicit reference to mathematical objects. Along the way, I discuss some questions about nominalism, about second-order logic, and about how structuralism understands the application of mathematics. Some difficulties of the eliminative program call for using modal notions, and their use in mathematics is a subject of Chapter 3. But in the end even the modal version of the eliminative program is rejected, and in §18 a version of structuralism is sketched that takes the language of mathematics much more at face value. Chapter 4 responds to an objection to the application to set theory of the version of structuralism I defend. Along the way, it considers other questions about the concept of set and the axioms of set theory.

The second main theme is a particular notion of mathematical intuition, which has its origin in the thought of Brouwer and Hilbert about the most basic elements of arithmetic but whose original inspiration comes from Kant. In Chapter 5, I lay out some basic distinctions concerning the notion of intuition, about which writers who use the notion (even to criticize it) are often unclear. But the main point of the chapter is to explain the particular conception of intuition that concerns me, develop some of its implications, and reply to some possible objections to it. Intuition so conceived offers part of the entry of mathematical thought into the infinite. The structure of natural numbers is shown to be witnessed by what can be called an intuitive model.

Chapters 6, 7, and 8 all concern the arithmetic of natural numbers, and it is in the first two of these that the work done by the conception of intuition of Chapter 5 is visible. Chapter 6 discusses the role of natural numbers as finite cardinals and ordinals and considers in a rather idealized way how language referring to natural numbers might originate. This genetic method is inspired by W. V. Quine's *Roots of Reference*. A conclusion of the chapter is that in the sense of Chapter 5, there is not intuition of numbers properly speaking. We also explore theories of finite sets and discuss the question of intuition of such sets, with the conclusion that the analogy with perception that such intuition requires would be too much stretched if it is claimed that the theory of hereditarily finite sets rests on intuition in the sense in which Hilbert and Bernays claimed that finitary arithmetic does.

The latter thesis is the main subject of Chapter 7, which assumes an interpretation of the language of arithmetic as referring to formal expressions and inquires how much arithmetic is intuitively known. Primitive recursion appears as an obstacle, and we are not able to conclude that exponentiation or faster-growing functions can be intuitively seen to be everywhere defined. The Hilbert school maintained that finitist arithmetic included primitive recursive arithmetic. Our conclusion is that it is quite doubtful that intuitively evident arithmetic extends that far.

Chapter 8 deals with some issues concerning the principle of mathematical induction, that any predicate that is true of 0 and is true of $n + 1$ if it is true of n is true of all natural numbers. As Poincaré pointed out a hundred years ago, this is the principle that makes arithmetic serious mathematics. I emphasize the open-endedness of "any predicate" in the principle. It is this that makes it possible to recognize the nonstandardness of a nonstandard model of formalized arithmetic and underlies Dedekind's proof that elementary axioms plus induction characterize the natural numbers up to isomorphism. The rest of the chapter discusses the uniqueness of the number structure and issues about impredicativity.

Chapter 9 turns to epistemological issues. After observing that mathematics has been characterized as rational knowledge, it introduces some issues about Reason and rational justification. Then it considers what can be said about the justification of principles in arithmetic and set theory.

A theme that appears in various places in the book,[1] but especially in Chapter 8, concerns schematic or second-order principles in mathematics, of which the most prominent examples are mathematical induction and the schemata of separation and replacement in set theory. As have other writers, I emphasize what is called the open-endedness of these principles, in that outside the context of specific formal systems they are not intended to be limited in their scope to particular formalized languages. But unlike some writers I do not see in this feature a convincing reason for regarding formulations in terms of second-order logic as canonical. Reasoning with second-order logic only moves the schematic character of principles into the logic. Furthermore, full second-order logic introduces a new assumption, that the instances of the relevant schemata are closed under second-order quantification, whatever one takes second-order variables to range over.

Certain issues that have been rather prominent in philosophy of mathematics in the last generation are commented on in the present work at most in passing. In one case, the question whether mathematical knowledge is a priori, the reader may find the omission surprising. The traditional affirmative view was vigorously attacked by W. V. Quine in some of his central writings. In more recent years it has been defended more than it has been attacked. I don't have a clear position to offer on this question. I am not convinced that the notion of a priori knowledge is as clear as is often assumed. It is quite obvious that experience and perception do not play the direct role in the justification of mathematical propositions that they play in natural science or in most factual statements of everyday life. The mathematician in proving a theorem does not appeal to experimental results or other forms of observation. Rigorous proofs can generally be represented as deductions from axioms. The question of the justification of axioms is a complex one, about which something is said in this work. Again there is no straightforward appeal to experiment and observation, but it is less easy to show that experience does not have some more subtle role that goes beyond the heuristic and motivating. In particular, that makes it harder to rule out the possibility that some unforeseen turn in

---

[1] And in places in *Mathematics in Philosophy*, for example, Essay 3.

the development of science might lead to the rejection as false of some assumption used in current mathematics.

But there is a consideration that makes this seem very unlikely.[2] That is that a mathematical theory of an aspect of the empirical world consists of taking a supposed actual system of objects and relations as an instance of a mathematical structure (where not every item in the structure is necessarily regarded as physically real). Then what confronts the "tribunal of experience" is the identification of a structure of this type with something in the world. That identification is falsifiable and has on many occasions been falsified. But then the resulting modification would consist of replacing the mathematical structure appealed to by another one, without abandoning the pure theory of the structure as mathematics. Thus, Euclidean geometry still plays a basic role in mathematics even though the view that space is Euclidean was questioned more than a hundred years ago and abandoned in the early twentieth century. It seems that some conceptual revolution of which we don't now have an idea would be required for us to abandon Euclidean geometry as mathematics. So it may be that much of current mathematics is "contextually a priori" in a sense proposed some years ago by Hilary Putnam.[3]

Another issue only glancingly commented on in this work is Benacerraf's dilemma. If it is put in terms of a causal theory of knowledge, according to which knowledge of certain objects requires causal relations of those objects and our minds, then I think the problem can be dismissed: mathematical objects are simply a counterexample to that theory of knowledge. But a more general form of the dilemma, expressed by W. D. Hart soon after Benacerraf's classic paper, is the difficulty of giving a naturalistic epistemology for mathematics.[4] That cannot be dismissed so easily. The more descriptive approach to mathematical

---

[2] I summarize here a point made in *Mathematics in Philosophy*, pp. 195–197.

[3] It would be hard to maintain this about the part of mathematics where there is uncertainty that might be serious, namely, the further reaches of set theory where very large cardinal axioms or other principles of high consistency strength are assumed. In this case, however, the mathematics makes no contact with actual natural science. If some of it is upset, it is far more likely that this will be a result of its internal development.

[4] See Hart, Review of Steiner, pp. 124–127. Hart's later paper "Benacerraf's Dilemma" corrects the exclusively ontological focus that many have given to the problem; in particular he points out that modal knowledge would raise similar questions. Insofar as Benacerraf's dilemma motivates nominalist constructions, that raises a question about the widespread use of modality in these constructions. Hartry Field may be influenced by a problem of this kind in seeking to limit his own use of modality to "strictly logical" modality. See "Is Mathematical Knowledge Just Logical Knowledge?" and "Realism, Mathematics, and Modality."

knowledge adopted in this work would probably not pass muster as naturalistic in the eyes of many contemporary philosophers. I don't see this as a serious problem for the foundations of mathematics. Naturalism as a philosophical tendency relies heavily on the authority of natural science. But modern science would be inconceivable without the application of mathematics. The actual methodology of mathematics, about which a descriptive approach aims to say something, has been found adequate (at least with the corrections arising in its own development, certainly influenced by applications) for the development of science over a more than two-thousand-year period. The absence of a naturalistic epistemology may mean that a kind of explanation or understanding of mathematical knowledge that would be desirable has not been attained. The search for it, like any enterprise in naturalistic epistemology, is on the boundary of philosophy and psychology. But even if it faces fundamental difficulties, they would not offer a convincing reason for abandoning current mathematics, or for reformulating it along some nominalistic or other lines, or for denying the claim of mathematical results to be true. To what extent we are still left with a challenge may depend on what counts as naturalistic, a matter that I leave to those who espouse naturalism to determine.[5] Furthermore, there are at least some reasons for doubting that what the naturalist seeks can be attained for our rational capacities in general, apart from the more special problems posed by mathematics.

Another omission is of any sustained discussion of the issues raised by constructivism in general and intuitionism in particular. Such a discussion would have been natural given the role played in this work by a conception of intuition that owes something to Brouwer. In fact, my original plan for the book called for a chapter on constructivism. The main reason why it is not there is that I have had my hands quite full with the other subjects I have taken on. It would be a large task to assess what the status of constructive mathematics and logic ought to be in the present day, or even to say accurately what it is. One thing, however, is clear: The use of classical logic in mathematics has survived Brouwer's attack on it, and mathematics obeying restrictions to ensure constructivity is a minority pursuit. A philosophical work that deals almost entirely with classical mathematics does not have to apologize for itself.

This work has been in the making for an unconscionably long time. During that time I have become indebted to a large number of

---

[5] For an argument that it is a challenge, and a proposal to answer it, see Linnebo, "Epistemological Challenges."

individuals, especially for intellectual stimulation and instruction, and to institutions for support. Columbia until 1989, and Harvard since then, have provided an academic home and an excellent work environment. Friends and colleagues in logic and philosophy of mathematics have been sources of stimulation, instruction, and questioning over many years. The late George Boolos, Solomon Feferman, Warren Goldfarb, Allen Hazen, Richard Heck, Geoffrey Hellman, Daniel Isaacson, Yiannis Moschovakis, Hilary Putnam, Michael Resnik, Stewart Shapiro, Wilfried Sieg, and William Tait deserve special mention, as well as Mark van Atten, Peter Koellner, and Agustín Rayo from more recent years. I still owe much to my longtime Columbia colleagues Isaac Levi and the late Sidney Morgenbesser. A consequence of the move to Harvard was more interaction with moral philosophers, which stimulated my interest in rational justification, dovetailing with an effort to understand ideas of Kurt Gödel. Without that, Chapter 9 of this work might not have been written at all. The late John Rawls provided the initial stimulus and helped me to understand his own views, and T. M. Scanlon has also been especially helpful. My fellow editors of Gödel's works, John W. Dawson, Jr., Feferman, Goldfarb, Sieg, Robert M. Solovay, and our Managing Editor Cheryl Dawson all contributed to my understanding of Gödel and his contemporaries.

None of my three principal teachers, Burton Dreben, W. V. Quine, and Hao Wang, lived to see the completion of this work. Some of the ideas were discussed with one or another of them, and their influence is no doubt more or less visibly present. I doubt that any of them would approve of what has finally come of the project.

Like many of us I have learned from my Ph.D. students. Most relevant to this work are R. Gregory Taylor, Richard Tieszen, Gila Sher, and Ofra Rechter at Columbia and Emily Carson, Michael Glanzberg, Øystein Linnebo, Michael Rescorla, and Douglas Marshall at Harvard. Tieszen especially has followed my writing over a long period and commented on earlier versions of many parts of this work.

In the long time that this work has been in progress, I have lectured on parts of it (sometimes as papers that have since been published) to many audiences. I am undoubtedly indebted to more members of these audiences than are mentioned here in connection with specific points. A special debt is owed to logicians and philosophers at the University of Padua, who twice invited me for extended series of lectures, which were accompanied by warm hospitality.

In the fall semester of 2006, the manuscript was discussed in a seminar at Harvard with both student and faculty participants. Questions

Preface

and objections have led to many changes, nearly all of which I hope are improvements. In particular, Koellner and Vann McGee saved me from mathematical errors. Doubtless other errors remain, for which I am responsible.

Mihai Ganea conscientiously examined an earlier version of Chapters 1–8 for bugs of various kinds, and Jon Litland has done the same for the final version of the whole work. Litland also has prepared the index and assisted with proofreading.

This work has had more institutional support than any single book deserves. Both Columbia and Harvard have provided sabbatical leaves. The project was begun when I was an NEH Fellow and a Visiting Fellow of All Souls College, Oxford, and work on it was done when I was a Guggenheim Fellow, a Fellow of the Netherlands Institute for Advanced Study, and later a Fellow of the Center for Advanced Study in the Behaviorial Sciences, the last with the support of the Andrew W. Mellon Foundation. I am very grateful to all these institutions. (The leaves they financed also encouraged other projects, especially the editing of Gödel's posthumous works.)

Terence Moore of Cambridge University Press welcomed the project and offered a contract on the basis of a very incomplete text. I regret that the work was not ready to submit before his untimely death. Two referees, one unmasked as Arnold Koslow, offered helpful suggestions. I am grateful to the Press for its continuing interest, in particular to the present editor Beatrice Rehl. I thank the production editor Laura Lawrie for her work and attentiveness to my concerns.

Much of the material in this work has appeared in articles with varying degrees of closeness to the form in which matters are presented here. More information with copyright acknowledgments follows.

I owe more to my wife, Marjorie Parsons, than I will venture to say here. I will, however, thank her for two specific things: for her unfailing support and assistance during two episodes of illness and for her patience in waiting so many years for a book to be dedicated to her.

Cambridge, February 2007

Sources and Copyright Acknowledgments

In Chapter 1 §§1–6 are an expanded version of my paper "Objects and Logic," *The Monist* 65 (1982), 491–516. Copyright © 1982 *THE MONIST*: An International Quarterly Journal of General Philosophical Inquiry. Peru, Illinois, U.S.A. Reprinted by permission.

Chapters 2 and 3 (except for §§14–15) draw on "The Structuralist View of Mathematical Objects," *Synthese* 84 (1990), 303–346, copyright © 1990 by Kluwer Academic Publishers, used with kind permission of Springer Science and Business Media.

§18 incorporates material from "Structuralism and Metaphysics," *Philosophical Quarterly* 54 (2004), 56–77, by permission of Blackwell Publishing on behalf of the editors of the *Quarterly*.

Chapter 4 is a modestly expanded version of "Structuralism and the Concept of Set," in Walter Sinnott-Armstrong et al. (eds.), *Modality, Morality, and Belief: Essays in Honor of Ruth Barcan Barcus* (Cambridge University Press, 1995), pp. 74–92, used here with the blessing of the editor.

Chapter 5 draws on "Mathematical Intuition," *Proceedings of the Aristotelian Society* N. S. 80 (1979–80), 145–168, reprinted by courtesy of the Editor of the Aristotelian Society, copyright © 1980 The Aristotelian Society. It also draws on "On Some Difficulties Concerning Intuition and Intuitive Knowledge," *Mind* 102 (1993), 233–245. An advanced draft of much of the rest was extracted and published as "Intuition and the Abstract," in Marcelo Stamm (ed.), *Philosophie in synthetischer Absicht* (Stuttgart: Klett-Cotta, 1998), pp. 155–187, included here by permission of the publisher.

Chapter 6 draws on "Intuition and Number," in Alexander George (ed.), *Mathematics and Mind* (Oxford University Press, 1994), pp. 141–157.

Chapter 7 overlaps considerably with "Finitism and Intuitive Knowledge," in Matthias Schirn (ed.), *The Philosophy of Mathematics Today*

(Oxford: Clarendon Press, 1998), pp. 249–270, and also draws on "On Some Difficulties."

In Chapter 8 §47 and §§50–51 draw on "The Impredicativity of Induction," in Leigh S. Cauman, Isaac Levi, Charles Parsons, and Robert Schwartz (eds.), *How Many Questions: Essays in Honor of Sidney Morgenbesser* (Indianapolis: Hackett, 1983), pp. 132–153, revised and expanded in Michael Detlefsen (ed.), *Proof, Logic, and Formalization* (Routledge, 1992), pp. 139–161. Material from the original paper is used by permission of the editors, who hold copyright. Material from the revision is included by permission of the editor and publisher.

§§48–49 rework the argument of "The Uniqueness of the Natural Numbers," *Iyyun: A Jerusalem Philosophical Quarterly* 39 (1990), 13–44. The later version appears in "Communication and the Uniqueness of the Natural Numbers," in the informally distributed *Proceedings of the First Seminar on Philosophy of Mathematics in Iran* (Shahid Beheshti University, Tehran, 2003).

Chapter 9 draws on "Reason and Intuition," *Synthese* 125 (2000), 299–315, copyright © 2000 by Kluwer Academic Publishers, used with kind permission of Springer Science and Business Media.

1 Objects and Logic

§1. Abstract objects

The language of mathematics speaks of objects. This is a rather trivial statement; it is not certain that we can conceive any developed language that does not. What is of interest is that, taken at face value, mathematical language speaks of objects distinctively mathematical in character: numbers, functions, sets, geometric figures, and the like. To begin with, they are distinctive in being abstract.

Roughly speaking, an object is abstract if it is not located in space and time and does not stand in causal relations. This criterion gives rise to some uncertain cases and would not be accepted by all philosophers.[1]

---

[1] This is close to what David Lewis calls the Way of Negation in his discussion of the abstract-concrete distinction in *On the Plurality of Worlds*, §1.7. Criticisms of the criterion are to be found there and in Burgess and Rosen, *A Subject with no Object*, §I.A.1.b. The conterexamples offered are for the most part not mathematical objects. However, these authors claim that some abstract objects of the kind called quasi-concrete in §7 below are located. There is one case relevant to mathematics, that of sets of concrete objects, which will be discussed there, in note 57.

A reservation about the causal aspect that is worth mentioning is the following. There has been some dispute about the kind of entities that enter into causal relations connected with the dispute about whether events are particulars or are proposition-like objects. Jaegwon Kim, a principal proponent of the latter view, also challenges the view that mathematical objects are "causally inert":

> Mathematical properties, including numbers, are no worse off than such sundry physical properties as color, mass, and volume, in respect of causal efficacy ("The Role of Perception," p. 347).

It is noteworthy that Kim talks of mathematical *properties*. The point could be put by saying that reference to mathematical objects is as likely to occur in a causal explanation as reference to other kinds of objects. This seems to me quite correct, but, as Mark Steiner had previously observed (*Mathematical Knowledge*, ch. 4), it can be put in the other framework, in which such reference will occur in the relevant descriptions of the events

1

It is not essential for our purposes that there should be a principled and exhaustive classification of all objects into abstract and concrete. Physical bodies and biological organisms, such as we encounter in everyday life, are concrete. If we assume that sense-perception necessarily involves a causal relation between the object perceived and the organism (the event or state of its perceiving), and that perception *locates* its objects at least in some rough way, then it follows that the objects of sense-perception are concrete. Thus it is generally assumed in discussions of abstract objects that abstract objects cannot be perceived by the senses.

In this they are not alone. It is not merely for this reason that abstract objects are thought to pose a general philosophical problem. Some presumably concrete objects, such as elementary particles, are far more recondite and no more perceivable than most "ordinary" abstract objects such as triangles and numbers. Nonetheless, some form of empiricism is a primary motive for finding abstract objects puzzling. We think there are elementary particles such as electrons, roughly because the assumption that they exist belongs to a theory that presents a picture of the objective world that accounts for the facts we can verify by perception. But why should a view of the world give place to objects that are not in it (not spatio-temporal) and do not interact with it? Is the existence of mathematical objects not a hypothesis for which we should have no need?

If the outcome of an investigation of such questions should be that mathematics is engaged in mythology, in speaking of remarkable configurations of objects that just do not exist, it would leave the existence of mathematics and its importance to science and practical life a much greater mystery than its objects are at the outset. This would suggest that what such questions should prompt us to do is to inquire more deeply into what a mathematical object *is*, what the "mode of being" of mathematical objects is. Or, in less ontological terms, we should inquire into the sense in which mathematics speaks of objects. This will be one of the main concerns of this book. It is, however, a serious question whether reference to specifically mathematical objects can be eliminated. This question will be addressed in Chapters 2 and 3 below. Assuming that a wholesale rejection of mathematics is not being considered, such an elimination can only proceed by not taking the language of mathematics entirely at face value.

---

standing in causal relations. This removes any temptation to attribute causal efficacy to the mathematical *objects*. By his emphasis on properties, Kim somewhat fudges the latter issue.

This question will prove to have relevance later; see §30.

§2. The concept of an object in general: Actuality

There is something absurd about inquiring, with complete generality, what an *object* is. The question may seem impossibly amorphous, but it is the extreme generality of the question, rather than vagueness in the ordinary sense, or something like it, that is the difficulty in putting it. If we ask what a gorilla is, an informative answer should distinguish gorillas from other animals. But from what can we "distinguish" objects? It is controversial whether it is meaningful to talk of "entities" that are not objects, and, even if it is, a distinction between objects and other entities is only part of what we are after in asking what an object is.

To the extent that philosophers discuss in general terms what an object is, the context is likely to be an inquiry into how thought or language can relate to objects, or, more amorphously, to "reality" or "the world." I will fix the way I wish to use the term "object" and simultaneously say what I think useful in such abstract discussions by saying that the usable general characterization of the notion of object comes from *logic*. We speak of particular objects by referring to them by singular terms: names, demonstratives, and descriptions. Since the unit of linguistic expression is the sentence, reference to objects will typically and standardly occur in the context of sentences, and, therefore, in the use of singular terms, by the application to singular terms of *predicates*. In its most general meaning, a "predicate" is a sentence with one or more empty places, called argument places, to be filled by singular terms.

Reference to objects may begin with singular terms, but it does not end with them, since the use of predicates is bound up with the use of those expressions which the formal logician regiments by quantifiers and bound variables. Thus, in English and many other natural languages, what I have called the "argument place" of a predicate can be filled by expressions like 'all men', 'every man', 'all ravens', 'everything', 'something', 'sometime', and others which serve to make the sentence general, or by pronouns that refer back either to one of these expressions or to a singular term. Thus, we have such sentences or discourses as:

All men are mortal. All ravens are black.
Everything is identical to itself.
Do something!
Come around to see me sometime.

> If anyone know just cause why this man and this woman should not be joined in holy matrimony, let *him* speak now or forever hold *his* peace.
>
> I saw Mr. Smith on January 10. *He* was in a bad temper *that day*.

Philosophers in the tradition of the modern theory of reference have disagreed about whether singular terms or quantifiers and bound variables are the more fundamental vehicles of reference to objects. It is quite evident that in practice we have to attend to both kinds of expressions. Singular terms may be dispensable in a final regimentation of the language of a rigorous science or mathematics, but, in order to understand reference and the notion of object, we have to look at language as it is before such regimentation, in which singular terms have an undoubted place. It is even more obvious that the use of singular terms[2] does not exhaust reference to objects in the sense that the objects whose properties and relations matter to the truth of what we say are not all referred to by singular terms, either in the statements themselves, or in any others. Our ontological commitments, that is, what is required to *exist* for our statements to be true, are not all expressed by the use of singular terms but are also embodied in general statements; indeed, a definite description (at least in its "attributive" use[3]) is a singular term such that the condition for being its *reference* is the truth of a statement involving a quantifier.

The phrase "concept of an object in general" comes, of course, from Kant's *Critique of Pure Reason*. It is used in connection with the categories, in particular with the thesis that the categories are the conditions under which what is given in experience can be thought of as an object. For example,

> The question now is whether *a priori* concepts do not also precede, as conditions under which alone something can be, if not intuited, nevertheless thought as object in general. (A 93/B 125)

Kant calls the categories "concepts of objects in general," but he also occasionally talks as if there were one concept of an object in general, that

---

[2] I assume that bound variables (and the pronouns of natural language that play the same role) do not count as singular terms. I don't think it matters where we draw the line, though with enough stretching of the category it might be made true, after all, that all objects of reference for our discourse are denoted by singular terms, since the semantics of quantifiers refers to assignments of objects to free variables or to interpretations in which the denotation of names is varied.

[3] In the well-known sense of Donnellan, "Reference and Definite Descriptions."

§2. The concept of an object in general: Actuality

is, a perfectly general concept of object.[4] But then it is the categories that give it its content, so that the difference with the plural use is merely verbal. Either way, Kant is thinking of objects of experience, and the proper application of the categories is to spatiotemporal objects of the sort that are given in perception or at least postulated in scientific theories.

In fact, when Kant talks of objects at this level of abstractness and generality, he is pulled in two directions. On the one hand, the categories are derived from formal logic and from the conceptual framework for talking of *any* kind of object, whether or not it is empirically known. The understanding, with its forms of judgment and categories, is more general than sensibility, with its forms of intuition, and thus there are "pure" categories whose *content* relates to objects "in general and in themselves," and are not limited by conditions of sensibility or experience.

On the other hand, it is of course central to Kant's theory of knowledge that the categories only have proper application giving rise to *knowledge* in application to the manifold given by sensible intuition, and thus they serve only for knowledge of the empirically given world. It is only in this context that he gives an informative account of consciousness of objects. And of course it has been difficult for many generations of readers to see how the application, even "problematic," of the categories to things in themselves is possible.

Even the pure categories, as a general conceptual apparatus for speaking of objects, envisage concrete rather than abstract objects. This is particularly true of the categories of relation: substance, causality, and community. We gave as one of the marks of an abstract object its not entering into causal relations. Although some marks of a substance may be possessed by abstract objects, the primary examples of substances in the philosophical systems of the past are paradigmatically concrete realities: organisms for Aristotle, God for Christian Aristotelians and the rationalists from Descartes on, and matter for Descartes and Kant. When Kant comes to the categories of modality, what in the original table (A 80/B 106) is called Existence (*Dasein*) and is later called actuality (*Wirklichkeit*) is explained in a way which clearly is meant to apply to concrete spatiotemporal objects:

> That which is connected with the material conditions of experience (of sensation) is *actual*. (A 218/B 266)

---

[4] A 51 / B 75, A 251. The latter passage seems to identify the concept of an object in general with that of the "transcendental object $= x$."

> The postulate for cognizing the *actuality* of things requires *perception*, thus sensation of which one is conscious, to be sure not immediate perception of the object itself the existence of which is to be cognized, but still its connection with some actual perception in accordance with the analogies of experience. (A 225/B 272)

By actuality, Kant means actual existence, and it therefore has the logical properties which, for existence *per se*, he appeals to in his famous criticism of proofs of the existence of God. It is not a property of objects; we could reconstruct it as what is expressed by the existential quantifier in empirical judgments. But the characterization quoted gives just what we have offered above as the marks of a concrete object.

In our modern logic, where we readily give the existential quantifier a more general sense, it is natural to render Kant's actuality as a restricted quantifier, so that 'Fs are actual' in Kant's sense would be rendered '$(\exists x)$ (x is actual $\wedge$ Fx)'. The German '*wirklich*' is less awkward as a predicate than the English 'actual'.

Just such a transmutation is carried out by Frege, who denies that numbers are actual (*wirklich*) in a sense which (probably consciously) echoes Kant:

> I heartily share his [Cantor's] contempt for the view that in principle only finite numbers ought to be admitted as actual. Perceptible by the senses they are not, nor are they spatial – any more than fractions are, or negative numbers, or irrational and complex numbers; if we restrict the actual to what acts on our senses or at least produces effects which may cause sense-perceptions as near or remote consequences, then naturally no number of any of these kinds is actual.[5]

Frege clearly understands *Wirklichkeit* as a property which some objects (such as bodies) have and others (such as numbers) do not.[6] In using the terms 'actual' and '*wirklich*', I will follow Frege's usage.

---

[5] *Grundlagen*, p. 97.
[6] As in the following striking passage:
> Some of what is objective is actual, other not. 'Actual' is only one of many predicates, and concerns logic no more particularly than, say, the predicate 'algebraic' applied to a curve (*Grundgesetze*, I, xviii–xix, my trans.).

Frege seems pretty clearly to have the same meaning of '*wirklich*' in mind when, near the end of "Der Gedanke," he questions whether thoughts are *wirklich*. The strict atemporality of thoughts is "annulled" by the fact that they can be now grasped by one person, at another time not. However, one might still recognize them as timeless on the ground that these changes affect only the "inessential" properties of thoughts. (We might now call them "mere Cambridge changes.") Moreover, he is prepared to say that a

§2. The concept of an object in general: Actuality

To return to Kant, I think it would be fair to interpret him as having a "concept of an object in general" which in its full sense connotes concreteness. This is even so in the less-than-full sense given by the pure categories. A problem in interpreting Kant is posed by the fact that, although he does make reference to mathematical objects, and is certainly not a nominalist, he does not have an explicit theory of them. There are indications that he would treat mathematical existence under the category of possibility; mathematical examples occur in the elucidation of that category in the Postulates.[7] There is some affinity between Kant's view and those now called modalist, which will occupy us in Chapter 3. This affinity is limited by the fact that Kant conceives possibility as "real" possibility, not certifiable by mathematical argument (construction in pure intuition) alone. It would take us too far afield to go into these problems, which I have discussed elsewhere.[8]

The expectation that objects as such should have the properties that Kant and Frege associate with *Wirklichkeit*, or other more rarefied versions that could be possessed by suprasensible concrete reality, is deeply rooted and motivates much resistance to mathematical objects and abstract objects generally. Taking the most general notion of object to have its home in formal logic (in Kantian terms, the forms of judgment rather than the categories) is intended to remove that expectation and serves to defuse the resistance it motivates. This perspective will underlie much of our subsequent argument.

Kant's refusal to apply the category of actuality to mathematical objects suggests another distinctive feature of such objects, that the distinction between potentiality and actuality, particularly between possible and

---

thought "acts" (*wirkt*) through being grasped and taken to be true. But because it is only in this way that thoughts enter into the causal order, the "actuality" of thoughts is very different from that of things:

> Thoughts are not wholly unactual, but their actuality is of a quite different kind from that of things. And their action is brought about by a performance of the thinker, without which they would be without effect (*wirkungslos*), at least as far as we can see. And yet the thinker does not create them but must take them as they are (p. 77; trans. from *Logical Investigations*, pp. 29–30, with modifications).

I am indebted to Mark Notturno for calling this tantalizing passage to my attention. In it Frege shows himself more properly "Platonist" than in his writings on the philosophy of mathematics.

[7] A220–1 / B268, A 223–4 / B271.
[8] "Arithmetic and the Categories," section I, and, in a briefer and more preliminary way, in the Postscript to Essay 5 of *Mathematics in Philosophy*, pp. 147–149.

actual existence, does not apply to them.⁹ This is an important point, which will be considered briefly in §4 below and more fully in Chapters 2 and 3.

§3. Intuitability

Although Kant would resist attributing to mathematical objects the properties summed up in the concept of actuality, he does apply in the mathematical context another characteristic of full-blooded empirical objects which will be important in what follows. Kant of course holds that a concept is empty unless it corresponds to *intuition*; intuition is necessary to establish the objective reality of a concept, that is, the possibility of instances. The forms of intuition, space and time, are therefore conditions to which all objects of experience must conform. The mathematical objects about which Kant is most explicit are geometric figures, which he calls forms of (empirical) objects. In proofs, they are constructed intuitively; in that sense they can be intuited. Intuitive representation also arises in mathematics for other mathematical objects, though for numbers in particular it appears that the relation to intuition is more indirect than for geometric figures. But arithmetic is according to Kant only applicable to sensible objects, that is, to objects given according to our forms of intuition.[10]

We will speak, generally and somewhat vaguely, of *intuitability* as a general condition on objects. The use of the term "intuition" rather than, say, "perception" is meant to preserve the generality of Kant's notion, which, in particular, is not meant to exclude the abstract. Kant means by "intuition" an immediate representation of an individual object. What is meant by "immediate" has been a matter of controversy.[11] When talking of intuition of *objects* in this work, I shall mean a mode of consciousness of individual objects that is importantly analogous to perception. But it is important to distinguish between intuition of objects and the different

---

[9] In 1790 Kant's disciple Johann Schultz wrote that "in mathematics possibility and actuality are one, and the geometer says *there are* (*es gibt*) conic sections, as soon as he has shown their possibility a priori, without inquiring as to the actual drawing or making them from material" (Review of vol. II of Eberhard's *Philosophisches Magazin*, in Kant's *Gesammelte Schriften*, XX 386 n.) (Cf. "Arithmetic and the Categories," p. 110 and n. 4.)

[10] Letter to Johann Schultz, November 25, 1788 (*Gesammelte Schriften*, X 555). This edition is hereafter referred to as *Ak*.

[11] See Essay 5 of *Mathematics in Philosophy*, especially the Postscript, the writings of Hintikka and Howell referred to there, and "The Transcendental Aesthetic," section I.

§3. Intuitability

but related notion of intuition as a propositional attitude. These notions and their relations will be explored further in Chapter 5. Clearly it is the former notion that is involved when we speak of the intuitability of objects.

Even if one takes the notion of intuition to be understood, the notion of intuitability poses two serious problems. As we have explained it, intuitability does not require that the object itself can be an object of intuition. But if it is, we might call the object *strongly* intuitable. We consider intuitable an object that can be "represented" in intuition, and we have not said what this means. Different relations will count as such representation; thus the notion of intuitability is to a certain degree schematic. In the relevant senses, however, representation of abstract objects by concrete objects or by objects relatively closer to the concrete is a pervasive phenomenon and of great importance for understanding abstract objects. An example is that a number could be represented by a set or sequence of that number of objects; such a representation occurs in Kant's argument to the effect that $7 + 5 = 12$ is synthetic. If, as in Kant's example, the elements of the set or the terms of the sequence are concrete, then the representation is what in §7 will be called "quasi-concrete." We will consider in Chapter 6 the question whether, if they are intuitable, the set or sequence itself is intuitable. If so, then the representation certainly qualifies as a representation in intuition; even if not, it is still a representation in intuition of a more indirect kind, since the concrete is reached in two steps. If such representation confers intuitability, then mathematical objects can count as intuitable even if no mathematical objects are strongly intuitable. This may seem to make the notion too weak to be useful; whether this is so will be considered in Chapters 5 and 6.

The second problem concerns the modal element in the notion of intuitability. We might call an object "perceivable" if it *can* be perceived, but this "can" can be understood in different ways; for example, to be precise we would have to determine how much to abstract from the situation and capacities of actual perceivers. Questions of this kind arise with particular force when we ask about the intuitability of mathematical objects. Is it true that all natural numbers can be represented in intuition in the sense that for any n, a set of n objects can be an object of intuition? That seems to require that in some sense we "can" take in an arbitrarily large finite collection, but, in practice, one will very quickly run up against limitations of time and memory. Classical conceptions of mathematical intuition found in Kant, Brouwer, and Hilbert, by contrast, appear to be committed to the view that in some way arbitrarily large, finite structures are representable in intuition. The viability of the idea of the "in principle" possibility of intuition's extending this far, and the question whether a conception of mathematical intuition along such lines requires it, will occupy us later.

One can ask how the notions of intuitability and actuality are related. The thesis that all intuitable objects are actual would rule out intuition of mathematical objects, but it is certainly maintained by many philosophers. It is interesting only if the notion of intuitability is either strong intuitability or based on a notion of representation in intuition that is not too liberal. The converse thesis that all actual objects are intuitable amounts to the thesis that those objects we postulate in a causal account of our perceptions are intuitable. Here, clearly, the question about modality is relevant, and also questions about the "theory-laden" character of observation.

§4. Logic and the notion of object

I have mentioned actuality and intuitability not just because of their centrality in Kant's conception of an object of experience. They represent requirements for being an object that have been applied by philosophers, in some cases not very consciously, to issues about abstract objects and inclined them to deny that there could be such objects, or, at least, to find them puzzling. The view that the most general notion of object has its home in formal logic is intended to reject such conditions as conditions for being an object.

This outlook seems to me a characteristically modern one and may not appear in full-blown form before Frege. Its most important advocates in more recent times are Carnap and Quine. Speaking of objects just is using the linguistic devices of singular terms, predication, identity, and quantification to make serious statements. We might in many cases be able to claim that the reference to objects in a discourse is only apparent. But this requires either a paraphrase or a special semantical explanation that removes it; one cannot, for example, just *say* that the objects in question are fictions. One might explain "fictions" as a *category of objects*, but in that case one has not fended off the interpretation that one is speaking of objects.[12]

---

[12] With reference to mathematics, one might also mean by such a claim that although the language of mathematics is to be interpreted to speak of objects, the mathematician is not committed to their existence because the statements in mathematics are not really

§4. Logic and the notion of object

Different writers in this tradition have stressed different parts of the apparatus of reference to objects. Frege placed particular stress on identity, Quine on quantification. But these differences are either differences of emphasis or concern peripheral issues. I shall assume that we do not have *objects* unless we can meaningfully apply the identity predicate. I hardly know how to begin in arguing for this. It is partly a terminological matter, since there may be other categories of entities, for example Frege's functions and concepts, to which identity does not apply and which are not objects. A more fundamental argument would have to be one concerning objectivity, such as is to be found in Kant's Transcendental Analytic and the tradition inaugurated by it. It is characteristic of an object that it can be represented in different ways, from different perspectives. But this statement hardly makes sense unless it means that the *same* object is thus represented. The possibility of representing the same object from different perspectives is necessary for our statements to have a content that is independent of the time of utterance and the speaker. The former mark of objectivity especially preoccupied Kant, the latter later philosophers who have been especially concerned with intersubjectivity.

The apparatus of reference to objects is very neatly regimented by first-order quantificational logic with identity. An influential view, forcefully maintained by Quine, is that questions of reference and ontological commitment are to be settled by a paraphrase of any discourse about which such questions arise into this first-order logic, based on classical propositional logic. This is an extreme view, which might be criticized on various

---

assertions but are rather to be compared with the occurrence of sentences in works of fiction. This is the view of Hartry Field (originally in *Science without Numbers*), who, indeed, insists on taking the language of mathematics at face value rather than reinterpreting it to eliminate mathematical objects, and then bites the bullet by holding that statements committed to mathematical objects are not *true*. His view then amounts to an instrumentalist view of mathematics.

Field's view that straightforward statements about numbers and sets are not true seems to me an extreme one; moreover, he has devoted more of his writings to elaborating his instrumentalist program and to dealing with objections than to working out the nominalist intuitions with which he begins.

Some particulars of Field's view are discussed in Chapters 2 and 3. Other versions of "fictionalism" have been proposed in recent years, but I do not propose to discuss them in this work. They tend to be even harder than Field's to reconcile with the claim of mathematics to be a science. I am sympathetic to the criticism in Burgess, "Mathematics and *Bleak House*." But it should be noted that Burgess is concerned there with versions of fictionalism that, unlike Field's, do not demand logical reconstruction either of mathematics or of its relation to science.

grounds. Because of its simplicity and power of expression, however, classical first-order logic does give an exemplary paradigm of discourse about objects. For the present, I only want to consider how the question what an object *is* is affected by possible reservations about taking this paradigm as a universal framework.

There are three questions that require some immediate discussion. The first is whether some intensional operators belong to logic. The second is whether there are "entities" that are not objects and, if so, how logic is to accommodate them. The third is whether reference to objects is necessarily reference to objects that *exist*. The second and third questions will be taken up in Sections 5 and 6, respectively.

On the first question, I confine myself to some brief remarks. The presence in our language of modal and intentional notions is obviously a serious obstacle to regimenting all reference to objects in terms of classical first-order logic or a similar extensional logic. Modal and intentional discourse gives rise to problems which are among the most tangled faced by a program of analyzing the "logical form" of expressions of natural language in terms of formal logic. Because a purely extensional logic is adequate for formalizing mathematical theories in the standard way, it might seem that problems of this sort can be avoided in the philosophy of mathematics. For a number of reasons, which we will not go into at this point, the matter is not so simple. Three considerations make it obvious that problems about modal notions have to be addressed in the philosophy of mathematics: the traditional view that mathematics is *necessary*; the idea, alluded to with reference to Kant, of explicating mathematical existence in terms of *possibility*; and the role, both in constructive mathematics and the foundations of set theory, of the conception of a potential totality. But this is not the place to explore the relevance of these considerations to a strictly Quinean conception of ontology.[13]

---

[13] See Essays 1, 7, and 10 of *Mathematics in Philosophy*. Epistemic notions play a central role in intuitionistic mathematics, but they are usually not incorporated into formalized intuitionistic theories. Recently it has been argued that epistemic operators should occur explicitly in the axiomatization of *classical* theories; this has resulted in the development of theories of "epistemic mathematics," in which an S4-like epistemic operator is added to the usual language. See the papers in Shapiro, *Intensional Mathematics*, and, for a forceful argument motivating such theories, Goodman, "The Knowing Mathematician."

§5. Is whatever is an object?

This question might be posed in a number of ways, with a number of possible "nonobjects" in mind. But the consideration that is most relevant to the ontology of mathematics is reflection on *predication*. In a simple predicative statement, which we can schematize as '*a* is *P*', it is tempting to say that the statement predicates *something* of *a*. But then that something seems to be *P*, and then we seem to have to say that '*P*' stands for something, just as '*a*' does. But if '*P*' stands for an *object*, then the simple predication seems to express a relation between two objects. But then does not the statement predicate something, expressed by 'is', of *a* and *P*? To admit this would threaten a regress. At all events, taking predicates to stand for objects has led into serious tangles.

The relevance of this problem to the question at hand is indicated by Frege's famous solution to it: according to him, what is predicated of *a* is a concept which is *not* an object; the apparently relational form is avoided because a concept is "unsaturated"; it has an empty argument place just as does the expression '( ) is *P*', which is applied to '*a*' to yield a sentence. Frege's position leads to taking as basic logic a second-order logic, as was the original Fregean logic in the *Begriffschrift* and its mature form in the *Grundgesetze*. Frege never showed any interest in disengaging the first-order fragment.[14]

By a second-order logic is normally meant an expansion of first-order logic by the addition of variables having the syntactic form of predicates, and quantifiers for them. The logic of the *Grundgesetze* is not precisely of this form because of its assimilation of sentences and singular terms. From this assimilation follows that predicates, for him (as for us) sentences with empty argument places, are to be treated like singular terms with empty argument places, which we will call *functors*. The latter have as their reference functions; thus concepts are a special case of functions. Indeed, it was partly on the strength of observed analogies between functors and predicates that Frege adopted a theory of reference that treats sentences like singular terms.[15] Although languages of this kind arise in

---

[14] He was not even tempted to do so by the paradoxes. On the contrary, his gradual abandonment of the concept of extension, which led finally to the "geometric" theory of number sketched in the last fragments of 1924–1925 (*Nachgelassene Schriften*, pp. 282–302), was accompanied by his holding firmly to his theory of functions and concepts and to the second-order logic that went with it as "fundamental logic."

[15] This analogy, it will be recalled, was one of the starting points of Frege's logic, before he formulated his theory of reference; as is well known, in the *Begriffschrift* he says he has replaced the concepts *subject* and *predicate* by *argument* and *function* (Preface, p. vii).

logical work even today,[16] we will usually assume that second-order variables are predicates. Much of what we say about second-order logic will apply to logics in which quantification of predicate places is subsumed under quantification of functor places.

Some form of generalization of predicate places is indispensable in mathematics, and it is for the sake of such generalization that we are most tempted to suppose that predicates have reference. Frege's arguments in *Funktion und Begriff* and elsewhere would imply the same about functor places. A very fundamental way in which generalization of predicate places occurs is in the statement of such general principles as mathematical induction and the axioms of separation and replacement in set theory. This might suggest that the underlying logic of much of mathematics must be second-order. However, the matter is not at all simple.

Frege held that predicates and functors have reference as such and that predicate and functor places can be generalized *directly*. This was essential to his thesis that functions and concepts have intrinsic argument places and cannot be objects. But it requires some explanation, because at first glance the general statements that express what I am calling "generalization of predicate places" are generalizations about *objects*. Consider for example the statements:

(1)　　Napoleon has all the qualities required in a general.[17]

(2)　　Metals are identified by such properties as malleability and ductility.

The first statement seems to be a general statement about *qualities*. If we ask for examples, we might cite bravery, decisiveness, and sense of timing. The expressions for these qualities are not unsaturated, and Frege would agree that they are objects. The same is evidently true of the "properties" cited in (2). But typically these qualities and properties are denoted by nouns or noun phrases that arise by nominalization of expressions of a

---

[16] Particularly type theories based on the λ-calculus or combinatory logic, beginning with Church, "A Formulation of the Simple Theory of Types." In such logics, the syntactic complication of introducing variables and quantifiers for what I am calling functors is avoided by λ-abstraction. Church regarded this as an abandonment of Frege's doctrine of functions as "unsaturated." See his "A Formulation of the Logic of Sense and Denotation," p. 4. The most significant examples of theories of this kind are intensional logics: Church's "logic of sense and denotation" and subsequent developments of it, and Montague's intensional logic.

[17] Cf. Whitehead and Russell, *Principia Mathematica*, I, 56. In the formalization (4) below, it might be more natural to render 'q is required in a general' as '$(\forall y)(y$ is a general $\to$ $\mathbf{R}(y$ has $q))$', and similarly in (3). This does not affect the points that concern us.

§5. Is whatever is an object?

predicative character. The quality of bravery is something one has if one *is brave*, one has decisiveness if one *is decisive*, and so on.

Now consider two formalizations of the first example (where **R** is a deontic operator meaning 'it is required that'):

(3) $(\forall F)[\mathbf{R}(\forall y)(y \text{ is a general} \rightarrow Fy) \rightarrow F(\text{Napoleon})]$

(4) $(\forall q)\{[q \text{ is a quality} \wedge \mathbf{R}(\forall y)(y \text{ is a general} \rightarrow y \text{ has } q)]$
    $\rightarrow \text{Napoleon has } q\}.$

(3) is the version by second-order logic; (4) appeals to a kind of objects called qualities. The sense in which (3) involves a direct generalization of a predicate place is that we can instantiate for '*F*' the predicate '( ) is decisive' and obtain

(5) $\mathbf{R}(\forall y)(y \text{ is a general} \rightarrow y \text{ is decisive}) \rightarrow \text{Napoleon is decisive},$

and, for example, from the assumption

$\mathbf{R}(\forall y)(y \text{ is a general} \rightarrow y \text{ is decisive})$

conclude that Napoleon is decisive. But to carry out this reasoning using (4) instead of (3), we need to be able to replace '( ) has decisiveness' by '( ) is decisive' and vice versa, the latter inside the (presumably intensional) operator **R**.

In much mathematical usage, as in the informal talk of qualities and properties, generalization of predicate and functor places is not done directly, but rather by way of generalizations about *objects* denoted not by predicates or functors themselves, but by terms obtained from them by what might be called nominalizing transformations, analogous to those by which nouns denoting qualities or properties might be imagined to arise from adjectives, verbs, or open sentences. Instead of qualities, properties, and relations, in mathematics one talks of sets, classes, functions, and relations, where the latter is generally meant extensionally.

Frege held that one will "recognize the same function" in occurrences of a functor with its argument place filled in different ways, such as in the cases '2. $1^3 + 1$', '2. $4^3 + 4$', and '2. $5^3 + 5$'.[18] Evidently he is claiming that the functor itself refers to the function. The usual present-day view would be that these terms do not, strictly speaking, contain reference to the function Frege has in mind, which would instead be denoted by a term such as '$\lambda x (2. x^3 + x)$'. The λ-abstraction transforms a functor into a singular term denoting a function. Confronted with the λ-notation, Frege would

---

[18] *Funktion und Begriff*, p. 6.

no doubt have agreed that λ-terms denote objects; his own notation for the graph (*Wertverlauf*) of a function has the same character. On the view that it is only objects that can be the reference of expressions or the values of variables, the sense in which it is true that one "recognizes the same function" in expressions where a functor occurs with different arguments, would have to be that one readily sees the equivalence of a term of the form '2. $t^3 + t$' to '$[\lambda x (2. x^3 + x)](t)$', where the reference to a function is explicitly made. Similarly, if one can "recognize" the statements

(6) Smith used to work for the man who murdered the second husband of his youngest sister.[19]

(7) Miss Brown used to work for the man who murdered the second husband of her youngest sister.

as placing Smith and Miss Brown in a common class, it is because one can recognize (6) as equivalent to

(8) Smith $\in \{x : x$ used to work for the man who murdered the second husband of x's youngest sister$\}$,

and likewise for (7).

We can, in effect, find in ordinary and mathematical usage different methods of generalizing predicate and functor places. We have followed Frege in taking second-order logic as embodying a direct method, in which predicates (and in some forms, functors) are taken as themselves having reference and as standing in quantifiable places; Frege relies on these assumptions in arguing that functions are essentially incomplete and cannot be objects. A second method, illustrated by (4), might be called the *method of nominalization*. This does not treat predicates and functors as themselves generalizable, but rather transforms the contexts in which they occur so that they are "nominalized" into names of qualities, properties, relations, sets, classes, or functions.

The existence of the method of nominalization and its frequent use both in ordinary language and in mathematics casts some doubt on Frege's view that predicates denote unsaturated entities which cannot be objects. It is, however, not always available or appropriate. Frege might have replied as follows to the above appeal to λ-abstraction in questioning his view that functors have reference: We are driven to the conclusion that functors denote functions, because we must accept functors in this

---

[19] Cf. Quine, *Methods of Logic*, 3rd ed., p. 111.

§5. Is whatever is an object?

role in order to apply the logical laws concerning functions. For example, for every function f and objects x and y, if $x = y$, then $f(x) = f(y)$. Thus, if $4 = 2 + 2$, then $2.\,4^3 + 4 = 2.\,(2+2)^3 + (2+2)$. The method of nominalization requires that the inference from the above generalization about functions to the conclusion be mediated by something like

If $4 = 2 + 2$, then $[\lambda x(2.\,x^3 + x)](4) = [\lambda x(2.\,x^3 + x)](2+2)$,

and the conclusion obtained by λ-conversion. But how are we to state the logical law which is applied in the λ-conversion? If it is a general law about functions (say that for any f and y, $[\lambda x\,(fx)](y) = f(y)$, then, if the generalization about functions contained in *it* rests on the method of nominalization, we will be in a circle. Here we are driven to assume laws in which functors occur *as such*, and not merely nominalized, because such laws are needed to get from nominalized generalizations to their non-nominalized instances.

Before I try to answer this Fregean argument, I shall point out another limitation of the method of nominalization. In the case of predication, the method works as follows: with a predicate 'Fx' we associate an object which (intending neutrality as to whether it is a set, class, property, or what) I shall designate '$(Ox)Fx$', and we have in the language a copula-like expression 'η' ('has', 'falls under', 'is an element of', or the like), such that for any singular term t, 'Ft' is (logically) equivalent to '$t\,\eta\,(Ox)Fx$'. But there must be some restriction, on pain of Russell's paradox. For if we let 'Fx' be '$\neg(x\,\eta\,x)$' and let t be '$(Ox)\neg(x\,\eta\,x)$', then we have '$\neg(t\,\eta\,t)$' equivalent to '$t\,\eta\,t$'. The supposition that $(Ox)Fx$ always exists is just the supposition that for any predicate 'F'

(9) $\qquad (\exists y)(\forall x)(x\,\eta\,y \leftrightarrow Fx)$,

which is just the inconsistent universal comprehension schema. Hence either $(Ox)Fx$ does not always exist (or at least does not always fall within the range of the quantifier $(\forall x)$), or the equivalence of 'Ft' to '$t\,\eta\,(Ox)Fx$' does not always obtain, or the term '$(Ox)Fx$' is not always well formed. Whatever "way out" we choose, it will mean that a generalization of predicates by the method of nominalization is failing to capture some predicates as "instances." The paradox is an argument for Frege's view that some generalizations are irreducibly about *functions* or *concepts* and cannot be reduced to generalizations about their object "surrogates."[20]

---

[20] Frege was of course unaware of this argument when he developed his theory of functions and objects, but he saw its force after Russell informed him of his paradox.

The limitations of the method of nominalization apply with particular force to an example we mentioned above for the point that generalization of predicate places is essential in mathematics: the axioms of separation and replacement in set theory. As Zermelo originally stated the axiom of separation, it is that for any "definite" propositional function P and any set a, there is a set b consisting of exactly those elements of a for which P holds.[21] Whatever he may have meant by a "definite" propositional function or property, it seems clear that it can be regimented by some kind of second-order logic, and we can state the axiom as

(10) $(\forall z)(\forall F)(\exists y)(\forall x)[x \in y \leftrightarrow (x \in z \wedge Fx)]$.

But in applying the axiom, one instantiates for the variable 'F' predicates that have not been determined to have sets as their extensions and in fact many predicates that can be proved *not* to have sets as their extensions. If we replaced the second-order quantifier by a quantifier over sets, (10) would reduce to the trivial statement that any two sets have an intersection.

Hence if we want to use the method of nominalization to express the generality of the axiom of separation, *sets* cannot serve as the objects associated with the predicates. Zermelo's idea of "definite properties" as another kind of object with which set theory is concerned has not found much favor in the form in which he presented it, but it would offer the possibility of saving the method of nominalization in this case. Historically, it is probably the ancestor of the more respectable notion of

---

[21] "Untersuchungen über die Grundlagen der Mengenlehre I," p. 263. Although the translation "definite property" is in general use, Zermelo actually uses the phrase "Klassenaussage $\mathbf{E}(x)$ definit für alle Elemente einer Menge M." The term *Klassenaussage* is more appropriately translated "propositional function" by Stefan Bauer-Mengelberg in his translation of the paper, van Heijenoort, p. 202. In fact, Zermelo's terminology shows some confusion; he does sometimes use the term *Eigenschaft* (property). In a later discussion of the notion in response to critics, he says his aim is to clarify axiomatically "the concept of 'definite' properties or statements (*Eigenschaften oder Aussagen*)" ("Über die Definitheit in der Axiomatik," p. 340). But then in his development he uses the term *Aussage*. I read him as being concerned with generalization of predicate places but vacillating, without making the distinction, between the method of nominalization and the method of semantic ascent.

I learned much about Zermelo's notion and its history from my student R. Gregory Taylor; see his dissertation, especially Chapter 6, and "Zermelo, Reductionism, and the Philosophy of Mathematics." See also Moore, "Beyond First-Order Logic."

§5. Is whatever is an object?

class in set theory, and we can interpret the variable '*F*' in (10) as ranging over classes.[22]

The more usual procedure in set theory, however, is to formulate the axioms of separation and replacement as *schemata*. This is in effect an application of a third method for generalizing predicate places, which I will call the *method of semantic ascent*. In axiomatizing set theory in the usual first-order way, the formulation (10) is not available; instead, of course, one assumes as an axiom *for each formula A (x)* of the language with the free variable x

(11) $(\exists y)(\forall x)[x \in y \leftrightarrow (x \in a \wedge A(x))]$.

This has the consequence that we cannot express in the language of the (formalized) theory the generality of the axiom in the formulation (10). In the metalanguage, however, we can express such a generalization by saying that all instances of (11) are *true*.

The method of semantic ascent enables us to express some generalizations for which the method of nominalization is either unsuccessful or inappropriate. Let us return to the Fregean argument against the claim that general laws about functions can be adequately stated as laws about function-*objects* denoted by λ-terms. The "law of logic" needed is the rule of λ-conversion, which might be stated as follows:

(12)   If $s(x)$ is a *functor* and t is a singular term, then '$[\lambda x.\ s(x)](t)$' is equivalent to $s(t)$; i.e., (as a matter of logic) '$[\lambda x.\ s(x)](t) = s(t)$' is true.

But stating this law in one of these metalinguistic forms, in which instead of talking of Fregean functions one talks of functors (and the other types of expression involved), serves the purpose. Because of the use of the truth predicate, it is not just a statement about language. ("Equivalent" could mean that one term can be substituted for the other, but as in other cases where one talks of logically permissible inferences, the understanding is

---

[22] Some of the axioms in the axiomatization of the notion of definite property that Zermelo attempts in "Über die Definitheit" are close to axioms assumed about classes in later work of Bernays and Gödel, for example the axioms of Group B in Gödel, *The Consistency of the Continuum Hypothesis*, p. 37.

In "What is Cantor's Continuum Problem?" Gödel refers to "property of sets" as "the second of the primitive terms of set theory" (1964 version, p. 260 n. 18). This suggests that something like Zermelo's original conception is not dead.

The concept of class in set theory is discussed in relation to the problem of generalizing predicate places in Essays 8 and 10 of *Mathematics in Philosophy*. In the latter essay, I consider the relation to intensional notions.

that truth will be preserved.) Such a metalinguistic form of statement of laws of logic is unavoidable; Frege himself needs it in order to state the rules of inference of his own system. According to Quine, it is the *only* way in which the generality of the laws of logic can be expressed.[23] We do not need to pass judgment on this. It is indispensable in logic, and we would naturally take many metalinguistic statements, such as that all statements of the form '$(\forall x)Fx \to Ft$' are true, as expressing logical knowledge, even though they are not logical truths according to the usual definitions.

The method of semantic ascent offers an alternative to the direct method for expressing the generality of the second-order principles of mathematics, as we have already seen in the case of the axiom of separation. This is of some importance for the axiom of induction in *number* theory, since we do not think of elementary number theory as involving commitment to properties, sets, classes, or Fregean concepts. The informal axioms have a generality that is not expressed by the statement, with respect to a particular formal language, that all instances of the schema in that language are true. The reason is that the informal schema need not be thought of as confined in its application to a particular, definite language. This is rather clear in the case of induction, where no one thinks of the validity of induction as limited to the language of, say, first-order number theory. If it were, then there would be nothing by virtue of which a nonstandard model of number theory is nonstandard. In practice, in any language in which we talk about natural numbers, we are prepared to affirm induction for *any* predicate of that language.

Stated in this way, the principle of induction has an inescapable vagueness. The same will hold for other informal second-order principles, so long as the method of semantic ascent is used to express them. We may of course express the principle of induction by the method of nominalization, as a generalization about sets, classes, or properties. This will in some contexts be more precise. But the same inescapable vagueness will appear again, as the application of such an axiom of induction will require minor premises to the effect that certain predicates have sets or classes as their extensions or that they express properties.[24]

The same inescapable vagueness arises for the laws of logic. The model-theoretic definition of validity of course allows a precise statement of a law of logic in the form of a statement to the effect that a certain schema such as '$(\forall x\, Fx) \to Ft$' is true under any interpretation, that is, for any

---

[23] *Philosophy of Logic*, pp. 11–12.
[24] Cf. Quine, *Philosophy of Logic*, pp. 53 ff.

§5. Is whatever is an object?

assignment of a set as domain for the quantifiers, a subset of the domain as extension for 'F', and an element of the domain as denotation for 't'. But the validity of the schema in this sense implies the truth of an actual statement of that form only on the assumptions that the manner in which the domain of the quantifiers is given makes it a set and that the predicate 'F' represents has a set as its extension. This pushes the question back to that concerning the principles of set theory, in particular the axiom of separation. The laws of logic have a certain dialectical character, in that the method of nominalization and the method of semantic ascent can both be used to state them, and neither can completely displace the other.

The systematic ambiguities which arise with respect to the notions of 'all sets' and 'all predicates (of any language we might come to formulate or understand)' have, however, no effect on the *extensional* characterization of the laws of classical first-order logic, provided we recognize the universal validity of the primitive rules of proof of some standard formalization. For then the completeness theorem informs us that any formula that is not provable has a countermodel of a very restricted sort, where the domain is the natural numbers and the sets which are the extensions of the predicates are arithmetical, in fact Δ_2. As G. Kreisel observed, this is true also if we envisage an "intuitive" conception of logical validity that takes in interpretations where the domain is not a set.[25] On common conceptions of discourse about proper classes in set theory, such a conception would involve the method of semantic ascent, and I would view that as the way to understand such a conception outside the specifically set-theoretic context. More needs to be said about the matter, but the present work is not the place to pursue it.

I should make clear that as I understand the method of semantic ascent, the objects talked about in semantic reflection are linguistic expressions. Properties do not come in by the back door in the formulation of general principles about the truth of sentences of a certain form. Where intensional entities may be needed in semantics is for handling explicitly intensional contexts, not for stating logical laws or second-order axioms for the extensional language of standard mathematics. But as our remarks about modality indicate, the kind of problems that give rise to intensions do arise in reflection on mathematics. In particular, they are involved in a treatment of proper classes in set theory that I have given elsewhere.[26]

---

[25] "Informal Rigour and Completeness Proofs," §2.
[26] *Mathematics in Philosophy*, Essay 10, sections III–IV.

These remarks should remind us that what has been given here is only the beginning of a treatment of generalization of predicate places. To begin with, just as Russell's paradox limits the method of nominalization, so the semantic paradoxes limit the method of semantic ascent. In the treatment of the concept of number, the second-order character of the principle of induction needs to be considered, as we will see especially in chapter 8. In the foundations of set theory, we need to look closely at all three methods, but especially the method of nominalization, since (in different contexts) in the practice of set theory both sets and classes serve as extensions of predicates. I have gone into some of these matters elsewhere.[27] I do wish to claim that the present discussion does show that considerations about predication do not lead inevitably to our taking second-order logic as our canonical framework and admitting, as values of our second-order variables, entities that are not objects.

In talking of the direct method and taking it to be captured by second-order logic, we may have given the impression that it already commits one to Frege's conclusion that what predicates designate cannot be objects. That is not quite true. The most that the appeal to Russell's paradox shows is that predicates of objects in a given domain cannot always have their reference *in that domain*, provided that the language can express the relation 'η' (Frege's 'falls under'). In the end, the force of the regress argument that we alluded to at the beginning of this section obviously depends on more about how one understands the relation of reference than we have gone into.

The syntax of second- and higher-order logic suggests a weaker view than Frege's, which many readers will find more plausible: that entities necessarily segregate into types such that expressions for entities of different types must be of different syntactic categories. If we admit the types to be different categories of *objects* (that is, subjects of predication to which identity applies), then we can carry out a "unification of universes" in which we suppress the differences of type and mark the domains of variables of different type by predicates. To be sure, some previously meaningless statements become meaningful, but their truth-values can be settled by stipulation in a way that does minimal violence to intuition. On this subject, I have little to add to the remarks of Quine.[28]

---

[27] In *Mathematics in Philosophy*. The parallelism of the difficulties posed by Russell's paradox and the liar paradox is a theme of Essay 9. On the methods of nominalization and semantic ascent in set theory, see also Essays 3, 8, and 10.

[28] For example *Ontological Relativity*, pp. 91–92.

§6. Being and existence

The reader will recall that the third point at which reservations about standard first-order logic as the universal measure of ontology can affect the notion of mathematical object is the ancient question whether reference to objects is necessarily reference to objects that *exist*. It is tempting to suppose that fiction, myth, and some propositional attitude constructions involve reference to objects that do not exist. 'Pegasus is a winged horse who was captured by Bellerophon' seems to be about Pegasus and Bellerophon, neither of which existed. We may not accept this as quite a genuine statement, at least not a true one. But then what of 'Pegasus is depicted in a statue above the entrance to the Columbia Law School' or 'Sherlock Holmes is more famous than any real detective, living or dead'?[29] And we can certainly say 'Pegasus did not exist; he is a mythical creature'. These statements, it will be said, involve reference to Pegasus and Sherlock Holmes, and are genuine truths.

This issue does not belong to the philosophy of mathematics, and we will not go into the merits of the claim that there is reference to nonexistent objects. I do want to point out how the meaning of 'object' and of the quantifiers might be affected. The most famous advocate of nonexistent objects, Alexius Meinong, held that *Sosein* is independent of *Sein*; that is, a singular proposition '*Fa*' can predicate '*F*' of an object *a*, perhaps truly, without imputing existence to *a*. 'The golden mountain is golden' says truly of an object, namely the golden mountain, that it is golden. 'The golden mountain does not exist' says truly of the same object that it does not exist.

According to Meinong, as thus represented, 'the golden mountain is golden' says of *an* object that it is golden; that is, Meinong is saying that some object is said here to be golden. Because that object does not exist, it seems that the quantifier-like expressions 'some object' and 'an object' do not imply existence, although we may regiment these locutions by the existential quantifier. Indeed, 'There *exists* an *x* such that *Fx*' would on this view be rendered 'Some *x* is such that *Fx* and *x* exists', so that existence is expressed by a *predicate*. Existence is not, as Quine would have it,[30] what the existential quantifier expresses. We will call the latter *being*. (The choice of this term does violence to Meinong; note that it is *Sein* (being) of which *Sosein* is said to be independent. There is, however,

---

[29] The second example is from Terence Parsons, *Nonexistent Objects*, p. 32.
[30] *Ontological Relativity*, p. 97.

no ready alternative, and we will adopt it provisionally and consider its difficulties later.)

This reading of Meinong's thesis is worth remarking on because it arises naturally in the construction and semantic interpretation of intensional logics.[31] Moreover, the distinction between being and existence is formally analogous to the distinction between existence and actuality, which arose in our comparison of Kant and Frege in §2. In fact, we might be tempted to turn on its head Quine's reason for holding that existence is what the existential quantifier expresses and conclude that this Meinongian *being* is what we were inclined to understand by existence, whereas 'existence' is a predicate that might better be called "real existence" or "reality." The idea that the concept of object has its home in formal logic is meant to imply that if we use the devices of singular terms, predication, identity, and quantifiers to make serious statements, then we are speaking about genuine objects. Nothing *more* can be demanded for the *existence* of objects than for the *truth* of statements of the form '$(\exists x)Fx$'. But if among our objects are the round square and Pegasus, we certainly do violence to ordinary language by taking the quantifier to express existence. It would follow that what the logical approach to ontology really envisaged was what we now call being.

It might be tempting to suppose that real existence is just Kant's and Frege's actuality and that what in mathematics is called existence is merely being. Matters are not so simple, however. The motives for admitting that some objects do not exist certainly apply in principle to mathematical objects. This is evident if we follow Meinong in thinking of empty definite descriptions as standing for nonexistent objects: the round square

---

[31] A straightforward semantics for modal quantificational logic involves a domain D of objects, which is independent of the choice of possible world. With respect to a given possible world an object in D may or may not exist. In what is called *fixed domain* semantics, the quantifiers range over D; to express existence, a predicate '**E**' needs to be added to the usual logical apparatus (whether or not '**E**' is counted as logical). Quantifiers in fixed domain semantics are often said to range over possible objects; but without some additional assumption there is no obstacle to allowing D to contain objects of which '**E**' is true in no world. A broader domain D is still relevant in the type of semantics (such as what, following R. H. Thomason, I call Q3) in which for each world, the quantifiers range over the objects existing in that world. Relative to a given world, singular terms may be interpreted to denote, and free variables may be assigned, objects that do not exist in that world.

The different semantic alternatives are presented briefly in Appendix 1 to Essay 11 of *Mathematics in Philosophy*. Thomason's "Modal Logic and Metaphysics" is still an instructive discussion of some general issues involved.

§6. Being and existence

is a nonexistent mathematical object; if there is such an object, then the sense in which there is cannot be mathematical existence. This kind of case, however, is the least persuasive for a Meinongian ontology. The other cases where it is argued that nonexistent objects are required are *de re* propositional attitudes and fiction. But on this view, surely fiction can speak of nonexistent mathematical objects, though of course the necessity of mathematics implies that the existence of such objects is *impossible*. Although fiction is doubtless constrained by considerations of coherence, it is certainly not limited to recounting the possible. Terence Parsons gives the example of a story in which a character named Atherton squares the circle.[32] But any reason to admit the squarer as an object applies with equal force to the construction by which he squares the circle. But no such construction can exist. Thus, if this conception of being were to be accepted, there would have to be a predicate of existence applicable to mathematical objects, which would differ from the predicate of actuality.[33]

As I have said, I do not intend to discuss the question whether we need to admit into the range of our quantifiers such objects as the golden mountain, the round square, Pegasus, and Sherlock Holmes. (My inclination is negative.) I do wish to point out some difficulties concerning the notion of existence which arise on such a view. A well-known one, arising from the treatment of existence as a predicate, is that it seems that such a predicate can be put into a description that designates an object; the round square is round, but it does not exist. Russell objected that 'the existent round square' should, on Meinong's grounds, stand for an existent object.[34] Meinong replied that the existent round square is indeed

---

[32] "The Methodology of Nonexistence," p. 657.
[33] However, this point does not rule out a conception of mathematical objects as nonexistent objects in the context of a Meinongian ontology. It might be held that the primary meaning of 'existence' is what we are calling actual existence, so that in the most basic sense, mathematical objects do not exist. The existence predicate of mathematical language would be different, perhaps more related to possibility. In fact, this seems to have been the view of Meinong himself; see "Über Gegenstandstheorie," §2 (*Gesamtausgabe*, II, 488).

The *incompleteness* of mathematical objects is a feature they share with the nonexistent objects postulated by Meinongian theories. Indeed, appreciation of the fact that incomplete objects seem to be forced on us by mathematics may lessen the resistance that is felt to Meinongian objects. By contrast, if the analogy is too much stressed, Meinongian objects may seem too much like abstract objects to serve the needs of the analysis of fiction.

[34] *Essays in Analysis*, p. 81.

existent but still does not exist.[35] Whether or not this makes sense, it seems to drive a wedge between the predicate of existence and the more proper sense of existence, which might be given by a quantifier.

We might try to formalize this interpretation of Meinong by means of a logic in which the quantifiers express being and there is a predicate, 'E', of existence.[36] The existent round square is $\iota x(\mathbf{E}x \wedge Rx \wedge Sx)$ (with 'Rx' for 'x is round' and 'Sx' for 'x is square'). But then if we admit that there is such an object, that is,

$$(\exists y)[y = \iota x(\mathbf{E}x \wedge Rx \wedge Sx)],$$

and that it is existent, that is,

$$\mathbf{E}[\iota x(\mathbf{E}x \wedge Rx \wedge Sx)],$$

then it follows that there exists an existent round square. Thus, Meinong's reply is not captured. He needs to make the distinction Russell could not see,[37] between existing, even as a predicate, and being existent. The predicate of existing must be disallowed in forming descriptions whose objects automatically satisfy the description. But a distinction between allowed and disallowed predicates can already be motivated by other considerations.[38]

---

[35] *Über die Stellung der Gegenstandstheorie im System der Wissenschaften*, p. 17 (*Gesamtausgabe*, V, 223).

[36] In its accommodation of singular terms not designating existing objects, such a logic resembles free logics, but in fact it could be just ordinary logic with the predicate 'E'. Typically in free logic, however, the quantifiers range over existents, and '$(\forall x)\mathbf{E}x$' is provable. The discussions of the relation of free logic to Meinongian ideas by Lambert (*Meinong and the Principle of Independence*, ch. 5) and Routley (*Exploring Meinong's Jungle and Beyond*, §§1.8, 1.14) make this assumption about what free logic is.

In the logic we have in mind, the truly "existential" quantifier is just the quantifier '$(\exists x)$' restricted to 'E'. The unrestricted quantifier might better, following Routley, be called the particular quantifier (see later). The old German terms *Seinszeichen* and *Seinsquantor* do not go naturally into English.

[37] "I must confess that I see no difference between existing and being existent; and beyond this I have no more to say on this hand" (*Essays in Analysis*, p. 93).

[38] Terence Parsons's neo-Meinongian theory of objects in *Nonexistent Objects* is based on a distinction, which can be traced back to Meinong, between "nuclear" and "extranuclear" properties. To any set of *nuclear* properties there corresponds an object having all those properties. Some restriction is needed to avoid paradox: if 'the square that is not a square' stands for an object *a* that is square and not square, '*a* is square' and '*a* is not square' would both be true, a contradiction. (Cf. *Nonexistent Objects*, p. 31. In "The Methodology of Nonexistence," p. 650, this is called the Paradox of Naive Object Theory.) Parsons avoids contradiction by concluding that *not* being square is not a nuclear property.

§6. Being and existence

Meinong had a deeper reason for resisting the interpretation of his theory of objects in terms of a being-existence distinction. He meant the independence of *Sosein* from *Sein* to be just that: that we can predicate things of the golden mountain does not imply that the golden mountain has being in *any* sense (except *Sosein* itself). So there really is no golden mountain; it isn't just that there is a golden mountain that fails to exist.

One might question whether this position would really help to analyze discourse about fiction, which is probably today the most plausible motivation for a theory of objects. But anyway, its coherence seems to me doubtful. Meinong needs to make general statements about objects: He professes to have a *theory* of objects. The language of such a theory will contain quantifiers. He may refuse to say that *there are* objects of a certain kind, except where such exist. But he wants to deny that all objects exist and gives such examples as the round square and fictional objects. But what is that but to say that *some* objects do not exist? The conclusion, "There are objects of which it is true to say that there are no such objects," was recognized by Meinong himself to be a paradox.[39] I think it is a *contradiction* unless 'there are' is admitted to be equivocal. One suggestion, not favored by Meinong's contemporary interpreters and apologists, is to regard the inner quantifier as objectual and as expressing being (no longer distinguished from existence in Quine's sense), whereas the outer quantifier is substitutional for a substitution class which includes what we would call singular terms lacking reference. This substitutional quantification might be relative to objects in the range of the objectual quantifiers.[40]

Our original suggestion resolves the paradox by taking the outer quantifier as the basic quantifier, the inner as really restricted to existing

---

Existence is also not a nuclear property (*Nonexistent Objects*, p. 25). So we can't assume that the existing round square exists. However, there is a "watered down version" of the extranuclear property of existence, which is nuclear and which might be called being existent (*ibid.*, pp. 43–44). If so, then the existent round square *is* existent but does not exist.

A similar distinction between "characterizing" and "noncharacterizing" predicates is used to meet this problem by Richard Routley; see *Exploring Meinong's Jungle and Beyond*, pp. 45–48, 86–91. In the latter passage, Routley considers the distinction between sentence and predicate negation, which complicates the picture: clearly one can admit the object *b* that is square and nonsquare, and conclude that *b* is square and *b* is nonsquare, if one rejects the inference from the latter to '¬(*b* is square)', even if one allows, as Meinong did, that nothing could *exist* that is both square and nonsquare.

[39] "Über Gegenstandstheorie," §3 (*Gesamtausgabe*, II, 490).
[40] In the sense of *Mathematics in Philosophy*, p. 214, also pp. 67–68.

objects.⁴¹ As an interpretation of Meinong's doctrine of *Aussersein*, it is certainly an anachronistic reinterpretation. The stress on quantifiers is not to be found in the text.⁴² In one crucial respect the interpretation makes Meinong's underlying ontological intuition not so different from our own (derived from Frege and Quine): In considering whether *there are* F's, we should not be guided too decisively by pictures derived from the case of actual concrete objects. I believe this was in fact part of Meinong's intent. But it clearly rejects Meinong's idea that nonexistent objects lack any kind of being. We can accommodate part of what the latter view apparently means, that not only do they not *exist*, but they do not have any analogue of existence such as the "subsistence" that Meinong attributed to universals and to facts.⁴³ But on our conception of objects and existence,

---

⁴¹ Parsons's version of Meinong follows this suggestion. He says that it is only a "terminological" question whether nonexistent objects have any kind of being (*Nonexistent Objects*, p. 10). The terminological issue would have to be whether 'there are Fs' implies 'Fs have being'.

⁴² Meinong says on p. 494 of "Über Gegenstandstheorie" that the paradox is *resolved* by the thesis that the object is by nature *ausserseiend*, in effect by the thesis of the independence of *Sein* from *Sosein* itself. Apparently this is to follow from the claim that "the whole opposition of being and non-being is primarily a matter of the Objective and not of the object" (p. 493). (An Objective is a state of affairs; see note 43 below.) I do not understand this, and I am not helped by Meinong's principal interpreters. It seems to say that the *being* of Fs is equivalent to the *truth* of 'there are Fs'. But then how is the implication avoided that the "objects of which it is true to say that there are no such objects" both have and lack being?

⁴³ We should mention another reason for admitting nonexistent objects which is certainly historically important. This might be called "fact metaphysics." If one holds that a true sentence stands for, or means, a *fact*, then for a false one there seems to be no fact for it to mean. Meinong held that a sentence designates an "Objective" (*Objektiv*) which "subsists" (*besteht*) if the sentence is true. Subsistence is the analogue for abstract objects of existence, which Meinong limited to the concrete, as Russell also did sometimes (e.g., *The Problems of Philosophy*, p. 100 of reprints). Thus a false sentence designates a nonsubsistent – we would say nonexistent – object.

One may be content to leave a false sentence denoting nothing and to take a locution like 'it is false that p' as simply another way of expressing the negation of 'p' and not as predicating something of a nonexistent "objective," state of affairs, or fact. But the matter becomes more serious when we consider propositional attitudes. To believe that p when it is false that p is still to believe *something*. If this something exists only when 'p' is true, then it seems to be something nonexistent. (Cf. Grossmann, "Meinong's Doctrine," p. 73.)

In spite of the central role of propositions in the logic of *Principia*, Russell objected to propositions on the ground that he could not see how there could *be* false propositions. For example:

> Time was when I thought there were propositions, but it does not seem to me very plausible that in addition to facts there are also these curious shadowy things going about such as "That today is Wednesday" when in fact it is Tuesday.... To suppose that in the actual world

§6. Being and existence

what is expressed by the quantifier 'there are' or 'some ...' is already an analogue of existence.

At this point it is instructive to consider the more uncompromising defense of Meinong's ontological intuitions by Richard Routley.[44] Routley discerns no difficulty in more comprehensive quantification than over existing objects, which he calls "neutral" quantification; although he advocates revisions of classical logic (particularly in the direction of relevance logic), he admits that standard first-order logic can have a neutral interpretation.[45] To be sure, he (reasonably enough from his point of view) thinks that even the notation '$(\exists x)$' for the "existential" quantifier is bound up with the reading of it as "there exists"; read neutrally, he prefers to call it the particular quantifier and to write 'Px'.[46] The English reading he prefers is "some" or "something"; "there are" he seems to regard as ambiguous between the neutral particular quantifier and the existential quantifier.

So far, nothing substantial distinguishes Routley's position from the version of Meinong we have emphasized. On that view, however, questions of reference and ontological commitment should be ambiguous: A term may lack reference to an existing object and still designate something, namely something nonexistent; we may be committed to F's, that is, to the conclusion that something is F, without being committed to, and even while denying, that F's *exist*. It seems to me that the logic-based notion of object implies that it is the broader rather than the narrower notions of reference and commitment that are the more fundamental ones.

That is emphatically not Routley's view. According to him, reference is to what exists; ontological commitment is commitment to existence. This is already suggested by his describing quantification encompassing

---

of nature there is a whole set of false propositions going about is to my mind monstrous ("The Philosophy of Logical Atomism," p. 223).

One might object that to suppose that there are such propositions is not to suppose that they are "shadowy things ... going about" or that they belong to the "actual world of nature." At all events, Russell sought to meet the difficulty by treating propositional attitudes not as relations to a proposition or fact but rather as relations to the *constituents* of what would be the proposition.

[44] I use the name under which the works under discussion here were published; as is well known, Routley later took the name Richard Sylvan. In "In memoriam: Richard Sylvan," Robert Meyer states that substantial parts of *Exploring Meinong's Jungle and Beyond* were written in collaboration with Valerie Routley, now Val Plumwood.

[45] *Exploring Meinong's Jungle and Beyond*, p. 81.

[46] *Ibid.*, p. 80.

nonexistent objects as "neutral." Routley follows Meinong in holding that mathematical objects do not exist, but then his position is more radical: He rejects Meinong's whole notion of subsistence and holds that mathematical objects and universals have no being at all, although he accepts theories that quantify over them.[47]

It is not easy to discern what issues other than terminological separate Routley's view from the version of the theory of objects outlined here. In spite of his strong criticism of what he calls the Reference Theory, his scheme of things allows a formally analogous relation between terms and objects that he calls aboutness, designation, or signification; and his objectual conception of "neutral" quantification preserves the conception of bound variables as ranging over a domain of objects, so that the question what that domain must contain for statements of a certain theory to be true will still arise.[48]

Where a more substantive issue emerges is with Routley's rejection of the classical conception of predication and the principle of substitutivity of identity that goes with it.[49] His main reason for this is that, with some other intensional logicians, he takes sentences of the form 'Fa' to be saying of a that it is F, even where 'F' is intensional and the sentence is not being read in a specifically *de re* way. In Routley's argumentation on this point, however, which consists mainly of criticism of alternative views about quantifying into intensional contexts and how to understand the well-known failures of substitutivity, he does not address the matter of ontological commitment or undertake to say why his conception of predication, if accepted over the classical one, should make possible "ontology-free" quantification over nonexistent objects. Evidently he thinks of the substitutivity of identity as bound up with what he calls the Reference Theory, but precisely on the point that now concerns us one has to distinguish two underlying ideas that form part of that "theory" as he understands it: the idea that (roughly) truth is a matter of reference, and the idea that reference is reference to what *exists*. It is the second that

---

[47] *Ibid.*, pp. 4, 851. Routley does not saddle his own position with the "fact metaphysics" referred to in note 43 and thus not with Meinong's identification of truth with the subsistence of "Objectives"; see particularly pp. 855–856. On his handling of "mathematical existence," see later.

[48] *Ibid.*, pp. 53, 81.

[49] *Ibid.*, pp. 96 ff. It seems clear that one can accept nonexistent objects and the independence of *Sosein* (true predication), at least from *existence*, without rejecting the classical conception of predication or the principle of substitutivity. Motives from intensional logic, however, can encourage both deviations from classical views, as Routley's writing illustrates. (He does not claim Meinong's authority on the second issue; see *ibid.*, p. 868.)

§6. Being and existence

bears most directly on ontological commitment, and the question I have raised is whether, once one entertains nonexistent objects, it is more than terminological legislation to continue to maintain that we have *reference* only where the object designated exists. It is the first idea that is bound up with the substitutivity of identity, and it continues to be so if one understands 'reference' as Routley's aboutness or designation. But so long as it is separable from the second, as it appears to be, it remains unclear why Routley's rejection of the latter, however well grounded, should make his views on ontological commitment themselves rest on more than terminological legislation.

One reason why Routley may believe otherwise is that he thinks of items as objects of *thought*. And of course it should be agreed that it should be possible to think of an object a, without either the thinker or someone reflecting on this situation being committed to there being such an object as a. But on every point except that of existence and being, Routley's conception of "items" is strongly realistic. They are not creatures of mind or language; his theory of them is not verificationist; they form a rich structure, at least as rich as the structures considered in mathematics, since items include the objects of mathematical theories. Of the latter objects, at least, Routley says that they "are in no way mind-dependent or tied to a thinking or perceiving subject or to human peculiarities or behavior or agreement."[50] Routley claims that the problems of platonism and mathematical existence disappear once it is conceded that the objects of mathematics do not exist. But in his own (admittedly sketchy) treatment of problems in the philosophy of mathematics, problems of the same kind reappear as problems about consistency and truth.

Where mathematical objects are concerned, however, Routley's view does differ from a weak Meinongian position, which holds that there are nonexistent objects, and that mathematical objects are among them, but otherwise (apart from the reading of the quantifiers) leaves logic and mathematics as it is. Consider, for example, the construction by which Atherton squared the circle, or the set of all objects that are not elements of themselves. Mathematics tells us that there are no such objects. The weak Meinongian position cannot stop with saying that they do not exist; it must be false that for some x, x is a construction that squares the circle, or x is a set containing all and only those objects that are not elements of themselves. It can avoid this conclusion by introducing a predicate to express what the mathematician means by existence, let us say (echoing

[50] *Ibid.*, p. 794.

Meinong and Russell) "subsists." Then what mathematics tells us is that the construction by which Atherton squared the circle does not *subsist*.[51]

Routley does not comment very explicitly on this weak Meinongian position, but implicit in his practice is a rejection of it. One reason may be that the introduction at this point of subsistence (however it is explained) would compromise the idea of nonexistent objects, unproblematic mathematical ones included, as altogether lacking being. If one is serious about the idea that the *ontological* status of mathematical objects is just that of objects of thought, no matter how fictitious or even inconsistently conceived, then there are no very clear grounds on which to reject Atherton's construction and the set of all non-self-membered objects while accepting the construction that proves the Pythagorean theorem and the set of all countable ordinals. This leads, however, into a revisionary attitude toward mathematics, which Routley in fact pursues in his interest in "paraconsistent" theories. With the objective of working out a position "which takes inconsistent sets as they come, as data, as objects of logical investigation, as objects which a satisfactory theory would let one talk about freely," he sketches a very ambitious program for revising the logical foundations of mathematics.[52] It is easy to see that such a program goes well beyond the weak Meinongian position. I do not wish to enter into discussion of radical revisions of existing mathematics, particularly when they are in large part merely projected. Routley's statement of his motivation, however, appears to run together the problem of formulating and clarifying existing mathematics (where any revision can be expected to have a mathematical motivation) and the problem of a logic for intentional discourse where mathematical objects are concerned. A revision prompted by the first problem can be expected to have a mathematical motivation. But it is the second that leads to allowing for the fact that we do talk of mathematical objects that do not exist. Without such a conflation, it is hard to see why a theory of mathematical objects should reject a notion that does

---

[51] Matters are more problematic in the Russell's paradox case, because '$(\exists y)(\forall x)[x \in y \leftrightarrow \neg(x \in x)]$' is *logically* false. There can be a set consisting of all *subsisting* objects that are not elements of themselves, but then it does not subsist. But there cannot be even a nonsubsisting set consisting of all objects, subsisting or not, that are not elements of themselves. The weak Meinongian can say that the property of being such a set is not a "nuclear" or "characterizing" property (see note 38). Dissatisfaction at this point is one factor driving Routley toward revision of logic.

[52] *Ibid.*, pp. 797–798. Paraconsistent logics allow contradictions to be true but by other restrictions, such as on implication, avoid the consequence that every formula becomes provable.

the work of that of mathematical existence, whether it is called existence, subsistence, or something else.[53]

If we decide in favor of the weak interpretation of the Meinongian view, with the concept of subsistence, we are still left with the question whether the "true" meaning of the existential quantifier is the permissive Meinongian one which we have called "being," existence that allows freely for abstract objects but that rules out *impossibilia*, or something like actuality. The logic-based concept of object does not decide between these alternatives, although, once it has been set forth, the case for the third is weakened. But in order to understand the notions of object and existence in mathematics we will have to put more flesh on the bare form given by formal logic. We need to fill out the logic-based conception by looking at cases. But as regards the three alternatives given above, it is mainly the choice between the second and third that will be affected. We have already indicated that considerations proper to mathematics will not lead us to favor the first over the second. General as the notion of object in mathematics is, there is still a constraint of possibility, coherence, or consistency that objects postulated in Meinongian theories are allowed to violate. But the case for such theories rests primarily on an analysis of intentionality, and particularly of discourse about fiction, which is not our concern.[54]

§7. Abstract objects and their concrete representations

We will close this chapter by making, provisionally, a distinction among abstract objects which is quite important for the philosophy of mathematics. Some abstract objects are distinguished by the fact that they have an intrinsic relation to the concrete; they are determined by their concrete embodiments. I shall call such objects *quasi-concrete*. Sense-qualities and shapes, among the objects prominent in traditional discussions of

---

[53] I may be attributing to Routley a more definite rejection of mathematical existence than he intends. In places he uses the term "consistent" with something like this meaning or at least extension. It is not clear to me whether Routley means consistency in this context to be a genuine property of objects (perhaps meaning something like "having consistent properties") or if an object is called consistent if it is talked about in the context of a consistent theory. In any event, the underlying idea is not so different from that of some earlier views about mathematical existence. For example, the idea that mathematical existence is to be understood in terms of consistency was current in the Hilbert school. See Bernays, "Mathematische Existenz und Widerspruchsfreiheit," who in the end understands it as a form of structuralism.

[54] But cf. the remarks about "fictionalism" with respect to mathematics in note 12 above.

universals, seem to count as quasi-concrete: They "occur" in the world as the qualities and shapes *of* whatever objects have them. This is true even on philosophical views according to which such qualities are ontologically more fundamental than, say, physical objects (or perhaps sensa) that might be said to instantiate them: On such a view, the most basic facts will still be of the occurrence of qualities at certain places and times or in certain relations to other qualities.

What makes an object quasi-concrete is that it is of a kind which goes with an intrinsic, concrete "representation," such that different objects of the kind in question are distinguishable by having different representations. The nature of the relation of "representation" will differ according to the kind of object. Thus if, for the moment, we adopt an empiricist picture according to which sense-qualities are qualities of sensa, that is immediate objects of sense, then the relation of representation is simply that of a sensum's having the quality in question. Two qualities are seen to differ by exhibiting a sensum that has one and not the other. In this case, we would expect that two qualities would be the same if just the same sensa have them, but perhaps we would need to allow the possibility that there are not enough actual sensa to make all distinctions among qualities. But even then we want to say that two qualities are the same if it is *impossible* that there should be a sensum having one and not the other.

An example that does not rest on empiricist philosophy and that will be important in the sequel is expression-types, whether of spoken or written natural language or of artificial symbolism. Consider written natural language, which has the advantage of being both familiar and straightforward. When we talk of words or sentences of a written language, we normally refer to types, that is, we treat as the identical expressions that are the same string of letters, spaces, and punctuations; and in determining the latter, we count a letter such as 'A' as the same in its many occurrences. Nonetheless, what a sentence *is*, is a matter of what physical inscriptions are or would be its tokens, and a sentence (if it is not impossibly long) can be pointed to by pointing to an inscription of it. Two expressions are the same if they have the same tokens, except in the case where neither has any tokens at all (that is, neither has been or will ever be written). To begin with, we can again say that two types are the same if, necessarily, any token of one is a token of the other. If one wishes to do so, the most obvious way of avoiding the modal operator is to identify the type with the sequence of its basic symbols; we can for example say that two types are the same if for every n, the nth symbol of the one is the

§7. Abstract objects and their concrete representations 35

same as the nth symbol of the other. (The symbols of the alphabet *can* be distinguished by actual tokens, as Quine has remarked.[55])

The most important examples of quasi-concrete objects are those in which the concrete representations are the sort of objects that can be perceived. However, the concept of a quasi-concrete object is not meant to be restricted in this way. Although sets are in general not quasi-concrete, it does seem that sets of concrete objects should count as such; here the relation of representation would be just membership. As I am understanding the notion, sets of physical objects that are inaccessible to observation would also count as quasi-concrete. But the empty set is already a doubtful case.

One might object, however, that an element of a set does not represent the set in the same way as a token represents its type or something concrete represents a sense-quality.[56] The reason is that one element can hardly represent the set as a whole. Take for example the set consisting of the prime ministers of all states of the European Union on, say, January 1, 1999. At most times its elements are widely distributed spatially; why should we accept one, say Gerhard Schröder, as representing it? Something like the mereological sum might be better chosen as representative, although for sufficiently complex sets one might question its concreteness. Although such a sum is unquestionably located, does it enter into causal relations? If one grants that it is concrete, that the same sum would represent different sets is not an objection, since the same is true of tokens and types. But I am not convinced by the objection; one can equally well take it to show that in this case the many-one character of the representation relation is essential to it.[57]

---

[55] *Philosophy of Logic*, p. 56.
[56] Such an objection was advanced by Christian Wenzel.
[57] This is an appropriate place to comment on the objection to our rough criterion of abstractness mentioned in note 1, the claim that sets of spatiotemporally located objects are located. (See Maddy, *Realism in Mathematics*, p. 59, earlier in "Perception and Mathematical Intuition," p. 179) Indeed, there seems no objection to saying that if the elements of a set $\{a_1 \ldots a_n\}$ (assuming for simplicity that it is finite) occupy the regions $R_1 \ldots R_n$ of space-time, then the set occupies the region that is the sum or union of these regions. It is not as natural a way of speaking as it appears, however, as one can see if one considers talk of such sets in a tensed language such as natural languages. A well-entrenched principle for talking of sets in modal contexts is that a set exists only if all its elements exist. (See *Mathematics in Philosophy*, p. 299.) If we think of sets as existing in time and apply this principle, we reach an anomalous result. Consider a set of objects no two of whose elements exist at the same time, say the set consisting of the Roman Empire, the Holy Roman Empire of the German Nation, and the German Empire proclaimed in 1871. If this set existed in, say, 1900, our principle would imply that the Roman Empire and

The kind of object with which quasi-concrete objects contrast most sharply is what I will call pure abstract objects. Pure abstract objects have no intrinsic concrete representation, and they are characterized not by conditions relating them to concrete objects of a specified kind but by conditions of a highly abstract character, involving objects in general. In this sense, it appears that the natural numbers are pure abstract objects, and also the *pure sets* that are the main object of study in set theory.[58]

It seems to be especially in mathematics that pure abstract objects arise.[59] The natural numbers are as elementary and indispensable an example of such objects as can be found. Their elementariness few would dispute. Whether they are genuinely indispensable we will consider in the next two chapters when we discuss some strategies for eliminating reference to mathematical objects. At this point, I shall simply explain what is meant by saying that they are pure abstract objects.

For a particular number, say the number five, what could be meant by an intrinsic concrete representation of it? A possible answer would be: Any configuration of five objects. However, something is concealed by the word "configuration." If by a configuration is really meant something concrete and spatial, then it does not determine the *kind* of object such that it consists of *five* of them. For example the inscription

Elizabeth has gone to school

is a configuration of five words, but of twenty-four letters (or of twenty-four letters and four spaces). This is a well-known point, much

---

the Holy Roman Empire existed at that time. We might avoid this result by saying that a set could change its elements over time; the same reasons supporting the rigidity of membership in modal contexts count against that. For example, our set, which intuitively has three elements, would at any time be identical either with a one-element set or (as now) with the empty set. (Cf. Sharvy, "Why a Class Can't Change its Members.") The conclusion we are left with is that regarding such sets as located requires the tenseless four-dimensional point of view. It is hard to argue that we *must* regard such sets as located, although the choice between rejecting such a conclusion and taking them as an exception to our criterion of abstractness is not determined by decisive considerations.

Temporal intuitions about sets seem to differ in this respect from modal ones. If a, b, and c are possible objects that are not compossible, then it seems quite reasonable to say that $\{a, b, c\}$ does not possibly exist, although each of its elements possibly exists.

[58] The phrase "pure abstract object" seems to have originated with Michael Dummett, who has the same examples in mind but gives a somewhat different explanation of the notion. See *Frege: Philosophy of Language*, p. 503.

[59] It may be that in ordinary thinking we do not find clear examples of pure abstract objects that are not mathematical. But we do find talk of such objects in philosophical theories. Perhaps the earliest clear reference to pure abstract objects is the discussion of the Greatest Forms in Plato's *Sophist*.

§7. Abstract objects and their concrete representations 37

emphasized by Frege in arguing, in the *Foundations of Arithmetic*, against the view that number is a property of physical things and in favor of his claim that a statement of (cardinal) number is a statement about a concept. There is no disputing Frege's claim that a physical configuration can only be treated as a configuration of a definite number of objects if a kind of object (i.e. a predicate) is given, and in a specific case there will be alternatives.

The plausibility of the suggestion that a configuration of five objects is a concrete representation of the number five comes from thinking of such a configuration, consciously or unconsciously, as a *set* or *sequence*. This does indeed avoid Frege's objection, since a set consists of determinate elements, and a sequence has determinate terms. But it seems clear that sets and sequences of concrete objects are not concrete but at best quasi-concrete. But then it might seem that numbers fail to be quasi-concrete only because of a certain second-order character; that is, we can think of a number as "represented" by a set of concrete objects and this set in turn by its elements, so that the concrete is reached in two steps rather than one. They might be called "quasi-quasi-concrete."[60]

In calling the numbers pure abstract objects, I mean to deny that this is the whole story. Here we should recall another point Frege uses in his polemical arguments, that numbers apply to objects in general, without any restriction. Thus we can number not only ships, stones, and pieces of sealing-wax, but also points, lines, numbers, and sets, and in addition gods, angels, or Platonic Forms. If the question how many angels can dance on the head of a pin is a silly or senseless one, it is not because *angel* is both a perfectly sensible concept and one such that it does not make sense to talk of this or that number of them. Where the concept of number does appear to cease to apply, the reason is that we are reaching or passing the limit of speaking of objects, at least in what I am taking as the primary sense. *Prima facie*, the form of words 'there are n Fs' is meaningful for any predicate or common noun phrase in place of 'F'. The schema already uses the plural formation, however, and this seems to rule out mass nouns. Other exceptions may arise where neither 'F' nor our understanding of the domain carries with it a principle of individuation. For example, the statement 'there are n things Reagan said in his speech' is at best very vague. We may attribute this to a difficulty of practice or principle in giving the identity conditions for things said. But in other cases, the lack of such conditions may be remedied by a contextual restriction either

---

[60] This term was proposed by an anonymous referee of this work.

on the range of the quantifier or on the intended interpretation of '*F*'. Thus, 'there are *n* red things' may make no sense if it is supposed to mean 'there are *n* red things in the universe', but it can of course be used in a context where a certain range of objects is in question and it is being said that *n* of them are red. Thus, although the applicability of 'there are *n F*s' presupposes individuation of the *F*s, it does not follow that '*F*' itself must be the sort of predicate that contains a principle of individuation, that is, a sortal. The fact that for a particular finite *n*, 'there are *n F*s' is equivalent to a formula of first-order logic with identity further reinforces the claim that number applies to objects in general.[61] But the question whether numbers are quasi-quasi-concrete deserves further discussion, which will take place in Chapter 6.

The concept of set has a similar generality, in that objects in general can be elements of sets.[62] In talking of sets, I will assume the iterative conception, according to which sets are obtained by (transfinitely) iterated application of the formation of a set from previously given elements, beginning with objects that are not sets (variously called individuals, *Urelemente*, or urelements).[63] It may be that for a set *no* individuals enter into its formation, that if one traces back from its elements, to the elements of its elements, and so on, every such chain ends with the empty set. Such a set is called a *pure set*.[64] It is evident that pure sets are pure abstract objects

---

[61] In these remarks, I have gone along with the thesis that the natural numbers are essentially *cardinals*, although I will question it in later chapters. The same considerations will emerge if one holds that they are essentially ordinals. On a more purely structuralist view, the possibility of interpreting the numbers so as to make them quasi-concrete or even quasi-quasi concrete is more remote.

[62] It is natural to say that absolutely *any* objects can be elements of sets. It may then be objected that proper classes are an exception. To this it might be replied that proper classes are not genuine objects. On this problem, see Essays 8 and 10 of *Mathematics in Philosophy*.

[63] This characterization is of course crude. Expositions of the iterative conception are to be found in a number of works on set theory, for example Shoenfield, "Axioms of Set Theory," Drake, *Set Theory*, ch. 1, or Krivine, *Introduction to Axiomatic Set Theory*, ch. 1. More philosophical discussions are Boolos, "The Iterative Conception" and "Iteration Again," Wang, *From Mathematics to Philosophy*, ch. 6, my *Mathematics in Philosophy*, Essay 10, and van Aken, "Axioms for the Set-Theoretic Hierarchy." Some of the issues about the iterative conception are discussed in Chapter 4.

[64] More precisely, we can define the transitive closure $TC(x)$ of a set x as the smallest set containing the elements of x, and such that the elements of every set belonging to it also belong to it. A set x is pure if $TC(x)$ contains no individuals.

It should be pointed out that writings on set theory by mathematicians characteristically assume a formulation that allows only for pure sets; even careful writers of introductory books will do this without any discussion.

§7. Abstract objects and their concrete representations

in our sense; otherwise, if some individuals are pure abstract objects, then a set formed ultimately only from such individuals will be a pure abstract object even if not a pure set.

It is easy to give examples of sets that, from what has been said so far, should count neither as quasi-concrete nor as pure abstract objects. A finite set some of whose elements are concrete and some of whose elements are pure sets would be an example, say the two-element set consisting of the von Neumann ordinal ω and my fountain pen; obviously such examples could be multiplied. It is reasonable to stipulate that a set is not a pure abstract object unless all its elements are, so that its transitive closure would have to contain only pure abstract objects (and so no concrete objects). There is no reason to count a set as quasi-concrete unless its elements are concrete. Thus the case of impure sets should make clear that the notions of pure and quasi-concrete abstract object do not provide an exhaustive classification of abstract objects, even the most basic mathematical objects. Of nonmathematical "universals," no doubt the most elementary are quasi-concrete, but those proposed in theories of morality and aesthetics, such as Plato's Justice and Beauty, are evidently neither quasi-concrete nor pure.[65]

---

[65] Our remarks about the null set might suggest that it is *both* quasi-concrete and pure. It's not obvious that this has to be ruled out, but I think the appearance results from two different ways of construing the null set. If we think of the sets whose elements come from a domain of concrete objects, then the null set is one of them; why should only it be excluded from being quasi-concrete? But the usual way of thinking of the null set is as the set that has no elements period, whatever the domain of quantification, perhaps taking it to be absolutely everything.

2 Eliminative Structuralism and Nominalism

§8. The structuralist view of mathematical objects

In order to focus on problems concerning mathematical objects, I turn now to what I call the "structuralist view" of them. By this I mean the view that reference to mathematical objects is always in the context of some background structure, and that the objects involved have no more by way of a "nature" than is given by the basic relations of the structure. The idea is succinctly expressed by Michael Resnik:

> In mathematics, I claim, we do not have objects with an "internal" composition arranged in structures, we have only structures. The objects of mathematics, that is, the entities which our mathematical constants and quantifiers denote, are structureless points or positions in structures. As positions in structures, they have no identity or features outside of a structure.[1]

Views of this kind can be traced back to the end of the nineteenth century, and the structuralist view in the sense I intend is an ontological conception that particularly fits the abstract mathematics that came into being in the late nineteenth century and has flourished in the twentieth. The statement that mathematics is about, or primarily about, structures is a frequently offered rough description of this character of modern mathematics. With reference to mathematics, the term 'structuralism' is often used to refer to no more than this rough description, or to a conception that expands on it in a different direction from what I have in view. One has

---

[1] "Mathematics as a Science of Patterns: Ontology and Reference," p. 530. This statement, as well as my own preceding it, could be read as meaning that the objects have no properties and relations other than those definable in terms of the basic relations. However, that reading ignores the fact that the objects stand in relations to other objects outside the structure in question. See, in particular, §14.

§8. The structuralist view of mathematical objects 41

in mathematics at least a pretty definite and agreed on concept of structure, so that the natural numbers, Euclidean and other spaces, groups, rings, and fields are paradigms of basic and relatively simple kinds of structure; and many more complex kinds arise in the development of mathematics. An example of a structuralist view of mathematics that is not, or at least not primarily, a structuralist view of mathematical objects is the view underlying the series of treatises of "N. Bourbaki," which covers a large part of modern mathematics. The emphasis of Bourbaki, and the organizing principle of the work, is that mathematics is built up starting with certain basic kinds of structure, for example, groups and topological spaces, which then combine and interrelate to give rise to the architecture of mathematics.

Such a view would suggest that structures themselves are the basic mathematical objects. As exemplified by the quotation from Resnik, however, the structuralist view of mathematical objects is the view that mathematical objects are "positions" *in* structures. Typically, structures are not taken as basic to mathematical ontology, and we shall see shortly that if one inquires what a structure is, one is led back to familiar kinds of mathematical objects. But the structuralist view is supposed to include such objects, in particular sets and functions, in its scope. Thus the statement that mathematics is about structures, or is the science of structure, falls short of offering the kind of view of mathematical objects that we are looking for. One fundamental branch of mathematics, category theory, might be read as taking structures as its basic objects (along with "arrows," maps between them). But that step by itself doesn't explicate what would be meant by the idea that reference to mathematical objects is always in the context of a structure, and that the properties and relations of such objects are only those that the structure in some way endows them with.

It would require some historical investigation to make clear how far mathematicians of the turn of the century had gone in formulating for themselves a structuralist view in the stronger sense that concerns me. The most developed and in some ways the clearest philosophical statements from before World War II are by Edmund Husserl, in explanations of "formal" mathematics and his conception of a "theory of manifolds."[2] In

---

[2] Although Husserl's clearest and fullest statement is in ch. 3 of the late *Formale und transzendentale Logik*, the essential ideas go back to the *Logische Untersuchungen* (see especially *Prolegomena*, §70). Husserl certainly would have repudiated structuralism as a complete account of mathematical objects, since certain basic objects, such as numbers, are according to him given in a more intuitive way. His views deserve a separate investigation.

1950, in his paper "Mathematische Existenz und Widerspruchsfreiheit," Paul Bernays articulated what could be called a structuralist view of mathematical objects in describing the existence of mathematical objects as typically "relative existence" (*bezogene Existenz*), that is as relative to a structure.[3] But one element is missing that belongs to that view as I conceive it: the idea just expressed that the properties and relations of these objects are only those that the structure in some way confers on them. But, at least for the case of natural numbers, that idea is clearly expressed by W. V. Quine somewhat later, in connection with his doctrine of "ontological relativity."[4] And it certainly underlies certain attempts to eliminate reference to mathematical objects, which will be discussed in this chapter and the next.

My aim will be to exhibit some difficulties that arise in stating the view more fully and to explore some directions in which we are led in dealing with them. As I have indicated elsewhere, I think that something close to the structuralist view is true.[5] It will turn out, however, that it does need some significant qualifications; in particular, some mathematical objects for which structuralism is not the whole truth must still have their place. I will consider in this chapter some other objections to the structuralist view. Moreover, in Chapter 4 we will consider a general objection to structuralism particularly in application to sets.

In spite of the problems I see in its precise statement, the structuralist view is familiar. I shall remind the reader of the main considerations but, because of their familiarity, not go deeply into them. Where it is most immediately persuasive is in the case of *pure* mathematical objects such as pure sets and numbers (in the broad sense, including the various number systems), just those mathematical objects that are pure abstract objects in the sense of §7. In these cases, we look in vain for anything else to identify them beyond basic relations of the structure to which they belong: for the natural numbers 0, S (successor), and perhaps arithmetic operations, for sets membership relations and perhaps whatever individuals might enter into their composition. A symptom of this is the problem of "multiple reductions," for a time much, probably excessively, discussed in the literature. The context in which it entered the philosophical literature is that of logicist or set-theoretic treatments of the natural numbers:

---

[3] For further discussion see my "Paul Bernays's Later Philosophy of Mathematics," §6.
[4] Quine is generally most explicit when speaking of the natural numbers. For a very explicit general statement, however, see *Ontological Relativity and Other Essays*, pp. 43–45.
[5] *Mathematics in Philosophy*, pp. 189–190, also pp. 20–22.

§9. The concept of structure

If one identification of the natural number sequence with a sequence of sets or "logical objects" is available, there are others such that there are no principled grounds on which to choose one.[6] The existence of multiple reductions was, however, long familiar from the "arithmetization of analysis" which gave rise to alternative constructions, on the basis of the natural numbers and set theory, of positive and negative integers, rational, real, and complex numbers.[7] The structuralist point of view is intended to dissolve this problem; I shall remind the reader how in §10.

Pure mathematical objects are to be contrasted not only with concrete but also with quasi-concrete objects in the sense of §7, such as geometric figures, or sets or sequences of concrete objects. A purely structuralist account does not seem appropriate for quasi-concrete objects, because the representation relation is something additional to intrastructural relations. Because they have a claim to be the most elementary mathematical objects, and also for other reasons, quasi-concrete objects are important in the foundations of mathematics. Their role leads, as we will see in §18, to the important qualification on the structuralist view alluded to above.

§9. The concept of structure

By a structure is usually meant a domain of objects together with certain functions and relations on the domain, satisfying certain given conditions. Paradigm examples of structures are the elementary structures considered in abstract algebra. For example, a *group* **G** consists of a (nonempty) domain G, together with a two-argument function on G, which we will write ∘, such that ∘ is associative, there is a (unique) element e that is an identity element for ∘ (i.e., for all $a \in G$, $a \circ e = e \circ a = a$), and every element a of G has a (unique) inverse with respect to ∘, that is, for every a there is a b such that $a \circ b = b \circ a = e$. One might regard the identity e along with ∘ as part of the structure.[8] (A distinguished element

---

[6] From Paul Benacerraf, "What Numbers Could Not Be" (1965) and my "Frege's Theory of Number" (1965), Essay 6 of *Mathematics in Philosophy*, esp. pp. 154–155.

[7] Philip Kitcher points out that multiple reductions also arise for the set-theoretic devices used in all these constructions. See "The Plight of the Platonist," p. 126.

[8] Often, however, one speaks loosely of the same structure when the basic objects and relations are not the same. For example, Euclidean geometry can be axiomatized with points, lines, and planes as basic objects (as in Hilbert's classic *Grundlagen der Geometrie*) or in a language in which points are the only individuals (as by Tarski in "What is Elementary Geometry?" and other work). There is a natural interpretation of one formulation in the other. Possibly there are reasonably precise notions of "same structure" that cover cases of this kind.

can be taken to be a function of 0 arguments.) The language in which we have set forth this simple example already refers to familiar kinds of mathematical objects: "domains," which one readily thinks of as sets, functions, and relations.

If we follow out this reading, then we arrive at the set-theoretic concept of structure, familiar, for example, from model theory, which provides a means of talking of structures as themselves mathematical objects, and offers the resources of set theory for reasoning about them. We take the domain of the structure as a set, and the functions and relations in the usual set-theoretic way. The structure itself is then a tuple, a set-theoretic object. For example, our group **G** will be the pair $\langle G, \circ \rangle$ or the triple $\langle G, \circ, e \rangle$. This way of talking about structures incorporates the dictum that mathematics is about structures into the familiar conception of set theory as the canonical language for all of mathematics, so that all mathematical objects can be construed as sets.[9]

Resnik's initial statement of his view seems to require cashing in in terms of the set-theoretic conception of structure or some variant of it. He talks freely of structures or "patterns" (his preferred term). It thus appears that *patterns*, or structures, are primary objects in his ontology, and one will naturally ask what these are. His statement of what a pattern is (*ibid.*, p. 532) is very close to the informal statement above of what a structure is. It is clear, however, that Resnik regards sets as on the same footing as other mathematical objects. It would take us somewhat afield to consider how Resnik reconciles the fact that in order to state his view, he has to talk of patterns as if they were a fundamental kind of object, while the view itself would make the patterns themselves "positions in patterns." We have, however, come up against the first difficulty in stating the structuralist view: It seems to require *structures*, and this concept seems to involve familiar kinds of mathematical objects, or perhaps to call for explication in "structuralist" terms that would threaten circularity. One answer to this difficulty is simply to use the set-theoretic conception of structure. This answer has an obvious difficulty: If we use it in order to give a structuralist account of some kind of mathematical object, then we will be assuming sets. The question will then arise how we are to give such an account of sets themselves. I shall, however, put that inadequacy aside for the moment and explore a version of the structuralist view that uses the set-theoretic conception of structure.

---

[9] Perhaps it is for this reason that Bourbaki relied on set theory as the axiomatic framework for mathematics in their work.

§10. Dedekind on the natural numbers

The set-theoretic conception of structure provides a very natural framework for giving accounts of particular kinds of mathematical object. We can illustrate it by the natural numbers, where an account of this kind was already given more than a century ago by Richard Dedekind in *Was sind und was sollen die Zahlen?* In this work Dedekind presents a development of arithmetic which is remarkably close to modern developments of arithmetic in set theory, although Dedekind's concept of *System* (set) is not axiomatized. His analysis of number, however, is presented as a characterization of the *structure* of the natural numbers.[10] This is given by the definition in para. 71 of a *simply infinite system*. A simply infinite system is a system (i.e., set) N such that there is a distinguished element 0 of N, and a mapping $S: N \to N - \{0\}$, which is one-one and onto, such that induction holds, that is:

(1) $(\forall M)\{[0 \in M \wedge (\forall x)(x \in M \to Sx \in M)] \to N \subset M\}$.

Let us abbreviate these conditions by $\Omega(N, 0, S)$. I will use Dedekind's term 'simply infinite system' for the *structure* $\langle N, 0, S \rangle$.[11] Dedekind has given an explicit definition of the kind of structure instanced by the natural numbers, although the presence of the quantifier $(\forall M)$ means that it is not first-order in N, 0, and S.

I want now to consider how Dedekind interprets talk of *the* natural numbers. So far, what we have been given is only a definition of a ω-sequence or progression. Although his intent is not entirely clear, his treatment of the natural numbers is meant to rest on two substantive claims: the claim that simply infinite systems exist (para. 72, resting on the argument of para. 66 that infinite sets exist), and the famous

---

[10] Thus the *aim* of Dedekind's analysis of number is quite different from that of Frege's *Grundlagen der Arithmetik*, where the main work is done by an analysis of the concept of cardinal number. The works are nonetheless often classified together, for good reasons: Frege's primitive notions of concept and extension are related to, though different from, Dedekind's primitive notion of system, and they rely on essentially the same analysis of mathematical induction, as Dedekind remarks in his preface to the second edition (p. x).

[11] Dedekind treats the natural numbers as beginning with 1; I follow contemporary usage in beginning with 0.

Note that Dedekind uses the notion of mapping, that is, function. He characterizes a mapping as a kind of *law* (para. 21). Once a reduction of functions to sets is available, however, multiple reductions arise again, as Kitcher pointed out (*loc. cit.*). Thus we have an additional reason for looking for a version of structuralism more thoroughgoing than the set-theoretic.

categoricity theorem (para. 132). What he actually proves in the latter section is, in our terminology, that any two simply infinite systems are isomorphic.

It is clear that Dedekind is not following the procedure, common in contemporary books developing the number systems in set theory, and which was followed by Zermelo, of presenting a simply infinite system and then identifying the natural numbers with that structure. The explanation he gives of talk about the natural numbers is somewhat awkward:

> If, in considering a simply infinite system N, ordered by a mapping φ, one abstracts from the specific nature of the elements, maintains only their distinguishability, and takes note only of the relations into which they are placed by the ordering mapping φ, then these elements are called *natural numbers* or *ordinal numbers* or simply *numbers*, and the initial element 1 is called the initial number (*Grundzahl*) of the *number series N*. (para. 73)

A reading of Dedekind that seems to me to accord reasonably well with this passage takes him as holding that statements about natural numbers are implicitly general, about *any* simply infinite system. A statement in the usual language of arithmetic will be expandable to one in which the arithmetical primitives are N, 0, and S, so that we can write it as $A(N, 0, S)$.[12] Let $\Omega(N, 0, S)$ be as above. When $A(N, 0, S)$ is taken as a statement about *the* natural numbers, we are to understand N, 0, and S as *variables*, and the statement is elliptical for:

(2)   For any N, 0, and S, if $\Omega(N, 0, S)$, then $A(N, 0, S)$.

The categoricity theorem implies that (2) holds if $A(N, 0, S)$ holds for a single simply infinite system $\langle N, 0, S \rangle$.

I will call this interpretation the *eliminative reading* of Dedekind. It clearly avoids singling out any one simply infinite system as the natural numbers and expresses the general conception I have in mind in speaking of the structuralist view. We shall soon see that it is probably not what Dedekind intended. But it is worth pursuing because it exemplifies a very natural response to the considerations on which a structuralist view is based, to see statements about a kind of mathematical objects as *general* statements about structures of a certain type and to look for a way of eliminating reference to mathematical objects of the kind in question by

---

[12] Eliminating functors introduced by recursion, such as addition, of course uses second-order means. But that is in accord with what Dedekind allows himself; in fact he was the first to show how to do this; see *Was sind?* §§9, 11–13.

§10. Dedekind on the natural numbers

means of this idea. The program is an instance of what I will call eliminative structuralism (see §11).

Dedekind's statement in the sentence after the above quotation, that one can call the natural numbers "a free creation of the human mind," raises doubts as to whether the eliminative reading is the best reconstruction of his intention. He explains this idea somewhat further in another text, a letter to Heinrich Weber of January 24, 1888, of which an extract was published in Dedekind's collected works. Weber had evidently suggested that the basic concept of number is that of cardinal; Dedekind argues for the priority of the ordinal concept, by which he means, I think, the structure of initial element and successor function. He then writes:

> If one wants to take your way... then I would advise not taking the number as the class (the system of all finite systems similar to one another) but rather as something *new*, corresponding to this class, which the mind *creates*. We are of a divine kind and possess, without any doubt, creative power not only in the material realm (railroads, telegraph) but most especially in the mental. This is just the same question of which you speak at the end of your letter in relation to my theory of irrationals, where you say that the irrational number is nothing at all other than the cut itself, while I prefer to create something new, distinct from the cut, which corresponds to the cut and of which I say that *it* brings forth the cut, generates it.[13]

What Dedekind proposes in the cases of cardinals and real numbers is well captured by W. W. Tait's notion of "Dedekind abstraction": Suppose one has described a structure of a certain similarity type, such as the finite von Neumann ordinals $\langle \omega, \varnothing, \lambda x\, (x \cup \{x\}) \rangle$, as a simply infinite system. Then one can introduce a new structure $\langle N, 0, S \rangle$ of that type, together with an isomorphism from the given one.[14] Clearly this does not square with the eliminative reading, although it is still structuralist in the general

---

[13] Dedekind, *Gesammelte mathematische Werke*, III, 489. I was led to look for this passage by W. W. Tait, who rightly questioned my earlier acceptance of the eliminative reading. In his published comment on the matter, Tait claims that "a new scandal will be created by taking Dedekind to be a structuralist" ("Critical Notice," pp. 590–591). He is right if he means (as I did in the place he comments on) the eliminative reading of Dedekind or some related eliminative structuralist interpretation. But the interpretation of Dedekind he himself recommends makes Dedekind's view of the number systems structuralist in the more general sense of §8; it would, of course, not be reasonable to attribute to Dedekind a structuralist view of sets and functions and thus a structuralist view of mathematical objects as such.

[14] Tait, "Truth and Proof," p. 87 n. 17 (n. 12 in original). I shall not pursue the question what meaning the term "abstraction" had for Dedekind and his contemporaries. At first sight, at least, Tait's conception is a reconstruction of Dedekind, not a direct interpretation.

sense with which we began. But I shall continue to explore the eliminative reading.

The eliminative reading avoids a difficulty faced by one crude attempt to deal with the "multiple reduction" problem, namely, saying that *any* simply infinite system can be, or can be taken to be, the natural numbers. Some such choices seem absurd, outside very special contexts; for example, suppose one has the natural numbers N and then takes as one's "natural numbers" the even numbers of N, with the appropriate successor function.[15] Again, suppose we have a certain simply infinite system $\langle N', 0, S' \rangle$, and suppose we take as "the" natural numbers the system $\langle N, 0, S \rangle$ obtained by replacing the 17th element of N' by Richard Nixon, and adjusting S' accordingly.[16] On this choice, the sentences 'Richard Nixon is a natural number' and 'Richard Nixon = 17' become true, which seems absurd. But it is clear that on the Dedekindian account neither can be true.[17] Another possible objection, that the *application* of arithmetic requires relations of numbers that are not internal to the structure of the numbers, will be considered in §14.

I want to consider another difficulty Dedekind's analysis faces. Dedekind thought it necessary to prove that simply infinite systems *exist* (paras. 66, 72). This seems intuitively necessary; otherwise it seems that one is not entitled to say (informally) that the natural numbers exist. A more troubling fact is that on the eliminative reading, if there are no simply infinite systems, then for any N, 0, S the statement (2) giving the "canonical form" of an arithmetic statement A is vacuously true. But then both A and $\neg A$ have true canonical forms, which amounts to the inconsistency of arithmetic. A version of this difficulty

---

[15] Cf. Quine, *Ontological Relativity*, p. 45.
[16] I.e., for m, n in N, $n = Sm$ iff $m = S'^{(16)}0'$ and $n =$ Richard Nixon, $m =$ Richard Nixon and $n = S'^{(18)}0'$, or $n, m \neq$ Nixon and $n = S'm$.
[17] This is not to say that they are false: One may wish to call a statement false only if the canonical form (2) of its negation is true; the example in the text shows that this does not hold for either sentence. It would be tempting to call such statements neither true nor false. I am indebted to Martin Stokhof for pointing out that clarification was needed here.

One might be worried because the sentence 'Richard Nixon = 17 ∨ Richard Nixon ≠ 17' is true, and therefore one or the other disjunct must be true. This is more harmless than it seems, since we are to think of '17' as containing implicit parameters for an unspecified simply infinite system. For each value of these parameters, one disjunct is true, but we can only *assert* what is true generally of simply infinite systems. The word 'true' is actually ambiguous in this context, since it might either presuppose particular values for the parameters, or mean "generally true," as it does in number-theoretic assertions and in the remark in the text.

§10. Dedekind on the natural numbers

is faced by eliminative structuralist accounts of natural number generally.

Dedekind sought to meet it by his famous, or notorious, argument for the existence of infinite systems (sets) in para. 66. This appeals to a kind of transcendental psychology. He argues that "the totality of things, which can be objects of my thought" is infinite; for given such an object s, we can let $S(s)$ be the thought that s can be an object of my thought, and this will be a new object of my thought. S is then a one-one mapping of the potential objects of my thought into themselves, whose range does not include objects of my thought that are not themselves thoughts. Thus, this totality is infinite.

With our more critical attitude toward set theory, we would question the treatment of the totality of objects of my thought as a set,[18] and the intrusion of nonmathematical concepts into Dedekind's argument would also give rise to objections. To translate Dedekind's development of arithmetic into a modern set theory like ZF, we would need the axiom of infinity, that is, the existence of an infinite set would simply be assumed. The question of how much better it is possible to do than Dedekind will then arise when we consider intuitive justifications of set-theoretic axioms. However, the problem of nonvacuity, roughly that of the existence of an instance of the structure, is a recurring one in structuralist treatments of objects of different kinds, and will continue to occupy us.

Dedekind's analysis deals neatly with the problem of multiple reductions, since no one simply infinite system is identified with the natural numbers. It appears to do so at the price of set-theoretic economy: The conventional development of arithmetic in set theory, for example by taking as the natural numbers the finite von Neumann ordinals, does not require the axiom of infinity. The difference is not quite so great as it appears: Dedekind did not make the distinction between sets and classes,

---

[18] Cantor, in his famous letter to Dedekind of July 28, 1899 (*Gesammelte Abhandlungen*, pp. 443–447), already mentions the *Inbegriff alles Denkbaren* as an example of an inconsistent multiplicity (therefore not a set) (p. 443). He does not explicitly relate this remark to Dedekind's argument. But the implied objection had already been communicated to Dedekind by Felix Bernstein. Bernstein's note on the matter appears in Dedekind, *Gesammelte mathematische Werke*, iii, 449, translated in Ewald, *From Kant to Hilbert*, II, 836.

Zermelo is quite explicit in criticizing Dedekind on this point, observing that the *Inbegriff alles Denkbaren* cannot be a set according to his own axioms, in "Untersuchungen über die Grundlagen der Mengenlehre I," p. 266 n. (trans. p. 204 n.). Fraenkel says that Dedekind recognized the justice of this criticism after the paradoxes had become known (*Einleitung in die Mengenlehre*, 2nd ed., p. 216).

and in fact his argument does not require that there be a simply infinite system whose domain is a *set*. Thus what is needed is just quantification over classes; however, in the absence of the axiom of infinity, common applications of induction and recursion require impredicatively defined classes.

The comparison of Dedekind's with the now conventional approach makes clearer the significance of set-theoretic constructions of number systems. These constructions should not be viewed as too literally *identifying* a certain system of objects as the numbers in question, but as proving that the concept of the structure in question is not vacuous.[19] If a set theory is assumed, this may not have much significance when the structure in question is the natural numbers; it is more significant when we are dealing with the positive and negative integers and the rationals, assuming only the natural numbers, or with the complex numbers, assuming only the reals (or natural numbers and sets thereof). Then, if we talk of numbers of these kinds as if they were objects *sui generis*, the construction insures that we avoid the problem of vacuity that threatened Dedekind's account of natural numbers.

A problem with which we began our exploration of structuralism was that the statement of the view seemed to require reference to *structures*, and we were then faced with the problem what kind of objects these are. One lesson we might draw is that we should distinguish what is required for a structuralist account of a *particular* kind of mathematical object, such as the natural numbers, and what is required to give a *general* statement of the structuralist view. It is the latter which at the outset required reference to structures. That the former requires such reference is not evident. Our Dedekindian account might be found wanting on the grounds that it builds general reference to structures into its canonical form for arithmetical statements. In our further discussion of eliminative structuralism, we will consider whether it is possible to avoid this result.

§11. Eliminative structuralism and logicism

I have described a certain reading of Dedekind's analysis as eliminative, in that it translates statements referring to or quantifying over numbers into statements in which such reference is absent. Although the analysis still assumes sets and functions, more recent developments of the same

---

[19] Earlier, I have appropriated Kant's term "objective reality," which means roughly real possibility. But that can have misleading associations, as Douglas Marshall pointed out.

§11. Eliminative structuralism and logicism

idea have sought to avoid such assumptions. By *eliminative structuralism* I mean a view or program that begins with the basic idea of the structuralist view of mathematical objects and develops it into an analysis in which reference or commitment to such objects, or to mathematical objects of a specific kind such as natural numbers, is claimed to be eliminated. Whether it is really eliminated can depend on questions of ontological commitment.

The underlying idea, although not yet the explicit program, is expressed by Paul Benacerraf when he concludes from structuralist intuitions that numbers are not objects and writes, "Number theory is the elaboration of the properties of all structures of the order type of the numbers."[20] The eliminative structuralist program has attracted a number of philosophers, and its tendency to be revived after attempts run into serious difficulties shows that the ontological intuition behind it exerts a powerful attraction.

Eliminative structuralism is an important type of structuralism, which has figured in classifications of structuralist views by several writers. We will contrast it simply with noneliminative structuralism, that is, structuralist views that do not claim to eliminate mathematical objects but hold that they are in some way only determined by the basic relations of a structure that is their home. Eliminative structuralism is what Dummett calls "hardheaded structuralism."[21] Rather tendentiously, Dummett contrasts it with "mystical" structuralism; because these are said to be the two types of structuralism, he seems to mean by mystical structuralism just what we mean by noneliminative structuralism. However, he attributes to mystical structuralism the thesis that the structures involved are free creations of the human mind, after Dedekind. The structuralist view of mathematical objects should surely be independent of any such thesis, however Dedekind is interpreted.

Stewart Shapiro uses the traditional distinction of *in re* and *ante rem* theories of universals to classify structuralisms. Eliminative structuralism

---

[20] "What Numbers Could Not Be," p. 291. This and some other remarks in Benacerraf's paper intimate an eliminative structuralist program, but he does not commit himself to it or indicate any way other than the set-theoretic of carrying it out. Note that the statement quoted quantifies over structures. On the same page, he talks of "systems of objects" and "relations." By contrast, it appears that he does want to avoid Dedekind's position of exempting sets from his structuralism; note his reference (p. 290) to Takeuti's reduction of Gödel-Bernays set theory to a theory of ordinals (see Takeuti, "Construction of the Set Theory").

[21] *Frege: Philosophy of Mathematics*, p. 296.

he regards as exemplifying an *in re* approach to structures; the idea is that there is "no more to structures than the systems that exemplify them."[22] The alternative is what Shapiro describes as *ante rem* structuralism. In part, this is characterized as noneliminative structuralism: Objects in mathematics so understood are "as bona fide as any objects are."[23] But something additional is involved, which is captured by the thesis that "structures exist whether they are exemplified in the nonstructural realm or not."[24] This again seems to me effectively to saddle noneliminative structuralism with a thesis that should be independent of it. As we noted at the end of the last section, it is not evident that a structuralist account of a particular kind of mathematical object should require reference to structures at all. And for a general statement of the structuralist view, different options about what structures are should be considered.[25]

So far, the only structuralist view we have formulated is the one extracted, with some violence, from Dedekind. Because of its assumption of sets and functions, however, we have so far not even stated a method of eliminating mathematical objects generally, whether or not we adopt a structuralist view of sets (and related objects such as classes and functions). Taken simply as an account of arithmetic, however, our Dedekindian analysis can be readily reformulated so that the role of set theory is taken over by *second-order logic*. We noted that the domains of simply infinite systems, although they need to be within the range of the variables, do not need to be sets. The same is true of other higher-order entities quantified over in Dedekind's account. His definition of a simply infinite system can be translated directly into second-order logic; this is simply a consequence of the fact that the structure of the natural numbers is second-order definable. The induction clause (1) is replaced by

(3) $(\forall M) \{[M 0 \wedge (\forall x) (Mx \to M(Sx))] \to (\forall x) (Nx \to Mx)\}$.

We write the new definition as $\Omega'(N, 0, S)$. Now suppose that A is provable in secondorder arithmetic. In second-order arithmetic, functions definable by the usual kinds of recursion (in particular, primitive recursion)

---

[22] *Philosophy of Mathematics*, p. 85.
[23] Ibid., p. 89.
[24] Ibid. Shapiro gives a similar explanation of the same classification in "Space, Number, and Structure," pp. 149–150. Bob Hale's distinction between "abstract" and "pure" structuralism is essentially the same as Shapiro's; see "Structuralism's Unpaid Epistemological Debts," p. 125.
[25] On the classification of structuralist views, see also "Structuralism and Metaphysics," pp. 57–59.

§11. Eliminative structuralism and logicism

are explicitly definable. We will suppose that in A expressions for such functions have been eliminated, so that A contains only 0, S, and logical expressions. If $A(N, 0, S)$ is the result of relativizing the quantifiers of A to N, then clearly the obvious reformulation of (2):

(4)   For any N, 0, and S, if $\Omega'(N, 0, S)$, then $A(N, 0, S)$,

is provable in pure second-order logic with identity (including comprehension). Moreover, if A is merely a truth in the language of second-order arithmetic, $A(N, 0, S)$ will hold in all standard models of $\Omega'(N, 0, S)$.

This simple translation of the language of arithmetic into that of second-order logic has been offered as a basis for a defense of the view that arithmetic is a part of logic. This form of logicism is a variety known as "if-thenism" or "deductivism." We will consider shortly a well-known defense of this view offered some years ago by Hilary Putnam,[26] which, however, used first-order formulations. Harold Hodes has vigorously defended the second-order form presented above, following Frege in interpreting second-order variables as ranging over functions and concepts.[27] Such a use of Frege's theory of functions gives this sort of if-thenism a claim to eliminate mathematical objects. A virtue of the second-order language, however, is the variety of interpretations to which it is susceptible; hence there may be other ways of using the above translation, or modifications of it that are still second-order, in an eliminative program. But, as Putnam's example illustrates, first-order versions have been pursued. Because they have tended to proceed by trying to approximate the second-order version, we will concentrate on the latter.

This version of logicism does not escape all traditional objections, for example to handling induction by means of a definition.[28] Nonetheless, it is worth some further exploration, if for no other reason because of the general applicability of the strategy. For example, two- or three-dimensional Euclidean geometry also has a categorical second-order axiomatization; we can then interpret Euclidean geometry along the same lines. We can also, it seems, meet the demand for a treatment of set theory, by formulating the axioms of separation and replacement as single statements in a second-order language, so that ZF becomes finitely axiomatizable.[29] This case illustrates that the second-order variables are

---

[26] "The Thesis that Mathematics is Logic."
[27] "Logicism and the Ontological Commitments of Arithmetic."
[28] I discuss this in *Mathematics in Philosophy*, pp. 167–170, 173–175.
[29] Relative, of course, to the logic: The full comprehension schema, which is here counted as logical, can of course not be replaced by a finite list of axioms, and even with it

doing the work of the set-theoretic notion of class rather than that of set.

This logicist eliminative program faces the same problem of the possible emptiness of the generalizations (4) as Dedekind's original formulation: if, no matter how N, 0, and S are interpreted, $\Omega'(N, 0, S)$ is false, then (4) is vacuously true, and is equally true if we replace A by $\neg A$. We can, as Putnam does, view the matter in terms of provability: If we can prove both (4) and its counterpart for $\neg A$, then $\Omega'(N, 0, S)$ will have proved inconsistent, and vice versa. The problem also arises on versions of the strategy that replace $\Omega'(N, 0, S)$ by a first-order statement.

What reason have we, on if-thenist grounds, to believe that things will not collapse in this way? We can in some cases appeal to another theory, within which it is possible to construct a model of the theory at hand or otherwise prove its consistency. Our proposal bears some resemblance to Frege's reinterpretation of Hilbert's idea that the axioms of geometry "define" the concepts of geometry.[30] Hilbert had made copious use of arithmetical and analytical models to establish consistency and independence.[31] But for the most elementary mathematical structures, something more than reference to another *mathematical* theory is needed to convince ourselves of the existence of structures of the relevant type, or even that their theories are consistent. This is particularly true of the natural numbers. Recall that this is just the point at which Dedekind found it necessary to appeal to transcendental psychology. We may not be able to avoid going outside the strictly mathematical, at least as understood by eliminative structuralism, in order to convince ourselves of something so basic.

At the time he defended if-thenism, Putnam was quite aware of this difficulty.[32] He replied that we do not need to assume the actual existence of structures satisfying the conditions in which we are interested, but only

---

the axiomatization of logic remains incomplete for the standard semantics, for which validity is not recursively axiomatizable.

[30] See Frege's correspondence with Hilbert and Korselt, in *Wissenschaftlicher Briefwechsel* and his two series of articles "Über die Grundlagen der Geometrie," in *Kleine Schriften*. This material is collected in translation in *Frege on the Foundations of Geometry and Formal Theories in Arithmetic*. It is instructively discussed by Resnik in *Frege and the Philosophy of Mathematics*, ch. 3.

[31] David Hilbert, *Grundlagen der Geometrie*.

[32] "The Thesis that Mathematics is Logic," pp. 26–33. For reasons that do not concern us, he saw the problem as arising in connection with the application of mathematics.

§11. Eliminative structuralism and logicism 55

that they *could* exist (*ibid.*, pp. 32–33).[33] This would presumably preclude the logical validity, in any reasonable sense, of the canonical versions, in the manner of (4), of both a statement and its negation. The formulation suggests the use of modality in explicating mathematical existence: What would be presupposed (in the case of arithmetic) is that it is *possible* that there are N, 0, S such that $\Omega'(N, 0, S)$ hold. This statement, however, is not interpreted in if-thenist terms, although it does not necessarily go outside the framework of eliminative structuralism more broadly conceived. Putnam made a version of this move shortly afterward,[34] as have many others since.

In the earlier paper, Putnam seems uncertain about the meaning of the modal term and says that whatever it means it implies syntactic consistency, which he thinks sufficient for the use he is making of the mathematical theory (in his case a version of the simple theory of types). One might question the adequacy of this in view of the deductive incompleteness of whatever axiomatization is involved.[35] To this it can be replied that we do not have better assurance of the "existence" of such a structure as the natural numbers than such consistency statements would

---

[33] Putnam admits that relying on our "intuitive conviction that certain infinite structures *could* exist" implies that mathematics is not fully reducible to logic (p. 41).

[34] "Mathematics without Foundations."

[35] On the second-order version of if-thenism that we have emphasized, this incompleteness resides in the logic; on Putnam's own first-order version, it resides in the axioms whose schematized versions form the antecedent of the conditional that is the canonical form of a mathematical statement.

Positions closely related to what we call if-thenism are instructively discussed in Rosemarie Rheinwald, *Der Formalismus und seine Grenzen*, Chapter II, under the name *Implikationismus*. The distinction she makes between the syntactic and semantic version there recalls ours between the "standpoint of provability" and that of semantic validity. As her name for the position suggests, however, there is an important difference between her interpretation and ours. She interprets a mathematical statement, made in the context of an axiomatic theory, as *expressing* logical implication of the "naive" statement by the axioms; it is for that reason that she says that if the mathematical statement is taken as an if-then statement, "if ... then ..." cannot be the material conditional (p. 49). Then the question arises whether by "implication" is meant logical derivability, semantic consequence, or something else; hence her distinction of syntactic and semantic implicationism.

In my view, Rheinwald's "implicationism" introduces an element of reflection into the interpretation of mathematical statements which is not essential to the idea of if-thenism or of eliminative structuralism. In the canonical forms we consider, such as (1), "if ... then ..." *is* the material conditional; the thesis is then (for example), that the *truth* of the arithmetical statement *A* consists in the *logical validity* of (1). Of course the syntactic/semantic distinction then arises for the interpretation of logical validity. But the reflection belongs not to the content of individual mathematical statements but to the account of that content.

give us. But for the if-thenist view, a problem arises about the interpretation of the consistency statement. A theory T (in this context, we can take Ω' or another antecedent as a theory) is consistent if there does not *exist* a logical proof of a contradiction from its axioms. Ordinarily we take this as a mathematical statement; a proof is an array of symbols satisfying certain conditions; these symbols are types and therefore abstract objects. This reading is at first sight outside the if-thenist framework; but note that symbols and proofs are not obviously pure mathematical objects and are plausibly quasi-concrete. It would, of course, be possible to regard the syntactical objects as themselves merely a structure, but then the appeal to consistency would be simply another form of reduction to another mathematical theory, or the consistency statement is actually being used in an *applied* context. What counts is what we can expect when we actually construct proofs. One way of taking the consistency statement, which bypasses both the structuralist understanding of syntax and the interpretation of it as about quasi-concrete but abstract symbol-types, is to take it as a statement of nominalistic syntax, where the objects are physical inscriptions. This offers a way of reconciling usual ways of understanding mathematical theories with nominalism: We interpret the theories in the if-thenist way, but deal with the problem of possibility by appealing to consistency, nominalistically interpreted.[36] This suggestion offers us the opportunity to discuss some issues concerning nominalism.

§12. Nominalism

I shall understand the term 'nominalism' in the sense most usual in contemporary philosophy of logic and mathematics, as the rejection (perhaps programmatic) of all abstract entities. It is this demand that would lead to the use of an interpretation of syntactical statements as referring to physical inscriptions. I shall take nominalism, moreover, to involve extensionalism, so that we are not allowed to use a modal language in talking about such physical objects as inscriptions. The view that allows modality but is otherwise nominalistic, which can be called modal nominalism, will be considered in Chapter 3.

---

[36] In "The Thesis that Mathematics is Logic," Putnam does not focus on nominalism and does not make this suggestion. The criticism of it made below is not meant to be a criticism of Putnam, even of his views of 1963 (evidently when the paper was written; see p. 17).

§12. Nominalism

It appears that in the eliminative structuralist program, the problem of the possibility of the natural numbers would have a simpler solution than Putnam's appeal to consistency: One would just describe (by a predicate) a domain of physical objects, a zero object, and a successor relation that would satisfy Ω'. The simultaneous refusal of abstract objects and modality means, however, that the quantifiers range over *actual* objects.[37] If we can give such a physical model of the natural numbers, then the physical universe is in some way infinite. Although recent nominalism has tended to be physicalist, the issue is the same if one seeks the model in some domain of mental objects, such as sensations, ideas, or thoughts.[38] Now, that the physical world is in some respect infinite is no doubt true. I am not sure whether according to current physical theory there are infinitely many physical particles or other objects that can claim to be "physical objects"; however, physical theories are based on a geometry in which space (or space-time) is infinitely divisible (whether or not it is infinite in extent). But should it be taken as a presupposition of elementary mathematics that the real world instantiates a mathematical conception of the infinite? This would have the consequence that mathematics is hostage to the possible future development of physics. Can we rule out the possibility that physics will abandon infinitely divisible space-time and replace it by some "quantized" conception? If not, and if this were to happen, then even accepting Hartry Field's claim that points and regions of space-time are physical and acceptable to the nominalist would not save the infinite in the physical world.[39] Would we be obliged to abandon the *mathematics* of the infinite, even the infinity of the natural numbers?[40] A great deal of the historically given mathematics would have to be jettisoned in this case.

---

[37] The second-order character of Ω' seems to pose a problem. This will be discussed later in this section and in §13.

[38] Note that Dedekind's model in para. 66 is not exactly of this kind, since he speaks of *objects* of his thought; moreover, the thoughts themselves seem to be not mental occurrences but rather thoughts in something like Frege's sense. (In fact Frege interpreted Dedekind in this way; see "Logik" (1897), in *Nachgelassene Schriften*, p. 147 n.)

[39] *Science without Numbers*, p. 31. Although Field is not an eliminative structuralist, in this work and subsequent papers (which importantly modify his position), he makes a number of moves that an eliminative structuralist might make, and some of the issues raised here arise for him. An adequate discussion of his views, however, would take us too far afield, and there is already a considerable critical literature on his work by others. We will discuss in §13 only one aspect of his position, his conception of second-order logic.

[40] Observe that the simple (first-order) statement that S maps N one-one into $N - \{0\}$ already has only infinite models.

The problem with the appeal to a physical model to deal with the problem of the possibility of a structure is that it makes mathematics presuppose an hypothesis that is stronger and more specific than needed, as I have argued elsewhere.[41] Any particular such hypothesis would be vulnerable to a refutation which would not upset the mathematics that is claimed to presuppose it. The nominalist might call attention to the fact that the world as we now understand it contains many different models of infinity, and then take the holistic position that no one of them is required for the infinity of the natural numbers.[42] In view of the dependence of all infinities in the real world on infinities in space and time, it is not clear that this position is really an advance over the Fieldian one just proposed; at any rate, it seems conceivable that *all* the hypotheses about infinities in the real world might be abandoned.

Let us now return to the suggestion mentioned at the end of §11, that the possible truth of the antecedent of the if-thenist's conditionals need amount to no more than formal consistency, interpreted as a proposition of nominalistic syntax. Such a syntax will have to be carefully formulated in order to avoid assuming that actual inscriptions are closed under the usual operations of logical syntax. But, no matter how this is done, the statement of consistency says no more than that there is no *actual* inscription that is a proof of a contradiction from whatever axioms are in question. This is of course not quite so weak as it seems, as future inscriptions are allowed: It says that no proof of a contradiction *will* ever be written down. Nonetheless, it cannot be stronger than the statement that no proof of a contradiction is *physically possible*, and such possibilities are constrained, for example, by the physical structure of space-time, in such a way as to make the consistency statement weaker than on its usual mathematical understanding. It is also hard to see what grounds other than inductive the nominalist can have for believing consistency statements for theories having only infinite models to be true. The most elementary mathematics used in proof theory, such as primitive recursive arithmetic, will not be known to be sound on any nominalistic interpretation,[43] and

---

[41] *Mathematics in Philosophy*, pp. 184–186.
[42] This possibility was suggested to me by Isaac Levi.
[43] If the inscriptions of nominalistic syntax do not satisfy the closure conditions of usual syntax, then the following kind of anomaly can arise: Suppose that for a given formal theory T we can give instructions for constructing a proof in T of a contradiction, so that we can give a strictly constructive proof (say, in primitive recursive arithmetic) that T is inconsistent. We might, however, be able to give a lower bound for the length of a proof of a contradiction in T, so large that on physical grounds it will be impossible actually to

§12. Nominalism

the intuitive idea that the axioms of such theories describe coherent possibilities is tainted with modality. That the consistency-based if-thenist form of nominalism is on stronger ground than other forms of strict nominalism is not evident. The other strategy suggested by Putnam's response to the problem of possibility, taking possible existence seriously, seems more promising but goes beyond strict nominalism.

I have assumed that the nominalist's objective is to interpret a reasonable amount of existing mathematics. It is for this reason that the conditionals of the if-thenist reconstruction have antecedents with no finite models. Some nominalist work has tried to dispense with any infinitistic assumption.[44] Such an attitude will of course avoid the kind of difficulty that I discern in the nominalist view. But it seems to me to have an even greater difficulty: At best, it suspends judgment about the truth even of the most elementary mathematics. For essentially this reason, work on the nominalist program in recent years has extended the allowed resources in one or two ways, either by allowing geometrical objects, as does Field, or by admitting modality. Discussion of any form of modal nominalism is deferred until Chapter 3.

Before we take leave of nominalism, however, I want to consider the possibility, more congenial to nominalism, that the *logic* of if-thenism should be first-order. In such cases as arithmetic and Euclidean geometry, it appears that the if-thenist has to jettison the intuition that these branches of mathematics describe a unique structure. In the if-thenist translation of an arithmetical statement, for example, $\Omega'(N, 0, S')$ will have to be replaced by a list of axioms that is no longer fixed once and for all but which must depend on the context of the statement. Consider, for example, the manner of doing this that disrupts the second-order formulation the least: Return to the set-theoretic formulation $\Omega(N, 0, S)$ (leaving S as a function symbol); add to it comprehension axioms corresponding to those of second-order logic with first-order comprehension, which must now, however, count as nonlogical axioms. They will correspond closely to those for classes in NB; it is thus best to think of the objects involved as classes rather than sets.[45] In this language, we can describe a finitely

---

write down such a proof. Then 'T is consistent' interpreted nominalistically will be true. If T contains primitive recursive arithmetic, we will be able to write down a proof in T of the (arithmetized) statement that T is inconsistent.

[44] For example Goodman and Quine, "Steps toward a Constructive Nominalism." I was reminded of this by Michael Resnik.

[45] It is simplest to have two sorts of variables, for individuals and classes; but of course a one-sorted formulation is readily available; then one needs the predicate 'is a class'.

axiomatizable conservative extension of the usual first-order arithmetic PA, related to it roughly as NB is to ZF; call the conjunction of its axioms Ω_1 (*N*, 0, *S*).[46] What one does, in effect, is to incorporate the second-order entities into the structure that is being considered only hypothetically.

Of course, if we use Ω_1 instead of Ω in our canonical form (2) the truth of the statement *A* amounts to its provability in this conservative extension of PA; because of the completeness theorem, it makes no difference whether we understand logical validity as semantic consequence or in terms of provability. Thus arithmetic truth becomes relative to the axioms assumed. Putnam was willing to accept this consequence.[47] The formulation we have chosen has the advantage that it makes the relativism arise in a natural place: Extensions of the axioms will be obtained by stronger comprehension axioms about classes or whatever other second-order entities one appeals to. These extensions will give rise to different translations of ordinary arithmetic statements.

The relativistic position is quite counterintuitive, not only because it contradicts the idea that the natural numbers are a unique structure (as are other number systems, and Euclidean space of a fixed number of dimensions), but because it makes the *meaning* of statements in these parts of mathematics depend on what axioms one is assuming, not only about the structure with which one is immediately concerned, but about classes and even about other objects, where reference to them may give rise to relevant comprehension axioms. But it would be difficult to refute categorically without more analysis of the intuition of the uniqueness of such structures as the natural numbers. That will be undertaken in Chapter 8. In the next section, we will consider whether the nominalist can after all use full second-order logic.

Concerning the problem avoiding vacuity in conditionals used to interpret mathematical statements, no solution has been found so far that preserves the purity of structuralism. Putnam's suggestion that the consistency of the antecedents is sufficient, if interpreted conventionally rather than nominalistically, compromises structuralism only by admitting

---

[46] A little delicacy is needed to obtain finite axiomatizability. Comparison with the axiomatization of NB shows that one must have pairing and projection functions. These can either be primitive or defined in terms of simple arithmetic functions, but in the latter case a few of these must be primitive, with recursion equations as axioms.

Some issues about the interpretation of theories of this kind and their relation to more conventional axiomatizations by schemata are discussed in Essays 1, 3, and 8 of *Mathematics in Philosophy*.

[47] "The Thesis that Mathematics is Logic," esp. pp. 22–25.

quasi-concrete mathematical objects, the expressions of formal syntax. (We have, however, not considered the problem of knowing such consistency presuppositions to be true.) But once one does that, one can take the expressions directly as an instance of the structure. Alternatively, one might give this role to expressions or other quasi-concrete objects but deal with them in some modal nominalist manner. That would solve the problem of nonvacuity in a natural way, based on the idea that it is sufficient if the existence of an instance of the structure is *possible*. Moreover, modal nominalism is a more promising account of syntax than strict, extensionalist nominalism. Although such an interpretation still leaves a dilution of the structuralist position, it suggests a general strategy for eliminating mathematical objects: Pure mathematical objects are dealt with in an eliminative structuralist way, while quasi-concrete objects are interpreted in a modal nominalist manner. This strategy will be examined in §15.

§13. Nominalism and second-order logic

The question still deserves discussion whether some way of interpreting second-order logic is available to the nominalist or, more generally, to the eliminative structuralist. It will be observed that in many of the cases to which the strategy we have been considering applies, number theory included, impredicative second-order logic is needed for the deductive development of the theory translated in one of the ways proposed.[48] Thus whatever entities comprise the range of the second-order variables will have to satisfy impredicative comprehension. An obvious candidate interpretation, advocated by Hodes (see note 26), is by Fregean concepts. Frege effectively embodied a comprehension principle for concepts and functions in his logic.[49] Although the question of predicativity had not been raised at the time (nor did Frege take note of it when it became an issue in the Poincaré-Russell debate), Frege's procedure is no doubt an expression of realism about the reference of predicates, a view that would have been shared both before and since. It is certainly not in the spirit of nominalism as we have been understanding it. In an eliminative

---

[48] However, some work on the question to what extent one can get around that will be discussed in Chapter 8, §50.

[49] It is worth noting how comprehension arises in Frege's logic. It is embodied in his axiom IIb of universal instantiation for function quantifiers and in his rules of substitution for Latin letters (free variables). See *Grundgesetze* I, §§25, 48 (rule 9). Behind these is of course Frege's conception of a function name; he took function names to be closed under quantification over functions.

structuralist program more generally, it is not ruled out of court, but what would be demanded is a justification of such realism *independently of mathematical objects* and compatible with the eliminative structuralist's reasons for wanting to eliminate them. Frege, whose conception of object was certainly meant to accommodate mathematical objects, did not take on any such burden himself. Moreover, the most visible internal difficulty of Frege's theory of concepts is that he is not in practice able to do without nominalizing predicates in talking about concepts. But this nominalization has to be rejected by the program of eliminating mathematical objects.[50] The history of the debate about Frege's theory does not encourage the view that concepts according to his conception are less problematic than mathematical objects.

Workers on the nominalist program have in fact preferred two other approaches. First, for many purposes mereology, otherwise known as the calculus of individuals, can do the work of second-order logic. Nelson Goodman played a role in developing the theory, and his motive was surely to have a logical tool for nominalist constructions.[51] It has been accepted as such a tool by nearly all more recent nominalists. Second, in two papers of the 1980s, George Boolos proposed an interpretation of *monadic* second-order logic as a logic of plurality and argued that theories formulated in this logic so interpreted have no ontological commitment beyond that to the objects over which their individual variables range.[52] Boolos's claim has also been embraced by nominalists.

Mereology is central to Hartry Field's construction. He interprets Hilbert's geometry, a second-order theory, by taking the second-order variables to range over regions, which in turn can be taken to be sums of points in the calculus of individuals.[53] Field exploits the fact that in a context where the individuals are atoms, that is, have no proper parts, a version of mereology will simulate monadic second-order logic. Field

---

[50] One can, of course, interpret this nominalization or the second-order language itself by means of semantic ascent. But then the theory one obtains is predicative relative to the underlying domain of objects, and does not serve the purpose at hand. (Cf. *Mathematics in Philosophy*, Essays 3 and 8.) The interpretation by semantic ascent is still highly relevant to structuralism; see later and §18.

[51] See *The Structure of Appearance*, Chapter 2, §4. For a full exposition of mereology see Simons, *Parts*.

[52] "To Be is to Be the Value of a Variable (or to be Some Values of Some Variables)" and "Nominalist Platonism." By "monadic second-order logic," I mean a restricted version of second-order logic in which all quantifiable second-order *variables* are monadic. But many-place predicates, and so all the apparatus of polyadic first-order logic, are allowed.

[53] *Science without Numbers*, pp. 37–38.

§13. Nominalism and second-order logic

describes the logic he accepts as "what might be called *the complete logic of the part/whole relation* or *the complete logic of Goodmanian sums*."[54] Platonistically, one can characterize this logic as postulating a domain of individuals such that for any nonempty *set* of individuals in the domain, there is an individual that is the sum of all of them.[55] No comparable nominalistic characterization of this logic is available, but it is at least clear that a form of full comprehension holds: Given a predicate of individuals that is true of at least one individual, there is a sum of just the individuals of which the predicate is true,[56] and moreover, the admissible predicates will be closed under quantification over all individuals, including these very sums. In the point-region case, one has, for any predicate of points, a region containing just the points of which the predicate is true (provided there are any), and these predicates will be closed under quantification over *regions*. As Field is well aware, this logic has properties parallel to those of standard second-order logic; in particular, it is not recursively axiomatizable.

Field introduces this logic for purposes somewhat different from those of the eliminative structuralist program we have been discussing, but still in the service of a program for eliminating mathematical objects. Its adaptation to our setting poses some problems, because it offers an interpretation of the usual second-order language only when the objects the individual variables range over are atoms in the sense of the calculus of individuals, that is, without proper parts. That would restrict the interpretation of the kind of translation of mathematical statements that we have been considering. Even with this restriction, however, given Field's assumptions, models of classical arithmetic and analysis are available. In any case it is worthwhile to consider Field's approach to second-order logic in its own terms.

We can understand Field's conception better without the help of set theory by adapting an understanding of second-order language that will concern us later. This is that the comprehension principle will hold for *any* predicate true or false of individuals. The point is that what will count as a predicate well-defined in this sense is entirely open-ended; in particular, it is not limited by the resources of any particular formalized language. It is most reasonably regarded as indeterminate at a given time, as it

---

[54] Ibid., p. 38.
[55] This amounts to saying that the individuals form a complete Boolean algebra, minus the zero element.
[56] This is the principle (C_S) of "On Conservativeness and Incompleteness," §4.

depends on what we will come to be able to express and understand. Since predicates can contain parameters (both first- and second-order) this allows the possibility that even a single instance of comprehension will, by generalization, yield more second-order entities than there are predicates in a particular (say, formalized) language.

So far, this understanding corresponds to the understanding of second-order principles in mathematical practice (cf. §5), and also in the design of new formal systems, where one expects principles such as mathematical induction or separation and replacement to be valid for a new formalized language. So long as the domain of individuals is treated as given, impredicativity enters only when we regard our stock of predicates as closed under quantification over second-order entities, in Field's case regions. This latter assumption is independent of the first one, that comprehension is to hold for any predicate that we recognize as well defined.

One version of this conception would regard the second-order entities as constituted in some way by the predicates themselves. Then the closure assumption just cited is very dubious; if it is indeed undetermined what predicates will or can be added to our language, then it is doubtful that a predicate quantifying over all such predicates will be definitely true or false of each individual. Appearances to the contrary arise from an implicit reducibility assumption; for example, to each such predicate corresponds an extension a; then in an extensional context we can replace the predicate 'F' by the predicate '( ) $\in a$', and thus replace the quantification over second-order entities by quantification over sets. This is of course not a procedure that Field could accept.

For Field, regions are to be physical entities. Then the assumption that the class of predicates is closed under quantification over regions is natural enough given a realistic attitude toward the physical world. If one were to question it, it would not be on the ground that it does not fit with the idea of regions as physical. What is more questionable is the transfer to this situation of the undoubted intuitive plausibility of comprehension principles. Field, as a nominalist, should reject the idea of an object associated with a predicate as something like its reference, as well as the "quasi-combinatorial" picture of "selecting" the elements of a given set that satisfy a given predicate that is often appealed to in motivating the axiom of separation.[57] It is not clear what direct argument for his principle Field can offer that would not trade on intuitions that

---

[57] For example Hao Wang, *From Mathematics to Philosophy*, p. 184.

§13. Nominalism and second-order logic

support non-nominalist principles, in particular, the existence of sets of points.[58]

The best position for Field would probably be to say that the comprehension principle is a hypothesis justified by its consequences in systematizing the geometrical basis of physics that constitutes, according to Field, the central part of true (as opposed to fictive) mathematics. The most direct use made of it, however, is logical, in Field's arguments for the conservativeness of platonistic mathematics over nominalistic theories.[59] Field's view, on this reading, puts him in a position in which we have found other formulations of nominalism: making the justification of mathematics turn on some hypothesis about the physical world, which is more vulnerable to refutation than the mathematics.

In §17 we will consider what can be accomplished by combining mereology with modality.

Let us now turn to Boolos's reading of monadic second-order language, although he did not advance it in the service of eliminative structuralism or any other program for eliminating mathematical objects in general. However, as the title of the second paper, "Nominalist Platonism," suggests, he did have an eliminative motive for advancing his interpretation. He claimed that a monadic second-order version of *set theory* is not, under this interpretation, committed to *classes* over and above sets, that is, not to proper classes.[60]

---

[58] If regions are construed as arbitrary nonempty sets of points, then, of course, it can be proved set-theoretically that regions satisfy Field's logic, that is, that they form a complete Boolean algebra without the zero element.
 In "On Conservativeness and Incompleteness," p. 137 n. 12, Field responds to the objection that "the intuitive basis of the complete theory of the *part of* relation is derivative from the idea of a set." He claims that "set theory has the theory of the *part of* relation as its main intuitive basis." It is certainly true that ideas of whole and part enter into the intuitive understanding of set-theoretic concepts and entered into the historical origin of the mathematical concept of set. Equally central to the origin of the set-theoretic approach to mathematics, however, were ideas about *functions*. But the crucial question for Field is what independent motivation he can give for logically strong principles of mereology such as his comprehension principles.
 Field attributes this objection to "one of the journal editors" (i.e., of the *Journal of Philosophy*). I confess to having been the editor in question. But I have no record of the wording that was communicated to him.

[59] *Science without Numbers*, Chapter 1, and "On Conservativeness and Incompleteness." It should be acknowledged that in the latter paper Field backs away from his earlier commitment to what he prefers to call the complete logic of the part-of relation. In the introduction to *Realism, Mathematics, and Modality*, p. 51, he describes his attitude toward the logical devices he uses as experimental.

[60] His skepticism about proper classes was of long standing; see "Reply to Parsons."

Boolos reads second-order formulae by paraphrasing monadic second-order quantifiers by plural quantifiers of natural language. If the first-order quantifiers of a second-order language range over, say, Gs, then the second-order quantifier '$(\exists F)$' is to be read 'there are some Gs' and an occurrence within its scope of, say, 'Fx', as 'x is one of them' (assuming that what 'them' cross-refers to will be unambiguous, but elaboration can take care of the problem if it is not). Thus, consider for example, in the second-order language of set theory, the comprehension axiom

(5) $(\exists F)\,(\forall x)\,[Fx \leftrightarrow \neg\,(x \in x)]$.

This will be paraphrased as

(6)   There are some sets such that each set is one of them if and only if it is not an element of itself.

This reading of monadic second-order logic has been criticized on the ground that there is not a natural reading of the universal quantifier, and Boolos in fact had to rely on the reduction of the universal quantifier to the existential. We can, however, do a little better than that, and render '$(\forall F)$' for monadic F as 'whenever there are some Fs ...'.[61]

Boolos appears to hold that the paraphrase of a monadic second-order sentence will not involve any ontological commitment to entities other than those that would be values of the individual variables. But at first glance, it appears that in a context of this kind a quantifier such as 'there are some sets' is saying that there is a *plurality* of some kind. Cantor's notion of "multiplicity" and Russell's of "class as many" were more explicit versions of this intuitive notion, both attempting to allow that pluralities might fail to constitute sets.

With respect to a number of nonfirstorderizable examples, Boolos argues that they do not involve commitment to *classes*.[62] In part, this rests on the absence of explicit mention of classes in the sentences in question (such as (6)). Now the same could be said of pluralities (a possible commitment Boolos does not explicitly address). The difficulty is that speaking of pluralities is already a kind of reduction of plural quantification to singular; to put (6) in terms of pluralities, one would replace 'there are some sets' by 'there is a plurality of sets', and then 'each set is one of them' by something like 'each set belongs to it'. If the result is taken as the

---

[61] David Lewis, *Parts of Classes*, p. 63. In §3.2 of this work Lewis argues in favor of the ontological claim of Boolos; see later.
[62] "Nominalist Platonism," pp. 73–78.

§13. Nominalism and second-order logic

canonical form, Boolos's intention seems to be violated: On his view, it is plural quantification that is canonical.

Still, there is in sentences making essential use of such plural quantification a form of generalization and cross-reference that one does not find in straightforwardly firstorderizable sentences. Thus, in (6) and other examples obtained by paraphrasing second-order formulae one has the pronoun 'them' referring back to 'some sets'; in effect, what follows 'there are some sets' has to say something about *some sets*: we would like to say, a certain (indefinitely indicated) plurality of sets; this is distinguished from saying something about *a set*, which would be indicated by an individual variable. The same is true in the example from Geach and Kaplan, more natural in English:

(7)   Some critics admire only one another.

Here the cross-reference is carried by 'one another'; again, that they admire only one another is said about *some critics*, not about an indefinitely indicated critic.

These observations express a discomfort not unlike one felt by Boolos himself, which he undertakes to deal with. He quotes the following statement by Harold Hodes: "Unless we posit such further entities [as Fregean concepts], second-order variables are without values, and quantificational expressions binding such variables can't be interpreted referentially."[63] In reply to this, Boolos offers an inductive definition of satisfaction for a monadic second-order language, which, he argues, obviates the notion of values of the second-order variables.[64] It is assumed that the metalanguage can express a pairing function for individuals (thus enabling a reduction of polyadic second-order logic to monadic) and indeed sequences of individuals, as well as the usual syntactical notions.[65] The inductively defined predicate is 'R and s satisfy F', where R is a second-order variable, s ranges over sequences of individuals, and F over formulae. All the clauses of the definition are the typical ones, except that for

---

[63] "Logicism and the Ontological Commitments of Arithmetic," p. 130.
[64] "Nominalist Platonism," pp. 81–83.
[65] The role of sequences is the usual in satisfaction definitions; thus finite sequences would be sufficient. Boolos is concerned with the case where the individuals are sets and the nonlogical apparatus is that of set theory, so that this assumption holds. It will not necessarily hold in applications for strictly nominalistic purposes; see later.

second-order quantification, which reads:

(8) If F is (∃V) G, then R and s satisfy F iff
(∃X) (∃T) {(∀x) (Xx ↔ T⟨V, x⟩) ∧
(∀U) [(‛U’ is a second-order variable ∧ ‛U’ ≠ ‛V’)
→ (∀x) (T⟨U, x⟩ ↔ R⟨U, x⟩)] ∧ T and s satisfy G}.

Looking at the matter "platonistically," R codes an assignment of second-order entities to the variables of the language; to the variable V is assigned λx R⟨V, x⟩. Then what the definiens of (8) says is that there is a (second-order entity) X and an assignment T such that T assigns X to V and agrees with R in what it assigns to other variables, such that T and s satisfy G.[66]

It is hard to agree with Boolos in finding that the treatment of second-order variables in this definition does not offer scope for the notion of a value comparable to what it offers to the notion of a value of an individual variable. The values of the individual variables, Boolos says, are just the terms of the sequences. Why should we not say, similarly, that X is a value of a second-order variable if it is λx R⟨V, x⟩ for some R and V, that is, if (∃R) (∃V) (∀x) (Xx ↔ R⟨V, x⟩)? The difference between the treatment of first- and second-order variables seems to lie just in the facts that in the second-order case functions are coded by predicates and that a function from individuals to n-argument second-order entities can be coded by an (n + 1)-argument second-order entity.[67]

It might also be pointed out that Boolos inductively defines a predicate with a second-order argument (the "assignment" R), to express satisfaction for a language in which there are no such primitive predicates, that is, in which second-order variables occur only predicatively (in atomic contexts of the form Vv). The straightforward way of eliminating this predicate would be by third-order logic; thus, whether one undertakes to eliminate it or not, one is stuck with atomic predication with second-order arguments. This is an enlargement of what Boolos has considered in his argument about ontological commitment. Although the English plural can still be used to read it, that does not cause the distinction between this sort of case and ones where predicates with second-order arguments are absent to disappear. Thus to read (8), let us call a *variable-pair* an

---

[66] This way of putting things assumes the second-order entities are extension-like; but if the language is not extensional, Boolos's definition will have to be modified in any case.
[67] An early criticism of Boolos's ontological claim is Resnik, "Second-order Logic Still Wild." References to further literature and an even-handed summary of the issues are to be found in Linnebo, "Plural Quantification."

§13. Nominalism and second-order logic

ordered pair whose first term is a variable. Then (8) might be read: There are some objects and some variable-pairs such that each object is among the former if and only if it is the second term of one of the latter pairs whose first term is V, and every other second-order variable U is such that for each object, the pair whose first term is U and whose second term is it is one of the latter pairs if and only if it is one of the R s, and the latter pairs and s satisfy G. This has the difficulty that nothing marks the final "the latter pairs" as a second-order argument; 'the latter pairs and s satisfy G' could say that all or most of the latter pairs are x's such that x and s satisfy G.

There are, however, cases where plural noun phrases unambiguously express second-order arguments; 'the latter pairs are infinite in number' would be an example. But what is the principle by which such cases are distinguished? Since Boolos's publications, it has been pointed out, notably by James Higginbotham,[68] that much simpler and more mundane English sentences have plural subjects where a rendering of them as second-order arguments, or at least an essential use of second-order logic, is tempting. These are the so-called collective readings of action sentences (and no doubt others) with plural subjects, as in:

(9)   The men lifted the piano.

(10)   Some men lifted the piano.

Each of these might be read as saying of certain men that each of them lifted the piano, but given the weight of pianos that is less plausible than to understand them as saying of the men, or some men, that they collectively lifted the piano. It might be tempting to render (10), for example, in second-order terms as

(11) $(\exists F)$[there are at least two Fs $\wedge$ $(\forall x)(Fx \rightarrow$ x is a man)
        $\wedge$ F lifted the piano],

thus understanding 'lifted the piano' as a predicate with a second-order argument, like 'are infinite in number' mentioned earlier. Apart from the fact that one might object intuitively to the idea of a second-order entity as lifting a piano, this reading raises the question how 'lifted' with a second-order first argument is related to the same verb with a first-order first argument. Higginbotham prefers a Davidsonian reading involving events,

---

[68] "On Higher-Order Logic and Natural Language," §§5–6.

so that (10) is rendered as:

(12)   ($\exists e$){Lifting(e) $\wedge$ Of(e, the piano) $\wedge$ ($\exists F$)[there are at least two Fs $\wedge (\forall x)(Fx \to x$ is a man) $\wedge (\forall x)$(Agent(x, e) $\leftrightarrow Fx$)]}.[69]

This reading still leaves open the question of interpretation for the second-order logic involved. The case for its ontological innocence might seem to be furthered by the fact that it arises in examples as simple as (9) and (10). In such examples, however, the second-order entities, if we admit them, have finite extensions, and finite sets arise fairly early in the understanding of arithmetic, plausibly still at the common-sense level.[70]

These considerations reinforce, it seems to me, an ontological intuition a little different from but complementing Quine's, according to which ontological commitment is carried by the expressions that indicate what one is talking *about*. We might say that they play the role of *subjects*, but here we must be careful not to revive the usage of traditional logic. But in logical terms they typically occur as arguments. In a primitive second-level predicate, second-order variables or expressions that can be substituted for them play that role. The same is evidently true of plural expressions in Boolos's paraphrases of second-order formulae, as I have indicated earlier. Boolos has, in my view, not made a convincing case for the claim that his interpretation of second-order logic is ontologically noncommittal. The great interest of his reading, in my view, is that it breathes new life into the older conception of pluralities or multiplicities. As a source of second-order logical forms, the plural and plural quantification are rightly distinguished from what was so much emphasized by Frege, predication and, more generally, expressions with argument places. In particular, if it is the idea of generalization of predicate places that we appeal to in making sense of second-order logic, then the most natural interpretations will be relative substitutional or by semantic ascent, and these will not license impredicative comprehension, and it is hard to see how that will be justified.[71] But if one views examples such as Boolos's as

---

[69] An alternative treatment develops a logical regimentation that takes the plural much more at face value. For different versions of this sort of treatment, see Byeong-Uk Yi, "Is Two a Property?" and Rayo, "Word and Objects."

[70] This issue will be explored in chapter 6; in §32 it will be argued that there is a possible genesis of notions of cardinality in which the concept of set plays no role; however, in §34 an alternative possible genesis will be developed that does involve finite sets.

[71] Cf. *Mathematics in Philosophy*, Essay 8 and elsewhere. Concentrating on natural language, Higginbotham (op. cit., §7) proposes replacing the introduction of second-order logic by admitting "classes as many" in the sense of the early Russell (see note 66), which

§13. Nominalism and second-order logic 71

involving "pluralities," they are more like sets as understood in set theory in that no definition by a predicate is indicated, so that one need not expect them to be definable at all. Thus one obstacle to the acceptance of impredicative comprehension is removed.[72]

An advocate of Boolos's interpretation in an eliminative structuralist setting could grant my claims about ontological commitment, but then take a position analogous to the Fregean: Second-order variables indeed have pluralities as their values, but these are not objects. Like Frege's claim about functions and concepts, this position would base an ontological difference, of objects and pluralities, on the grammatical difference between the singular and plural terms that refer to each. It does not seem to me to have the same intuitive force as Frege's position, since there is no analogue to the regress argument that can be made if one views the reference of a predicate as an object. There will still be, just as with Frege's concepts, the irresistible temptation to talk of pluralities as if they were objects, as we have already noted above. The only gain this interpetation offers over the Fregean is the removal of an objection to impredicativity.[73]

Suppose, however, that one does not accept our view of the matter. Intuitions about ontological commitment are disputable, and those nominalists who have applied second-order logic and Boolos's reading of it have not been persuaded by the arguments against Boolos's "meontological" claim. The disputable character of such intuitions was surely one of Quine's reasons for insisting on regimentation in terms of first-order logic and resisting not only higher-order logic but also modal logic or

---

are very close to what I call pluralites. He sees this as a rejection of second-order logic rather than an alternative interpretation of it because he assumes that second-order logic will be so interpreted that second-order entities are "predicational," that is, in some way derived from predicates. Given the entrenched character of the understanding of second-order entities as sets, this assumption seems questionable. But the issue may in the end be terminological.

[72] It would be going too far, however, to hold that on this interpretation impredicative comprehension is obvious in general.

[73] In reflecting on set-theoretic paradoxes, both Cantor and Russell entertain the idea of pluralities that are not "unities": Cantor, in the letter cited in note 18, uses the term "inconsistent multiplicity"; Russell speaks of objects that form a class as many but do not form a class as one, such as "the classes which as one are not members of themselves" (*Principles of Mathematics*, p. 102). In explaining the distinction of class as many and class as one, Russell appeals to the difference between plural and singular reference (*ibid.*, pp. 68–69); his use of the plural in deploying the former notion resembles that in Boolos's examples and in his reading of second-order formulae.

It is of course an inference from certain pluralities failing to be unities to their failing to be objects. On this issue with reference to Cantor, see *Mathematics in Philosophy*, pp. 280–286.

generalized quantifiers in the apparatus of a theory whose ontological commitment is to be assessed. Whatever else may be said of Quine's procedure, it at least makes such questions reasonably definite questions. One also might respond by saying that just because of its disputable character ontological commitment should not be a central issue in the foundations of mathematics.[74] Boolos would hardly dispute that if monadic second-order logic is added to a theory that admits pairing of individuals, then the power of the theory is greatly increased; one can go from a theory whose proof-theoretic analysis is quite straightforward, as is the case with first-order arithmetic, to one that it is beyond the most powerful resources of contemporary proof theory to analyze, full second-order arithmetic. It is somewhat telling that there are significant intermediate stages given by admitting the comprehension schema

(13) $(\exists F)(\forall x)[Fx \leftrightarrow A(x)]$

with restrictions on the formulae $A(x)$ that are admitted. Thus, we may admit only first-order comprehension, that is, not allow A to contain bound second-order variables, or any of a number of weaker restrictions designed to insure predicativity. If we read the second-order variables as ranging over sets, then the most natural readings of them order them in a straightforward way so that each logically stronger one admits more sets. Does that not create a burden of proof for someone who wants to argue that, on the reading by means of the plural, the differences are not ontological differences? If there is no enlargement of ontological commitment as one passes to less restricted versions of the comprehension schema, then perhaps that speaks against the importance of the notion. But that is an objection to nominalism: Of all the main trends in the foundations of mathematics, it is nominalism that gives greatest weight to ontology.

One consequence of the attention given to the plural by logicians from Boolos on is that plural logic has come to be a logic in its own right.[75] The ontological issues canvassed above then come to be issues about the ontological commitments of theories in plural logic. But a consequence of this development is that many issues concerning the plural and its logical force can be discussed independently of controversial questions about ontology. And, as John Burgess has illustrated, it becomes possible

---

[74] Cf. Wang's remarks many years ago, in "What Is an Individual?"
[75] Two somewhat differing formulations are given in Rayo, "Word and Objects," and Yi, "The Logic and Meaning of Plurals." Linnebo, "Plural Quantification," follows Rayo.

to show what plural logic can accomplish in set theory while leaving for separate discussion the issue about commitment to classes or the like that motivated Boolos.[76] But I don't think the issues discussed in this section would be materially changed by putting things in the language of plural logic rather than monadic second-order logic.[77]

We can sum up our discussion of second-order logic as follows: If the eliminative structuralist uses it, he will not be able to avoid ontological commitments more uncomfortable on balance than that to mathematical objects, either to Fregean concepts or to multiplicities that are not "unities." If he does not, he is faced with the relativistic consequences that have made if-thenism an unpersuasive view for some time. Although we will look at this issue again in §16 in connection with "modalist" versions of eliminative structuralism, this picture will not change.

§14. Structuralism and application

Any view of the nature of basic mathematical objects has to provide for the application of the relevant mathematical theories. For this reason we have to consider objections to a structuralist view of numbers that raise this issue. The objections can be discussed with a version of the eliminative reading of Dedekind as our paradigm, whether formulated in terms of set theory as in §10 or in terms of second-order logic interpreted in one of the ways we have discussed.

The structuralist view implies that the application of numbers as cardinals is not directly reflected in their identity, as it was in the constructions of Frege and Russell. They seem themselves to have regarded this fact as an objection. Frege rightly held that any view of the foundations of arithmetic and analysis must make provision for their applications. As he famously remarked, "It is applicability alone that raises arithmetic from the rank of a game to that of a science. Applicability therefore belongs to it of necessity."[78] But as Dummett has emphasized (p. 61), he criticized others' accounts of real numbers on the ground that "the principle governing all possible applications of the real numbers should be displayed in their definition." The same idea presumably lay behind his procedure in the foundations of arithmetic, where the conception of numbers as cardinals

---

[76] Burgess, "*E pluribus unum.*" Cf. *Fixing Frege*, §§3.6–3.7.
[77] I hope to pursue the questions further elsewhere.
[78] *Grundgesetze* II §91, translation from Dummett, *Frege: Philosophy of Mathematics*, p. 60. All quotations from Dummett in this section are from this work, which is cited by page number in the text.

was a driving idea of the analysis. The analysis of cardinality is used to buttress the thesis that numbers are objects by leading Frege to his criterion of identity for numbers, that the number of *F*s is the same as the number of *G*s just in case there is a one-one correspondence of the *F*s and the *G*s.[79] One can be led into confusion here: An analysis of the locution "the (cardinal) number of the *F*s" should certainly lead to the consequence that Frege's criterion is true. But Frege seems to have demanded more, that the sense of arithmetical propositions, including those of pure arithmetic that on their surface do not refer to cardinality, should in some way incorporate their role as cardinals. Once he introduces his explicit definition of the number of *F*s as the extension of the concept 'equinumerous with the concept F', the proof of his criterion uses only the definition, the basic property of extensions (i.e., the notorius axiom V), and simple logic.

One might object to Frege's view on the ground that the ordinal conception of number is more fundamental than the cardinal; indeed, Dummett offers just this criticism of Frege (p. 293). But the structuralist view has a more general quarrel with the attitude we are attributing to Frege. Once we have the natural numbers with their structure, then that is sufficient for applications. For the cardinal use of the natural numbers, one needs some understanding of cardinality, and for this we can for the moment follow Frege (who himself followed Cantor). But as Dedekind seems already to have seen, a structuralist understanding of what the numbers are does not stand in the way of a reasonable account of their cardinal use.

It is true that the application of arithmetic requires relations of numbers that are not internal to the structure of the numbers themselves or even to some larger universe of mathematical objects. We can see that the structuralist view should not be formulated so as to hold that mathematical objects have no properties or relations other than those definable in terms of the basic relations of the structure in which they reside. For example, according to a widely accepted analysis of counting, in counting a group of objects a one-to-one correspondence is established between the *numbers* from 1 to a certain number *n* and the objects of the group. Such relations in which the elements of a structure may stand will be called

---

[79] *Grundlagen*, §§62–68; cf. *Mathematics in Philosophy*, p. 153. This criterion has been much discussed; see in particular Wright, *Frege's Conception of Numbers as Objects*, and many of the essays in Demopoulos, *Frege's Philosophy of Mathematics*. More will be said about it in Chapter 6. But I do not attempt in this work to deal seriously with Wright's "neo-Fregean" program. On that, see now also Hale and Wright, *The Reason's Proper Study*.

§14. Structuralism and application

external relations. Evidently, the structuralist view has to allow numbers or other objects it treats to have external relations.

Clearly, those relations between elements of a simply infinite system and other objects will only count as relations of *numbers* if they are in a certain sense invariant under choices of realization of the structure. Let $\langle N, 0, S \rangle$ and $\langle M, 0, R \rangle$ be simply infinite systems in the sense of §10. By Dedekind's theorem, they are isomorphic; let h be the isomorphic mapping of N onto M. If, now, for $n \in N$, f is a one-one correspondence of some objects, say the Fs, and $\{m: m \in N \wedge m \leq_S n\}$, then, of course, if we set $g(x) = h[f(x)]$, g is such a correspondence between the F's and $\{m: m \in M \wedge m \leq_R h(n)\}$.[80] Thus if, using N as "realization" of the numbers, one concludes on the basis of f that there are n Fs, using M one would conclude on the basis of g that there are $h(n)$ Fs, which is the right result.

This treatment of external relations is generally applicable; suppose $S = \langle S, \ldots \rangle$ and $T = \langle T, \ldots \rangle$ are structures, and h is an isomorphism of S onto T. Then if R is a relation of elements of S to elements of another set U, and we set

$$xR'y \text{ iff } (\exists z \in S) [x = h(z) \wedge zRy],$$

then for any x in S, y in U, xRy obtains if and only if $h(x)R'y$, and R' can do the same work as R. This statement is of course vague and would have to be filled out in particular cases, as we have done for the example of counting. It should also be clear that it does not depend specifically on the set-theoretic conception of structure, and this treatment of external relations will be applicable to other structuralist views of such objects as numbers. If the eliminative structuralist view is formulated in terms of second-order logic, then in this argument only simple cases of first-order comprehension are used.

The idea is simply that the relations of numbers to other objects that are needed for applications are essentially invariant under isomorphism, so that the form of structuralism that denies that there is a more intrinsic characterization of what the numbers are can accommodate these relations. It is of course a further demand that the general principle for the application of the natural numbers or the real numbers should be *explained*. In the case of the natural numbers, one can well question whether there is a single such general principle, as the reduction of the cardinal use to the ordinal or vice versa could be questioned. The analysis of cardinality that Cantor made mathematically usable and general,

---

[80] $\leq_S$ is the order relation induced by S as successor; similarly for $\leq_R$.

which was then used by Frege, certainly offers an account of one very general such principle. But it is far from evident that its work requires that their role as cardinals be determinative of what objects the natural numbers are.

Some other objections to a structuralist view of the natural numbers are more technical. A simply infinite system or progression, as we defined it, is a structure with a distinguished initial element and a unary operation satisfying Dedekind's conditions. This does not tell us whether the initial element is to count as 0 or as 1; if we have a progression $\langle N, a, f \rangle$, then the structure $\langle N - \{a\}, f(a), f \rangle$ is also a progression and isomorphic to it. But this influences how we interpret any Arabic numeral with respect to the progression; if the initial element is 0, then 3 is the fourth element; if it is 1, then 3 is the third. Also, given our progression, we can give two definitions of finite cardinality. Let n be an element of our progression $\langle N, a, f \rangle$. Then we have:

(9) $(Nx)Fx = n$ iff the Fs are equinumerous with the terms preceding n.

(10) $(Nx)Fx = n$ iff the Fs are equinumerous with the terms up to n.

(9) is the correct definition if we take the initial element a as 0, but (10) is correct if we take it as 1.

Dummett takes this observation to be a serious objection to a structuralist view of the natural numbers (pp. 52–53). He holds that it cannot be right to say that 0 is just the initial element of a progression, or that 3 is the third element; and in fact these are incompatible. This claim is undoubtedly right: '0' does not mean 'the distinguished (initial) element of a progression supposed given', and the translations considered in §§10–11 of arithmetic language cannot just presume that the distinguished element is 0. But how fatal is this observation to structuralism?

One might at first reply that what is needed is simply more structure; although the numbers beginning with 0 and those beginning with 1 are isomorphic, this ceases to be so if we expand the structure to include addition, since $n + 0 = n$ while $n + 1 \neq n$ for any natural number n. Dummett replies that the structuralist must have in mind the second-order characterization, since it is categorical. At all events, one also might reply on his behalf that, given a progression, there are two possible ways of introducing addition, say by primitive recursion:

(11) $m + a = m; m + fn = f(m + n)$.

(12) $m + a = fm; m + fn = f(m + n)$.

§14. Structuralism and application 77

(11) gives a the role of 0, while (12) gives it the role of 1.[81] Which is the "real" addition?

It seems to me that the correct reply for the structuralist is that if we have a progression, and we want to go on to develop mathematics, then whether we take the initial element to be 0 or 1 is a *convention*. In the versions of eliminative structuralism that we have considered, this will be reflected in the manner in which arithmetical language is translated, and in the choice of second-order explicit definitions of addition, whether to mirror the recursion (11) or (12), and then other operations. Frege gave a good reason for preferring the convention that the initial element is treated as 0, as otherwise the progression would not contain a cardinal number for empty concepts, but this reason is relevant only when cardinality is at issue, and even then one can postpone introducing 0 until, for example, one is also introducing negative numbers.[82]

If this seems unnatural, it is because a structuralist understanding of arithmetic belongs to sophisticated mathematics; it is not part of the layman's understanding of arithmetic or even the mathematician's before a certain amount of foundational reflection has been undertaken. That, as Dummett says, "we take them [the natural numbers] as too intimately connected with certain immediate applications of them to regard them as identifiable solely through the internal structure of the natural-number system" (p. 52) is true. It does not follow that such a connection is what individuates the numbers as objects, or even that they have such individuation at all.

Dummett makes a further objection in the following eloquent passage:

> The identity of a mathematical object may sometimes be fixed by its relation to what lies outside the structure to which it belongs; what is constitutive of the number 3 is not its position in any progression whatever, or even in some particular progression, nor yet the result of adding 3 to another number, or of multiplying it by 3, but something more fundamental than any of these: the fact that if certain objects are counted 'One, two, three', or, equally 'Nought,

---

[81] This observation is made by Tait, "Frege versus Cantor and Dedekind," p. 225.
[82] An illustration for this conventional character might be the following: In the ordinal use of number concepts in everyday language, the initial term of an ordering or sequence is described as "the first" and thus made to correspond to the number 1. It is not clear that this everyday usage gives any place to 0 as an ordinal number. But in mathematical usage, in particular in mathematical logic, it is quite common to assign 0 to the initial element of a sequence, and that is in keeping with the very entrenched practice of regarding the ordinals as beginning with 0.

one, two', then there are 3 of them. The point is so simple that it needs a sophisticated intellect to overlook it; and it shows Frege to have been right, as against Dedekind, to have made the use of the natural numbers as finite cardinals intrinsic to their characterisation. (p. 53)[83]

In our terms, Dummett claims here that 3 is characterized by an external relation rather than internally to the structure of numbers with just an initial element and the successor function or relation. The point really comes back to the question whether the initial element of a progression is understood as 0 or 1, as 3 is characterized as $S2$ and 2 as $S1$, so that once the convention regarding 0 and 1 has been made, 3 is after all characterized in intrastructural terms, without expanding the structure.

Dummett attributes to Frege the view that numbers are "specific objects," and he reasonably enough regards Benacerraf as denying this. Eliminative structuralism will in general deny this, for it, '3' might be taken to signify rather a role. A difficulty for the alternative that Dummett does not address is that on the standard constructions of the number systems, there is the positive integer 3, the rational number 3, the real number 3, and the complex number 3, and these are not strictly identical, although there is a canonical mapping of each system into the next in which the next 3 is the image of the preceding one. In other words, if one considers the full use of '3' in mathematics, it's far from clear that being 3 is not a role of some sort rather than being a "specific object."[84]

Given the above treatment of external relations, we might answer the demand that some stronger constraint than the structuralist allows on what objects the natural or real numbers are is needed for application as follows, taking this time the real numbers as our example: Let $\mathbf{R}_0$ be the real numbers, so understood that they obey the constraint; let $\mathbf{R}_1$ be some copy of the structure of the real numbers. If we have got the right structure, then $\mathbf{R}_0$ and $\mathbf{R}_1$ are isomorphic. Let g be the isomorphism of $\mathbf{R}_0$ onto $\mathbf{R}_1$. But now $\mathbf{R}_0$ provides for applications, by hypothesis. In particular, there are truths such as that the ratio of the magnitudes m_1

---

[83] The "sophisticated intellect" is presumably that of Paul Benacerraf, whose view in "What Numbers Could Not Be" is the primary target of the objections by Dummett we have been discussing.

[84] This is before one considers that a 3 might arise in still other structures, for example, fields of characteristic p for $p > 3$.

To be sure Dummett does not claim that the observation in our quotation shows that 3 is a specific object; one option about the natural numbers he considers left open is "that they are not specific objects at all, even though capable of being characterised by reference to their application, rather than by pure structure" (p. 54).

§14. Structuralism and application

and m_2 is r, for $r \in R_0$.[85] But surely we could provide for the same applications by, for $s \in R_1$, giving to 'the ratio of the magnitudes m_1 and m_2 is s' just the truth-conditions presumed already given for 'the ratio of the magnitudes m_1 and m_2 is r' where $r = g^{-1}(s)$. Furthermore, suppose that the general principle concerning applications (such as the Fregean one we have presumed) that governs our understanding of statements about real numbers should prove inadequate; applications arise that are not covered by the principle. By virtue of what are these applications of *the real numbers*? I do not see any answer to this question other than that the mathematical structure involved is that of the real numbers.[86] I conclude that at this stage there is no reason to conclude that concern for applications should lead one to reject a structuralist view.

---

[85] Concern for applications led Frege to undertake to define real numbers as ratios of magnitudes. R_i (i = 0, 1) is of course the domain of $\mathbf{R}_i$.

[86] Quine, commenting on views of Russell, remarks, "Always, if the structure is there, the applications will fall into place" ("Ontological Relativity," p. 44). Much of the argument of this section is an elaboration of that remark.

3 Modality and Structuralism

§15. Mathematical modality

In §§11–12 the use of modal notions appeared attractive as a way of formulating eliminative structuralist views of mathematical objects so as to deal with the vacuity problem that such views faced. Before I pursue that idea, I shall consider modal notions more directly in relation to mathematics. Although this provides necessary background for continuing the previous line of argument, it is also of interest in its own right and will be relevant also to other parts of the present work.

Modality is what is expressed primarily by the words 'necessary' and 'possible' and by the modal auxiliaries 'can' and 'must'. These words have such a variety of uses, some of quite different character, that we could not hope to survey them in a brief compass (and probably to do so at all would be a major linguistic undertaking). However, two distinctions made between uses of modal words or between modalities are of general importance and important for our purposes: (a) that between epistemic and nonepistemic and (b) that between absolute and non-absolute.

(a) In *Naming and Necessity,* Kripke stresses the difference between the epistemological notion of the *a priori* and the notions of necessity and possibility with which he is concerned, which he says belong to metaphysics (pp. 35–36). In this discussion, he does not emphasize the idea of epistemic *modalities* or epistemic uses of modal words, but of course such can be found. In everyday language, the modal auxiliaries especially are often used in relation to the speaker's or the community's knowledge; for example 'it can rain tomorrow' is most likely to mean 'for all I (we) know, it will rain tomorrow'. Kripke acknowledges such an epistemic use when discussing our inclination to say that "the answer to the

§15. Mathematical modality

question whether Hesperus is Phosphorus might have turned out either way" (p. 103). He observes that "the four-color theorem might turn out to be true and might turn out to be false" (*ibid.*, evidently as of 1970). "Obviously, the 'might' here is purely epistemic."[1]

In mathematics, modality naturally arises in two ways: First, mathematical truths are thought to be necessary, and second, mathematical existence is intimately related to possibility. Whether or not the modal notions arising in these two ways are identical, they are certainly prima facie nonepistemic.

Nonetheless, there is a notion that has been treated as a modality in logical work, namely *provability*, which is certainly prima facie epistemic. There are two types of studies in which modal logic is used with $\Box A$ having the intended meaning 'It is provable that A'. In the first, provability is provability in a given formal system. In that situation, if the system satisfies the conditions of Gödel's incompleteness theorem, the modal logic cannot be an aletheic modal logic of the usual kind. For the axiom schema T ($\Box A \to A$) would amount to a "reflection principle," the principle that what is provable is true. A well-known theorem of Löb's states that for any formula of arithmetic one can prove that if the corresponding instance of the reflection schema is provable, then the formula itself is already provable. Thus, on the provability interpretation, the schema L:

$$\Box(\Box A \to A) \to \Box A$$

---

[1] It is not so clear that Kripke means by this to acknowledge epistemic *modalities*. With respect to the statement 'Gold *might* turn out not to be an element', he seems to regard 'might' as a sort of shorthand for something purely epistemological (p. 143 n. 72). He goes on to say, "If I say, 'Gold *might have* turned out not to be an element,' I seem to mean this metaphysically and my statement is subject to the correction noted in the text." That is, it is not really correct but is, rather, a loose way of saying that it might (metaphysically) have been the case that there is a substance with all the properties originally known of gold but not an element.

In these cases, the idea of epistemic modality has the difficulty that one seems to acknowledge the epistemic possibility of states of affairs that are not metaphysically or "logically" possible, contradicting the idea that the latter is the most general kind of possibility. In applying modal logic to epistemic notions, for example in the standard epistemic logics, epistemic possibility is identified as truth in some world that is an "epistemic alternative" for the knower. But then this implies truth in some possible world or other. Hence either epistemic possibility implies possibility *tout court* (and the sort of statement Kripke considers is not epistemically possible, at least on his view of what is possible *tout court*), or else for the analysis of the epistemic notions some more generous notion of possible world is needed, so that some epistemic alternatives are not "really" possible worlds.

is sound. If to the minimal normal modal logic K (which is sound on the provability interpretation[2]) we add both T and L, obviously we can infer $\Box A$ for every A; on the provability interpretation, this amounts to the inconsistency of the system. A modal logic sound for the provability interpretation, then, would be one in which one replaces the schema T by the schema L. (Following Boolos, we call the resulting logic GL.) This logic turns out to have considerable mathematical interest; in particular, it turns out to be *complete* in the following sense: If a propositional formula is not provable in GL, then arithmetical formulae can be instantiated for its sentence letters such that, on the interpretation of $\Box A$ as 'A is provable in first-order arithmetic', the resulting instance is not provable in first-order arithmetic.[3]

In another type of study, the necessity operator has as intended meaning something like "informally provable" or "provable by arbitrary correct means." In this case $\Box A \to A$ is sound, and the reasonable modal logic will at least contain S4. A certain amount of work has been done developing mathematical theories in which such an operator is added to a classical theory; the resulting "epistemic mathematics" has interesting relations to intuitionistic mathematics.[4] Another interesting line of research studies in an epistemic-logical way the issues surrounding the connection, real or alleged, between Gödel's theorem and mechanism.[5]

On its face, the necessity of mathematics is not epistemic; however, it may be related to the fact that mathematical knowledge is characteristically obtained by proof. The fact that Goldbach's conjecture is not known to be true or false does not alter the fact that if true, it is necessary; if false, it is necessarily false. I need not go into this point because it has been vividly illustrated by Kripke.[6]

---

[2] This soundness is closely related to the first two of the "derivability conditions" proved by Hilbert and Bernays in their proof of Gödel's second incompleteness theorem. See *Grundlagen der Mathematik*, II, §5 (d). But the second condition amounts to the soundness of the schema 4
$$\Box A \to \Box\Box A$$
characteristically assumed in systems of epistemic logic. That schema is derivable in GL. For the modal-logical treatment of these matters see Boolos and Jeffrey, *Computability and Logic*, ch. 27, and, more extensively, Boolos, *The Logic of Provability*.

[3] This rather difficult result is due to R. M. Solovay. See Boolos, *The Logic of Provability*, ch. 9.

[4] An introduction to the subject is provided by the papers in Stewart Shapiro (ed.), *Intensional Mathematics*. For an informal treatment with philosophical motivation see Goodman, "The Knowing Mathematician."

[5] Reinhardt, "Epistemic Theories."

[6] *Naming and Necessity*, esp. pp. 36–37.

§15. Mathematical modality

Whether the possibility related to mathematical existence is epistemic may seem more controversial, since the constructivist view that existence involves construction has been developed in a manner in which explanations in terms of truth-conditions are replaced by those in terms of proof-conditions. For a structuralist understanding of classical mathematics, however, such a conception is inappropriate. If, for example, we think of the existence of a structure of pure mathematical objects in a rough way as consisting in the possibility of an instance of it, the realist point of view about mathematical objects and truth will require this possibility to be independent of our knowledge. In the further development of the structuralist view, I will assume that the modalities involved are nonepistemic.

(b) What necessity or possibility in some absolute sense would be contrasted with is what is necessary or possible at some point in time, in some situation, or by virtue of some aspect of how things are. Thus everyday use of modal words, especially the auxiliaries 'can' and 'must', often indicates necessity or possibility of the latter kind. Thus today it is possible for me to go to New York tomorrow (assuming I have no pressing obligation preventing it); but on, say, January 1, 1936, it was not possible for me to go to New York the next day, because a child less than three years old cannot undertake such a journey. Though it is possible for me to go to New York tomorrow, under a number of easily conceivable conditions it would not have been possible. Thus, it seems that the possibility depends not only on the time but on a variety of contingent facts.

The idea of an absolute modality has a very simple model in terms of possible worlds: Given a domain of possible worlds, there is a natural necessity operator $\Box$ such that $\Box A$ is true if and only if A is true in *all* the worlds in the domain; $\Diamond A$ is true if and only if A is true in *some* world. Then the resulting modalities satisfy S5; in particular, $\Diamond A$ implies $\Box \Diamond A$. In terms of possible worlds, the difference between absolute and nonabsolute modalities is represented by the difference between unrestricted and restricted quantification. Interpretations of normal modal logics weaker than S5, where $\Box A$ is true at a world i if A is true at every world j that is "accessible" from i, clearly give rise to nonabsolute modalities (except in the trivial case where every world is accessible from every other).

The "metaphysical" or "broadly logical" modalities central to much recent writing on modality clearly seem to be understood as absolute; it is generally taken as clear that they satisfy S5 and, in particular, that they give rise to a class of possible worlds such that necessary truth is truth in all possible worlds. In fact, the term 'absolutely necessary' is often used to mean 'necessary in the strictest sense'.[7] That the strictest necessity should be absolute in the above sense is to be expected, for to obtain such a strict necessity one would generalize over all the conditions on which a nonabsolute necessity might depend.

Among nonepistemic modalities four should be distinguished (roughly in order of stringency of necessity):

(i) "Physical or natural"; this would be the necessity that the laws of nature possess. What is allowed by the laws of nature, and is not otherwise contradictory or incoherent, is physically possible.

(ii) Metaphysical or "broadly logical."[8] Several different writers seem to aim at the same notion, which Kripke generally calls simply necessity, or necessity *tout court*. Generally it is regarded as obvious that necessity in this sense is stricter than physical or natural necessity, and that many stories that conflict with the laws of nature nonetheless describe logical possibilities. Thus Alvin Plantinga writes:

> Unlike Superman, furthermore, the rest of us are incapable of leaping tall buildings at a single bound or (without auxiliary power of some kind) traveling faster than a speeding bullet. These things are impossible for us; but not in the broadly logical sense.[9]

The prevalence of "science-fiction" examples in many philosophical discussions suggests a widespread belief on the part of philosophers that the philosophically relevant sense of possibility is much more liberal than the physical or natural.

Both Kripke and Plantinga refer to the necessity that concerns them as strict necessity, necessity in the highest degree. When we turn to modal ideas associated directly with mathematics and logic, we shall, however, allow that some truths are not necessary in one of the latter senses that are yet metaphysically necessary. We can reconcile this apparent

---

[7] I myself recommended the term 'absolute necessity' for strict necessity in the sense I took Kripke to intend (*Mathematics in Philosophy*, p. 177). I now prefer to use the term 'absolute' in the more formal way indicated in the text.

[8] Clearly this notion is meant to share some of the properties of the notion of "logical" necessity prominent in analytical philosophy before Quine's critique of the analytic had wide impact. Neither this earlier notion nor that in modal realist literature was meant to be particularly tied to formal logic; hence the term 'broadly logical'. This latter term is the one favored by Plantinga to distinguish the notion he works with from other notions of necessity; see, for example, *The Nature of Necessity*, p. 2.

[9] *Ibid.* Kripke, however, declines to commit himself as to whether there really is a difference between physical necessity and strict necessity (*Naming and Necessity*, p. 99).

§15. Mathematical modality

contradiction by saying that the more liberal possibility allowed by the logico-mathematical notions is not "real" possibility. This traditional term is, however, far from clear.

Whether or not it is what this traditional term envisaged, we can see that metaphysical modality obeys a constraint that in considering mathematical and logical modality we will violate. To make out a statement as expressing a metaphysical possibility, one describes a situation in which it is true. This description will not serve its purpose unless it does no violence to the meanings of the terms in the statement. Take, for example, the old chestnut, 'Something is both red all over and green all over'. It is generally thought that one cannot make that out as possible in the relevant sense. In order to do so, one would have to describe something that is both red all over and green all over, and it would have to be clear that it was red in just the same the sense in which everyday things are red, likewise green in just the sense in which everyday things are green, and what may be the more delicate aspect, the relevant sense of 'all over' also has to be respected. For example, it can't be something just covered with red and green spots. Regarding an example in the above quotation from Plantinga, if we ask whether it is possible that he should leap over a tall building in a single bound,[10] we are to describe a situation in which someone does this (which appears easy enough to conceive because it is involves only a great enlargement of scale of capacities that some real athletes possess) and furthermore that this person is *Alvin Plantinga*. Particularly assuming names to be rigid designators, this latter possibility is perhaps not so easy to make out; one gets into questions about just what is necessary to be the particular individual that one is.

Whatever the idea of logical modality connected with formal logic might turn out to be, it will not be constrained in such a way as to give rise to such delicate questions. To make out a statement as logically possible, it should be sufficient to construct a model in which it is true. But in such a model the name 'Alvin Plantinga' can stand for *anything* satisfying the condition which, in the model, represents leaping over a tall building in a single bound. Thus, although it may not be clear whether it is metaphysically possible that Plantinga leaps over a tall building in a single bound, its logical possibility in the narrower sense seems to be a triviality.

---

[10] Possibly we should distinguish between the possibility, considered by Plantinga, that *he* leaps over a tall building in a single bound and the possibility, whether weighed by him or by another, that *Plantinga* does this. I assume that it is the latter that is in question.

It may be claimed that the constraint we have discerned in the notion of metaphysical modality is one that any reasonable modal notion should obey; otherwise, how is the possibility that Alvin Plantinga leaps over a tall building in a single bound the possibility of *that* state of affairs and not some other (say that a "counterpart" of Plantinga performs this feat)? I shall consider this question further in connection with the "formal" modalities that violate the constraint.

(iii) Mathematical modality, which I defer until I have taken up

(iv) Logical modality in the strict sense. The idea here is that formal logic carries with it its own notion of necessity. The extent to which such a notion has been developed in the literature is remarkably slight; the more usual conception of "logical modality" would make necessity coincide either with analyticity or with metaphysical necessity. We have already briefly indicated the difference between the latter and formal logical necessity; this difference should be still clearer if we consider the range of necessary truths claimed in *Naming and Necessity* or other recent modal realist works.

The idea of a formal logical notion of modality is that logical truths, or logically valid statements, should be necessary, but only such statements should be necessary. Thus, a sentence should express a possible truth if it cannot be refuted by pure logic, or if it has a model. Because we have not yet specified what "logic" includes, clearly we have to allow the possibility that these will not be equivalent.

Let us assume first that nonmodal logic is just classical first-order logic.[11] Then a sentence in this language (thus without modalities) will express a *necessary* truth just in case it expresses a logical truth in the usual sense. Suppose we now extend the language by the usual modal operators thus interpreted. We know the truth-conditions for sentences of the form $\Box A$, where A contains no modal operators; namely $\Box A$ is true if and only if A is true in all models. Because this no longer depends on the model, for a given model we can make it the condition for $\Box A$ to be true in that model. Then the stipulation determines the truth-conditions for all sentences of the extended language. Consider, for example, a given (usual first-order) model **M** and a sentence A with no nested modalities, in which $\Diamond$ has been eliminated in favor of $\neg\Box\neg$. Consider a subformula of A of the form $\Box B$. This is true, regardless of **M** and the assignment to the free variables of B, just in case B is logically valid. Replacing $\Box B$ by T

---

[11] My earlier discussion of logical modality in Essay 7 of *Mathematics in Philosophy* also made this assumption.

§15. Mathematical modality

or F according as B is or is not valid, we obtain an equivalent of A without modalities. But then we can determine the truth-value of $\Box A$ in **M** by the same criterion. This interpretation amounts to a possible worlds model structure in which $\Box$ is absolute in the sense mentioned earlier, and the possible worlds are a collection of models rich enough so that every non-modal formula of the language that is not valid has a countermodel in the collection.[12] The interpretation is "absolute" in another sense: If D is such a collection of models, and D' is an extension of D, then the truth-value of a modal formula at a model in D is the same if the necessity operator is interpreted over D' as if it is interpreted over D.

By adding such an operator of "logical necessity," however, we lose nice properties of first-order logic. Clearly the absolute character of the operator implies that a closed and modally closed formula is *true* just in case it is logically necessary. But if A is a nonmodal formula, $\neg\Box A$ is true just in case A is *not* valid in the usual first-order sense, and thus it is "logically true" just in that case. But such A are not recursively enumerable. It follows that there can be no proof procedure for logical validity in the extended language.[13]

So long as one thinks of logical necessity as having the properties of first-order logical truth this will be an unacceptable result. The matter looks different in the context of our discussion of eliminative structuralism, since there we have used second-order logic and have considered analogues of the standard semantics, where the valid formulae are already not recursively enumerable.

There is, however, a more philosophical objection to the idea that formal logic gives rise to absolute modalities. That a statement (say in first-order logic) is not logically valid is made out by a *mathematical* construction, typically of a countermodel but possibly a more proof-theoretic argument showing the impossibility of a derivation. The idea that a logical possibility obtains with logical necessity, if it obtains at all, implies that arguments of this kind themselves yield *logical* necessities. Many

---

[12] This is not quite sufficient, as a formula B may contain free variables to which objects are assigned before one searches for a world that might falsify it. It would be sufficient to have models whose domains are sets of natural numbers, such that for any satisfiable formula A with n free variables, for any sequence of natural numbers of length n the collection contains a model that satisfies A with the given sequence assigned to its free variables.

[13] Two alternative definitions of logical modalities are more briefly considered in *Mathematics in Philosophy*, pp. 180–181. They are in my opinion less attractive than the one discussed in the text.

arguments concerning logicism have taken the existential commitments of mathematics to show that it is not part of logic (or that it would be only if such commitments could be eliminated).[14] There is thus a conflict between the idea that mathematical existence goes beyond logic and the treatment of logical modality as absolute.

Because with this interpretation statements of the form $\Diamond A$ are equivalent to consistency statements, some will be quite strong from a proof-theoretic point of view. For example, von Neumann-Bernays-Gödel set theory NB can be formulated as a finitely axiomatizable first-order theory. Hence there is a single first-order sentence A such that $\Diamond A$ is equivalent to the consistency of NB. Because this consistency could only be proved with the help of strong abstract principles, which from some philosophical points of view could be rejected or held to be unintelligible, the assumption that, if true, it is logically necessary, is incompatible with one possible conception of logic, namely, that logic codifies the principles that are necessary for all reasoning about objects in general.

If we allow to logic richer means of expression, such as second-order quantifiers or generalized quantifiers, then possibility statements have correspondingly greater expressive power. For example, considering ZF as a second-order theory (i.e., replacing the schema of replacement by a single second-order statement), it is finitely axiomatizable.[15] Its standard models are structures $\langle V_\alpha, \in \rangle$ where α is a strongly inaccessible cardinal; hence there is a single second-order sentence A whose standard models are just these structures.[16] $\Diamond A$ is not quite equivalent to the existence of a strongly inaccessible cardinal, because the possible truth of A does not imply that there is a model of it whose domain is a *set*. Still, it amounts to the possibility of a structure (class) of sets where the power set is the full power set, and the axioms of ZF hold. Of course, to A large cardinal axioms can be added. Because much stronger conditions on α can be expressed by second-order sentences, possibilities in standard second-order logic can imply the possible existence of very large cardinals.[17]

---

[14] I endorsed this view in *Mathematics in Philosophy*, pp. 131, 165–166.
[15] Relative to the underlying logic; see note 29 of Chapter 2.
[16] V_α is the set of sets of rank $< \alpha$, that is, the result of iterating the power set operation α times beginning with $\varnothing$, or, in the more general situation where urelements are admitted, beginning with the set of urelements.
[17] This is true even for second-order indescribable types of cardinals, such as measurables. (See Drake, *Set Theory*, pp. 281–282.) For let $B(x)$ be a formula of first-order set theory saying 'x is a measurable cardinal'. If A is the conjunction of the axioms of second-order ZF, as in the text, then a standard model of $A \wedge (\exists x) B(x)$ will be a structure $\langle V_\alpha, \in \rangle$ where α

§15. Mathematical modality

These considerations, together with the well-known ones about delineating the class of logical constants, make it in my opinion very doubtful that a generous notion of logical possibility would be distinguishable in a principled way from the mathematical possibility appealed to by other writers.[18] What we can do is to single out a more restricted case of mathematical modality, such as the one based on first-order logic considered earlier, and describe this as logical modality. If it is to do more than to identify the specific manner in which the restriction has been defined, however, the label 'logical' is not of great significance.

To return to the issue of absoluteness, statements of logical possibility of the kind we have been considering are made on the basis of mathematics, and the considerations in favor of regarding mathematical truths as necessary also favor regarding them as necessary – in whatever sense mathematics is necessary. These doubts about the absoluteness of logical necessity have force if one regards the difference between logical and mathematical modality as a principled one. Otherwise, as statements of logical possibility are anyway mathematically necessary, taking them to be logically necessary is simply imposing a constraint on the specific notion of logical necessity being introduced.

(iii) To return to mathematical modality, the notion arises in two contexts: first, as possibly the appropriate interpretation of modality for the thesis that mathematical truths are necessary, and second, as an explication of the idea that what mathematical constructions establish is a kind of possibility.

If one takes the view about mathematical existence and logic mentioned in the discussion of logical modality above, the necessity of mathematics will not be strict logical necessity. Otherwise, if one does

---

is a strongly inaccessible cardinal such that there is a measurable cardinal $\kappa < \alpha$. Thus $\Diamond [A \wedge (\exists x) B(x)]$ implies that the existence of a measurable cardinal is possible.

[18] In this I appear to differ from Field in "Is Mathematical Knowledge Just Logical Knowledge?" He regards the notion of modality he uses as strictly logical (pp. 84–85). In note 7, he contrasts it with mathematical modality as Putnam described it in "Mathematics without Foundations." In my opinion, he misinterprets Putnam here. According to Putnam, it is surely mathematically possible, say, that there should be no sets of uncountable rank, although it is a theorem of ZF that there are such sets.

Field does apparently regard his logical modalities as first-order; see the remarks on NB, p. 88. Putnam's treatment of set theory has the modal operator do some of the work of second-order apparatus. In the text, at least for the purpose of giving eliminative structuralism the best possible run, I have used modal statements where second-order quantifiers are assumed to have a standard reading. Of course this is a further step than Field takes down the slippery slope toward mathematical modality.

hold that the necessity of mathematics is strictly logical, one's idea of logical necessity will not be much different from mathematical necessity as here understood.

Mathematical truths are no doubt necessary in the "metaphysical or broadly logical" sense considered earlier; hence, one might ask why a particularly mathematical modality should be appealed to for an interpretation of the necessity of mathematics. Questions of metaphysical modality have to do with the nature of objects of this or that kind. In practice, the kind of objects is generally one that we know from experience or from some other relation of the objects to ourselves and represents some aspect of ourselves or the natural world (or the supernatural if there should be such). There are necessities considered in the literature on these matters where specific kinds of object seem not to be involved. An example would be the necessity of true statements where the identity predicate is flanked by names. But these also would hold if the necessity is understood in a logical or mathematical sense. Because of the generality of the content of mathematical statements, they resemble the necessity of identities rather than such cases as that gold has atomic number 79. Indeed, the structuralist view of mathematical objects in effect denies that mathematical objects have "natures" at all. The necessity of mathematics should have the same formal character as the content of mathematical statements themselves.

When we turn to possibility, moreover, in the case of mathematical possibility we do not have the problems noted earlier with metaphysical possibility. The friends of metaphysical possibility, where it is to go beyond physical or natural possibility, have to make judgments on the basis of intuitions presumably expressing their understanding of the notions involved in the statements whose possibility is being assessed. This must in general be done without the help of a developed body of theory. Intuitive judgments of mathematical possibility are also made, for example concerning structures of sets that satisfy large cardinal axioms or determinacy assumptions; and in such cases there has been uncertainty and controversy. In contrast to the metaphysical case, it is possible to attack the question by developing mathematics on the basis of the axioms questioned and seeing how the theory that emerges coheres with the rest of mathematics. Thus there is a richer and more exact theoretical setting in which intuitions can be tested. Questions of consistency can also sometimes be settled by less problematic methods, as, for example, with those set-theoretic statements

§15. Mathematical modality

that follow either from Gödel's axiom of constructibility or from Martin's axiom.

Another relevant observation has been made by Geoffrey Hellman.[19] This is that we might express some mathematical necessities as counterfactuals: "If $\langle N, 0, S \rangle$ were a ω-sequence, then ..." In counterfactuals in everyday language, there is a background of relevant conditions that is not explicitly stated. This is not true in these mathematical cases, and the problems that are notorious in the theory of counterfactuals do not arise. But it is not so clear that they are absent with metaphysical modalities.

Let us now return to the question which of the different modal notions considered above is the appropriate one for stating eliminative structuralism on modal lines. Physical or natural possibility is too restrictive, as we indicated earlier in our remarks on nominalism: It demands too much to ask that the structures considered in mathematics be physically possible; indeed, in the case of higher set theory, there is every reason to believe that they are *not* physically possible.[20] At the other end of the spectrum, formal logical modality was found either to be an awkward notion generally or not in the end to differ from mathematical modality. Thus, mathematical and metaphysical modalities are the survivors.

It should be clear from this discussion that, in my view, mathematical modality avoids some difficulties that metaphysical modality is subject to and that it is altogether more appropriate to our purpose. But it may seem that in an eliminative structuralist account of mathematical objects we are dealing with statements that contain only logical expressions essentially and therefore that the two kinds of modality will not differ. This is indeed true on certain understandings of mathematical modalities. But if one takes the language of mathematics at face value, then the usual intuition that mathematics is necessary has the consequence that any pure mathematical objects that there are exist necessarily. If we take metaphysical necessity to be a unitary notion, it should certainly have this property. However, thinking of mathematical existence as potential seems to underlie eliminative structuralist proposals, and it has its attractiveness outside that context. That is a reason for interpreting the necessity in (3) and (4) below as mathematical necessity, where this is to be distinguished

---

[19] *Mathematics without Numbers*, p. 36.
[20] *Mathematics in Philosophy*, pp. 191–193.

from metaphysical necessity in particular in this respect.[21] But many of the remarks later will be independent of this choice.

However, it is not evident that mathematical modality is a unitary notion. In commenting on Hartry Field, I attributed to Putnam a notion of mathematical possibility that allows it to be mathematically possible that there should be no sets of uncountable rank (see note 18). I consider that the natural notion for many purposes. But it will be replied that the axioms of ZF are true, and it is a theorem of ZF that there are sets of uncountable rank. Since mathematical truths are necessary, it is necessary that there are sets of uncountable rank. Since that's a contradiction, it seems that there have to be different standards of necessity and possibility at work. Different points of view will offer different ways of resolving this difficulty. One will be presented in the next section. But our preferred solution will be offered in §18.

§16. Modalism

By "modalism," I mean the program of eliminating mathematical objects in favor of modalities, or the thesis that mathematical objects can be eliminated in this manner.[22] The simplest version of modalism would simply expand the resources of the if-thenism considered in the last chapter by claiming the possible existence of the structure in the framework of which mathematical statements are interpreted; for example, statements about natural numbers could still have the canonical form

(1) $(\forall M)(\forall x)(\forall f)[\Omega'(M, x, f) \to A(M, x, f)]$

---

[21] Cf. *Ibid.*, pp. 327–328. It should be clear to the reader of that essay that I allow mathematical modalities to be *de re*. I mention this here because Timothy McCarthy points out some technical difficulties for Putnam's modalist interpretation of set theory unless his modality is allowed to be *de re* ("Platonism and Possibility," pp. 288–289). McCarthy, however, seems to suppose that if that is conceded, then the modality is metaphysical in the above Kripke-derived sense. I do not agree. Moreover, in spite of remarks that give his interpretation a basis in the text, I think McCarthy is probably wrong in attributing to Putnam an intended interpretation of mathematical modalities close to what above is called strictly logical. It is this that gives rise to the difficulty he points out.

[22] The term has been used in roughly this sense by other writers, for example Rheinwald, *Der Formalismus*, ch. III. The inventor of modalism was perhaps Putnam, who, however, seems to repudiate it at the outset ("Mathematics without Foundations," p. 57). But what he is denying is not that mathematical objects *can* be eliminated in favor of modalities but that this account of them is in all respects clearer or more fundamental than taking mathematical language at face value as referring to mathematical objects (the "mathematical-objects picture").

§16. Modalism

(in effect (2.4) of §11). But a presupposition for number-theoretic statements in general would be the possibility of a simply infinite system, i.e.,

(2) $\quad \Diamond (\exists M)(\exists x)(\exists f)\, \Omega'(M, x, f).$

Once we have introduced modality, however, it would be more in keeping with the idea to hold that the mathematical truth of A consists, not in the *logical* truth of (1), but rather in its *necessary* truth. This can readily be incorporated into the canonical form by replacing (1) by its necessitation. There is then a simplification, because the mathematical truth of A will consist simply in the truth of its canonical form, which in the case of a number-theoretic statement will be:

(3) $\quad \Box (\forall M)(\forall x)(\forall f)\, [\Omega'(M, x, f) \to A(M, x, f)].$

It is also possible to take the necessity operator as obviating the quantifiers; that is, we can think of the canonical form as simply

(4) $\quad \Box [\Omega'(N, 0, S) \to A'(N, 0, S)].$[23]

On the basis of the discussion in the previous section, I will assume that the modal operators are understood either in the sense of mathematical modality or of metaphysical modality. Modalism seems to give a satisfactory solution to the problem of nonvacuity. (2) demands no more than that a simply infinite system should be possible, in a sense that may require some further explanation, but which (at least if one of (i)–(iii) of §15 is chosen) is weaker than physical possibility. An analogous claim would hold in the case of other structures. There may be an epistemological problem about how we know a statement such as (2) to be true. But its truth seems to follow from the supposition that theories in physics describe coherent possibilities, and perhaps it can be seen in more direct and intuitive ways, as I will argue in §29.

The modal version of eliminative structuralism offers a resolution of the apparent ambiguity of the notion of mathematical modality noted at the end of the last section. It allows seeing the ambiguity not as one in the modality but in the mathematical statements said to be necessary or possible. Let A be the set-theoretic statement that there are sets of

---

[23] (4) is obviously equivalent to the result of replacing '$\to$' by the strict conditional in the result of dropping the outer quantifiers in (1). Whether the difference of (3) and (4) is significant depends on the interpretation of the modal operators and on refinements of modal logic.

uncountable rank. Considering it as a theorem of set theory, the modal translation might render it as of the form

$$\Box(\forall R)[\text{ZF}_2(R) \to A(R)],$$

where R is a binary predicate symbol, and $\text{ZF}_2(R)$ and $A(R)$ are respectively the conjunction of the axioms of second-order ZF and A, with $\in$ replaced by R. Assuming that the modal logic satisfies at least S4, the translation *is* necessarily true. The possibility of the nonexistence of such sets would be the possibility of a structure, say of the hereditarily finite sets, that does not admit them. This can be expressed directly in the second-order modal language, and it is not incompatible with the translation given above. Thus the "modalist" way of rendering mathematical statements allows for a unitary notion of mathematical modality.

The translation of arithmetic might be objected to on constructivist grounds; in the context of classical logic, which we have not questioned, (2) seems to state the possibility of an "actual" infinity. Indeed, a more thoroughgoing modalism might replace (2) by a weaker statement, to the effect that any structure of the type of an initial segment of the natural numbers can be extended.[24] I will not pursue or reply to this sort of objection here, since even the weaker statement allows various means of interpreting classical arithmetic, and constructivism is a large subject in itself.

There are two types of objections in the literature that I do wish to consider. The first questions the faithfulness to mathematical discourse of the canonical forms the modalist offers. The second questions whether, given the apparatus that the modalist's translations must use, there really is a reduction of reference to mathematical objects.

Thus our first question is whether it is at all reasonable to suppose that the modalist's canonical forms represent what the mathematician actually means by a mathematical statement of the relevant sort. Turning in particular to arithmetic and thus to (3) or (4), one obvious objection

---

[24] In the logical context we are assuming, this would follow from the sort of statement claimed to be intuitively evident in Chapter 5, §29. It should perhaps be pointed out that a "modalist" treatment of arithmetic using such a weaker statement seems to require quantifying into modal contexts and thus an interpretation of the modal operators that goes with that; see for example *Mathematics in Philosophy*, p. 47 n. 10. (In line 3 from below of the portion of this note on p. 47, 'members' should be 'numbers'; in line 17 of the portion on p. 48, '$(\forall x)Zx$' should be '$(\exists x)Zx$'. I am indebted to the late J. M. B. Moss for pointing out these and a number of other such misprints, which are corrected in the 2005 paperback edition.)

§16. Modalism

can, I think, be discounted. When the number theorist talks of numbers, surely he does not refer to objects of a merely hypothetical domain; although one ought perhaps not to make too much of what is involved in the idea that number theory involves genuine reference to objects called numbers, genuine reference there is. But the emphasis we have placed on the problem of nonvacuity should make clear that the views we have considered do not regard the reference involved as reference to purely hypothetical objects. The canonical form (3) or (4) arises in a context in which there is the presupposition (2); (2) plays a role analogous to that played by existential presuppositions in the use of singular terms.

Nonetheless, there is a related discomfort, which perhaps will be felt more strongly about (3) (as well as its ancestor (1)) than about (4): Do the intuitions behind structuralism really support the interpretation of arithmetic statements as of this *explicitly* general second-order form? The generality also appears in the presupposition (2), since that the possibility of which is presupposed is an existential generalization. The vagueness of Dedekind's talk of abstraction perhaps had its point. Even if we cannot single out *one* simply infinite system that is the numbers *par excellence*, perhaps it is falsifying the sense of discourse about natural numbers to take the step we took in interpreting Dedekind, of taking arithmetical statements to be really about *every* simply infinite system.

It should be observed that an objection of this kind can be made to any of the canonical forms proposed by different versions of eliminative structuralism. What force is granted to it depends on the importance of eliminating mathematical objects relative to other objectives. Given a domain of mathematical objects that, like the natural numbers, forms a second-order definable structure, statements about them will implicitly have the same kind of generality, by the counterpart of Dedekind's isomorphism theorem. Putting this generality into the explicit content of the statements will for some be a small price to pay for avoiding an ontology of mathematical objects. A noneliminative version of structuralism should, however, be able to avoid it.[25]

---

[25] As does the position of Field, who insists on taking the language of mathematics at face value and then claims that statements involving reference to pure mathematical objects are not true. This view is reiterated in "Is Mathematical Knowledge Just Logical Knowledge?" where Field enlarges the logic he admits to include modalities and adopts a position having some kinship with what I am calling modalism.

§17. Difficulties of modalism: Rejection of eliminative structuralism

Let us turn now to the second type of objection to modalism, whether the modalist's apparatus really does offer an elimination of mathematical objects. A first question would touch any reductive program making use of modality: Does not the modal operator itself involve us in an ontology of possible worlds? Although possible worlds are not paradigmatic *mathematical* objects, they are in one way or another problematic enough, so that eliminating mathematical objects in favor of possible worlds is a dubious gain.

There is no need to dwell on this objection, since the ontology of possible worlds arises not directly from modal discourse but from a semantical treatment of it that was expressly designed to use conventional (extensional) mathematics. If we simply take modalities at face value, there is no reason to take a statement, say, of possibility as involving the existence of anything. Commitment to possible worlds only becomes a problem in a framework in which the modalities are primitive if the modal logic is powerful enough to simulate quantification over possible worlds. Such a logic may be needed for some modalist treatments of set theory,[26] but in this case there is another more intuitive objection to the eliminative program, as we shall see.

Second-order logic, however, is still a problem for the modalist. The discussion of §13 did not leave him in a significantly better position than the nominalist. In our discussion of arithmetic, it appeared that the only alternative to using second-order logic in eliminative structuralist reductions was a relativism that we found unacceptable. Putnam still takes that position in his modalist treatment of arithmetic in "Mathematics without Foundations" (pp. 47–48). When he turns to set theory, however, he claims to do better. He seeks to capture the notion of a *standard* model of set theory by means of first-order modal logic; he asserts that this notion "can be expressed using no 'non-nominalistic' notions except the '□'" (p. 57). Thus the difficulties posed by second-order logic would be bypassed.

The models in question are to be certain graphs; Putnam writes:

> The model will be called standard if (1) there are no infinite-descending 'arrow' paths; and (2) it is not possible to extend the model by adding more 'sets' without adding to the number of 'ranks' in the model. (p. 57)

---

[26] *Mathematics in Philosophy*, pp. 324–325.

§17. Difficulties of modalism

This definition is not on its face first-order. But it appears that Putnam intends the graphs in question to be thought of as individuals, so that the modalized quantification over graphs in (2) is first-order. Moreover, it seems that both the "paths" of (1) and the "ranks" of (2) are to be internal to the model. How the definition is supposed to work, without assumptions about graphs that would defeat the purpose, is still not clear to me; it is not clear to me how *any* definition of the sort he proposes can work given the non-finite axiomatizability of Zermelo set theory. Neither Putnam nor anyone else has published a working out of the idea.[27]

The problems concerning second-order logic may not seem fatal to everyone. We also might point out that one serious problem we raised, that of nonvacuity, seems to be solvable on a modal nominalist basis at least for arithmetic, and it will also be solvable for classical analysis if (like Field) one is willing to grant that the physical world instantiates in some way or other one of the standard Euclidean or non-Euclidean geometries. Even without this hypothesis, a lot of mathematics can be rescued: Because, as we pointed out earlier, many basic results of analysis can be proved in rather weak theories, the modal nominalist solution to the problem of nonvacuity is available for a lot of everyday analysis. Moreover, the eliminative structuralist who admits second-order logic of course has second-order arithmetic at his disposal, that is, classical analysis with the real numbers interpreted as second-order entities rather than as objects.

The question remains how from this point of view one might convince oneself of the possibility of more elaborate structures, either defined in higher than second-order terms or involving domains of objects of higher cardinalities. Any structuralist approach to set theory involves admitting the possibility of such structures; conversely, set theory allows rich possibilities for constructing models of theories and therefore of showing the possibility of structures where no more nominalist way of doing so is available.

Modalist eliminations of set existence assumptions have been proposed, such as that of Putnam in "Mathematics without Foundations" just discussed. At least with the help of second-order modal logic, the technical part of such a program can be carried out. But then, it seems

---

[27] In understanding Putnam's definition, I have been much helped by Timothy McCarthy's exposition (see note 21), in spite of the disagreements noted above. But I do not see how his own version would work without ordinals external to the model to serve as ranks. In correspondence, McCarthy has proposed a new reconstruction which, however, does not purport to give the kind of eliminating translation that the use of Putnam's idea for the eliminative structuralist program would require.

to me, any such elimination faces a very simple and fatal difficulty. The nonvacuity assumption, of the form of (2) or perhaps some weaker alternative, will still have to be made. But then the concept of object in terms of which objects satisfying the required conditions are possible is a very general one; there is no reason to believe that structures of the required kinds are possible where the objects involved have any of the characteristic marks of concreteness. In cardinality they will outstrip anything that can be represented in the physical world. One may, like Putnam, begin with a spatial conception that models some aspects of the concept of set, and then conclude that a *space* is possible whose cardinality is high enough so that it can contain a standard model of some set theory.[28] Here, I think, there is a slide between the notion of physical space and the notion of a space as a structure of the general sort considered in geometry. Putnam is, to be sure, not claiming that such a space is *physically* possible. What he claims is the mathematical possibility of a physical space satisfying certain conditions. If one goes beyond conditions of a geometric character, and the cardinality, what is to be added to make this possibility that of a *physical* space? And, more to the point, of what relevance would it be to the acceptability of set theory as *mathematics*?

It might be thought that matters are different if we suppose that there is possibly some structure of the *mind* that satisfies whatever conditions are required, or that there is some plausible weaker modal condition on operations of the mind whose truth is sufficient for an eliminative modalist interpretation of set theory. Many writers on the concept of set, who could appeal to remarks of Cantor himself, have given a central place to a mental operation of "collecting" in their account of the concept. If the operation gives rise, by a kind of synthesis, to the set as a new object, it does not help in the present context.[29] But the role of the object can perhaps be played by the operation itself.[30] Then, the idea goes, quantification over sets can be interpreted in terms of possibilities of collecting. Such a view, however, meets the same kind of difficulties. It is not only

---

[28] "Mathematics without Foundations," p. 57; cf. *Mathematics in Philosophy*, p. 192 n. 32.

[29] This was the proposal of Husserl in *Philosophie der Arithmetik*, which, in spite of its psychologistic guise, is of particular historical interest because Husserl was Cantor's colleague at the time. Husserl evidently thought later that the basic idea was not essentially bound up with the psychologism that he had rejected, since it reappears in the late *Erfahrung und Urteil*, §61. The most immediate difficulty with Husserl's proposal seems to me to be not psychologism but how to make it apply to infinite sets. The idea is discussed further in §34.

[30] As is proposed by Philip Kitcher in *The Nature of Mathematical Knowledge*, ch. 6, for whom, however, the operation of "collecting" is not mental but an overt action.

§17. Difficulties of modalism

that we are asked to accept as mathematically possible iterations of mental operations that go well beyond the actual capabilities of the human mind. Such an extension of human capability, whether or not it is thought of mentalistically, seems to be involved in the much more entrenched idea of computability or constructibility "in principle."[31] Computations, and at least the more elementary constructions, however, can at least be thought of spatiotemporally. But for the possibility of "collectings" of a large transfinite number of objects, iterated through a large transfinite sequence, we need at least the possibility of a "time" that would be much richer than actual space-time or any possible space-time that physical considerations would lead us to consider.[32] The concept of mental operation is thus extended by rather extravagant analogies. We may see the concept of object that allows such objects as large transfinite sets as an analogical extension of a concept of object that doubtless begins with the physical objects, organisms, and events of everyday experience. But particularly if it is understood so as to be purged of some of the pictures traditionally associated with platonism, this extension is limited to what the mathematics actually requires. It is hard to see what problem with mathematical objects is really helped by assuming the possibility of a physical space, or a system of mental operations, with the kind of structure higher set theory requires.[33] It is probably only the (perhaps inchoate) sense that there is some scandal to human reason in supposing that there are (or even that there possibly are) *objects* that stand in *relations* that the conception of a particular mathematical structure such as a model of set theory calls for, with perhaps nothing more to them than that (or only something more of a very abstract character) that leads philosophers to entertain such ideas. In my view, such an attitude should be turned on its head. We are able to understand higher set theory and to have enough of an intuition concerning the principles of set theory to create a highly developed theory, which shows no sign of being inconsistent or incoherent, and about which there is too much agreement for it to be *ad hoc*. If

---

[31] I discuss the question in what sense this is so in "What Can We Do 'In Principle'?"
[32] Compare the remarks on Wang in Essay 10 of *Mathematics in Philosophy*, pp. 272–280, and on Kitcher in my review of *The Nature of Mathematical Knowledge*, pp. 133–134.
[33] In recent years, a favorite reason for seeking to eliminate mathematical objects has been the dilemma posed by Benacerraf in "Mathematical Truth." It is easy to see that the kind of assumption considered here is of no help with Benacerraf's problem. First, only what is actual stands in causal relations to our knowledge; second, once we consider the sort of space or mental operations that is required, we no longer have any idea of what a causal theory of knowledge involving them would be like. More generally, we do not have an idea of what a naturalistic explanation of such knowledge would be.

a "formal" concept of object is the only one that can interpret set theory without assuming extravagant possibilities, that is a strong argument in favor of such a concept of object.

§18. A noneliminative structuralism

Eliminative structuralism begins with the observation that there is in the end nothing to the pure abstract objects of mathematics other than their being related in certain structures, and infers that talk of objects with so little of an intrinsic nature must be only a *façon de parler*. The conclusion of our discussion is that, at least when we come to higher set theory, we have no reason to suppose even the possibility of objects making up the structure that are more "concrete" than pure mathematical objects. A structuralist view of higher set theory would then oblige us to accept the idea of a system of objects that is really no more than a structure. But then there is no convincing reason not to accept it in other domains of mathematics, in particular in the case of the natural numbers. It would be highly paradoxical to accept Benacerraf's conclusion that numbers are not objects and yet accept as such the sets of higher set theory. One might, indeed, conclude that set theory should be an exception to a structuralist conception of mathematical objects. In §19, we will indeed consider an argument that would imply that conclusion. First, however, I wish to set forth a version of the structuralist view that does not undertake to eliminate reference to mathematical objects.

If we abandon eliminative structuralism, we need no longer seek a canonical form for the languages of various mathematical theories that reinterprets them in the manner of the proposals we have been discussing. We can take the language of mathematics more at face value. We will not regard talk of natural, real, and complex numbers, sets and functions on them, or various spaces and their points and mappings as a *façon de parler* for something else that would be more acceptable from a nominalistic or other reductionist point of view. If so, how is our position still structuralist? The main point is that taking the language of mathematics at face value does not require us to take more as objectively determined about the objects about which it speaks than that language itself specifies.

Let us concentrate for the moment on the natural numbers. I assume that we have at our disposal a language that contains '0', 'S', '=', and probably additional predicates or functors, such as '+' or '×'. I will also assume that it includes the apparatus of quantificational logic, with either

§18. A noneliminative structuralism

a predicate N read as 'natural number', or the natural numbers as the intended domain, or the natural numbers segregated as a type.

To see how our view of the natural numbers is structuralist, it is useful to consider another view that is sometimes advanced, according to which natural numbers are *sui generis*; in particular, no natural number is identical with an object given independently. Natural numbers are just what they are; in particular they are not sets or other objects that might be introduced to cash in Frege's idea that numbers are "logical objects." The view says nothing about whether other objects might be reduced to numbers, so that the qualification "given independently" is important. A classical version of the view, at least, would say that natural numbers are nonspatial and nontemporal and do not stand in causal relations. So this much could be said about their "nature." It's not clear what such a view would say about the role of numbers as cardinals and ordinals; of course all would agree that it is necessary that they should be able to play that role, but versions of the *sui generis* view might differ as to whether either the cardinal or ordinal role is essential to what objects they are and whether one is prior to the other.

Such a view is not obviously incompatible with the aspect of the structuralist view stressed by Bernays, that existence for mathematical objects is in the context of a background structure. But the *sui generis* view implies that for the natural numbers that structure must be *the natural numbers* in some distinguished sense, and it must distinguish natural numbers from all objects given otherwise, either by making such identities false or ruling them out of the language by singling out a type of natural numbers.

One further step that the structuralist view takes is to reject the demand for any further story about what objects the natural numbers are, except in certain special contexts, in particular, to reject the idea that they are essentially cardinals, or, for that matter, essentially ordinals. I will put aside for the moment the question what to say about their abstract character.

But that seems to imply that if an isomorphic copy of the natural numbers can be constructed in another structure, then it is a question whether the copy is not just as good as the original, an equal claimant to be *the* natural numbers. Some copies do not have a good claim, such as progressions constructed within the natural numbers or in some other way that obviously presupposes the natural numbers. But others are not so easy to rule out, the most prominent being the number sequences constructed in set theory and higher-order logic.

Suppose we accept in some context a progression constructed in some larger structure, most likely in set theory, as being the natural numbers. That seems to be what is done when someone developing arithmetic in set theory defines 0 as a certain set, and Sx by some operation on sets (say as $x \cup \{x\}$), and Nx in such a way that induction (as a generalization about sets or classes) will hold. Why should we not accept this progression as being the natural numbers? One may have the worry that it is not really a progression but a nonstandard model. If (as we can reasonably expect) induction can be proved for any predicate of the larger theory, we should be able to prove that our structure is not a nonstandard model. But the existence of nonstandard models of the larger theory might lead to doubts as to whether this result can be taken at face value. We will not consider such doubts now but will address them in §§48–49.

Now the question is, what reason is there not to accept our progression as the natural numbers? It cannot be a difficulty in developing pure number theory in the theory of the larger structure, at least if the latter is well chosen. In particular, set theory would offer the resources for developing the relations to other structures that are needed to prove more sophisticated results in number theory, even those that are in the end provable in PA. For reasons developed in §14, the difficulty cannot be a concern about the application of arithmetic. Thus it seems that it has to be some philosophical concern, in a broad sense metaphysical, about the identity of the natural numbers as objects.

Structuralist views have been encouraged by the once much debated "multiple reductions" problem, that is, that in the construction of the number systems in set theory (or higher-order logic, on the model of *Principia Mathematica*), the constructions do not give a unique identity to the numbers involved. An example that goes back to the nineteenth century is the two classic constructions of the real numbers, as Dedekind cuts or as equivalence classes of Cauchy sequences. But what has dramatized the matter for philosophers is the existence of different progressions of sets that offer usable models of arithmetic, in particular the finite von Neumann ordinals and the so-called Zermelo numbers, where 0 is defined as $\varnothing$ and Sx as $\{x\}$.

Different reactions have been expressed to this situation. (For simplicity, we will consider only the natural-number case.) Roughly, they take the form of saying that all identities of numbers and sets are false, that there is no fact of the matter as to whether a natural number is identical to a set (even if the candidates are only those that arise in constructions actually used), and that such identities are context-dependent. There is

§18. A noneliminative structuralism

a fourth reaction, of a different nature from the first three, that numbers are not objects at all.

The first reaction fits well with the *sui generis* view of the natural numbers that we discussed above. But it is actually otherwise quite unnatural. It requires (as talk of natural numbers in more comprehensive first-order theories would) regarding such identities as meaningful, but rejecting identities actually used in mathematics, in the construction of arithmetic in set theory, and in other contexts such as representing numbers in the λ-calculus.

The second would fit well a canonical language in which there is a syntactically segregated type of natural numbers, which is also congenial to the *sui generis* view. Other ways of realizing it will be discussed after we consider the third. The latter seems a plausible formulation of the structuralist intuition and allows that someone working within a set-theoretic construction is nonetheless talking about the natural numbers.

This third reaction is tempting in that it seems in accord with the idea of structuralism and is maximally accommodating to the actual use of mathematical language.[34] But it faces a serious difficulty. It seems to have as a consequence that in one context, call it c (perhaps a lecturer developing arithmetic in ZF), '2 = $\{\varnothing, \{\varnothing\}\}$' is true, while in another context c' (perhaps the development of the Zermelo numbers in §12 of Quine's *Set Theory and its Logic*), '2 = $\{\{\varnothing\}\}$' is true. But now it is surely the case that '$\{\varnothing, \{\varnothing\}\} \neq \{\{\varnothing\}\}$' is true in both contexts. But this implies that either the numeral 2 has changed its reference from c to c', or one of the two set terms has. The second hypothesis is rather implausible.

There are, it seems to me, good reasons for saying that it is the numeral that has changed its reference. It is the only plausible response consistent with the idea of the third reaction. And it seems called for by the general idea that mathematical existence, and with it reference to mathematical objects, is in the context of a structure. Although there is no change in background structure between identifying natural numbers with von Neumann numbers and identifying them with Zermelo numbers, other changes of background structure do motivate the idea that numerals change reference, for example, from considering the natural numbers as a stand-alone structure and taking the background structure as consisting of sets.

---

[34] It is at least suggested by a remark of mine in "Structuralism and Metaphysics," p. 76.

In §14, considering the case of the natural number 2, the rational number 2, the real number 2, and the complex number 2, it seemed forced on us to say that across these contexts 2 is more a role than a definite object. The difficulty we have raised for the third reaction to multiple reductions implies that already the natural number 2 has this character. That would be a way of understanding the apparently metaphorical talk by other writers of numbers and other mathematical objects as "places in structures" or "positions in patterns." But before drawing this conclusion we should revisit the second reaction.

Each of the other reactions implies that there is an element of fiction in the development of arithmetic in set theory as well as in traditional logicism, though in the latter case unintended by Frege or Russell. Either the identity sign is used without the speaker or writer quite meaning it, or he does not quite mean it in saying that in this context numerals designate natural numbers and that the objects satisfying the defined number predicate are the natural numbers. They would be surrogates for the natural numbers, as surrogates for many objects (not necessarily mathematical) arise in the construction of models. The latter choice is probably the more workable one, since the former would make the identity sign ambiguous within set-theoretic usage.

This very moderate degree of fictionalism about mathematical reference does not trouble me much as such. But it seems to presuppose that there is genuine, nonfictive reference to them, which will be present only when the background structure is the natural numbers. A manner in which this is realized that fits the second reaction is by treating the natural numbers as a type, so that identities with other objects cannot be stated. This is, of course, the way they are treated in some attractive formal theories, for example, the typed theories of constructions used by Martin-Löf for intuitionistic mathematics and by Tait for classical mathematics. But is it really evident that this is what reference to natural numbers in informal mathematics aims at, as opposed to a first-order theory in which such identities can be stated? And are we really justified in privileging in this way one framework for talking of numbers?

In this connection we might consider the introduction of natural numbers by Dedekind abstraction, as described by Tait.[35] Suppose we have a structure $\langle M, a, f \rangle$ that is a progression, that is, it satisfies Dedekind's definition of a simply infinite system or some other condition doing the same work (such as the treatment of induction in §47, if the

---

[35] "Truth and Proof," p. 87 n. 17 (p. 369 n. 12 in original publication).

§18. A noneliminative structuralism

progression is explicitly given). It may be a system of sets of the sort we have been discussing, or it may consist of quasi-concrete objects as proposed in chapter 5; there are no doubt further alternatives.[36] Then we introduce a new structure $\langle N, 0, S \rangle$ and an isomorphism onto the given one. This will evidently also be a progression. But it conforms to the basic structuralist intuition in that the number terms introduced do not give us more than the structure. This procedure provides a very suggestive model of how reference to natural numbers might be introduced.[37] But it gets its force from the use of a typed language. Thus, the question arises what is to prevent us from later, for some specific purpose, speaking of numbers in a first-order language and even affirming identities of numbers and objects given otherwise. If it is taken as the last word about reference to natural numbers, it faces the problems discussed earlier.

One might ask on what grounds we can say that Julius Caesar is not a natural number. The conclusion that natural numbers are in the end roles rather than objects with a definite identity might imply that we cannot rule it out. That would be implied by the view that *any* progression can be treated as the natural numbers. But although there is evidently scope for convention with respect to what progression is treated as the natural numbers, it does not follow that anything goes. A convention in which a certain number (say 15) is Julius Caesar is not a very good convention, for example, because it either makes the existence of Julius Caesar necessary or that of 15 contingent, or because it would upset the relation of at least one of them to space and time. That may seem question-begging. If we do not assume it false that Caesar $= 15$, how can we be sure that the one contingently exists and the other necessarily? I would reply that this difference is a feature of our informal talk about Caesar and about numbers, the "grammar" of these expressions before we offer a more explicit account of what objects 'Caesar' and '15' designate. There is no prior deep metaphysical reason why Caesar and 15 cannot be treated as identical.[38]

---

[36] For example the finite cardinals of a neo-Fregean introduction of arithmetic.
[37] These remarks indicate what I would now defend of my earlier remark, "Tait's idea of Dedekind abstraction . . . can be adapted to give a convincing description of discourse about *the* natural numbers" ("The Structuralist View," p. 336).
[38] For further discussion see the Appendix to "Structuralism and Metaphysics." However, it is not made clear there that the context-dependence of some identity statements involving numerical expressions can imply context-dependence of what these expressions designate.

Whatever one's stance toward the specific problems we have been discussing, the considerations underlying structuralist views of mathematical objects have sometimes been expressed by saying that mathematical objects are "incomplete," meaning by this that there is only a certain specific range of predicates such that there is a fact of the matter as to whether they are true of the object in question. In view of the fact that there are different background structures relative to which we can use the natural number vocabulary, a definite range of predicates can be given in advance only when that is specified. But the description is itself incomplete, for it neglects the fact that the *relations* of a structure are themselves given only by formal conditions, and in different realizations of a type of structure, not only will there be different domains of objects, but these relations will have different realizations. The "incompleteness" of the pure abstract objects of mathematics is for this reason more radical than that of another kind of object often said to be incomplete: fictional or nonexistent objects (assuming, for the sake of argument, that there are such objects). Fictional objects are taken to be undetermined with respect to properties and relations whose holding or not cannot be reasonably inferred from the story; a more drastic incompleteness obtains for Meinongian objects like the golden mountain, since what we have to go on is only that it is a mountain and that it is golden. In these situations, however, we do not envisage any reinterpretation of the predicates applied to the objects; Sherlock Holmes is a detective in a sense that we can take to be fixed, also when we consider other detectives (real or fictitious). We have, independently of the story, an understanding of notions such as that of a detective, of a murder, of London, of Baker Street (because these are real places). There is at least some level of understanding of this kind of simple mathematical notions like addition, multiplication, or set membership, and of more complex ones such as curve or surface or computation. On a purely structuralist view, however, none of these notions is fixed in a way in which the nonfictional vocabulary used to describe a fictional situation is. Their role in mathematics, rather, is in the genesis and motivation of mathematical conceptions, and in the application of mathematics.

Another difficulty that will be raised is whether we are still entitled to reject Benacerraf's claim that numbers are not objects. The view we have defended implies that they are not definite objects, in that the reference of terms such as 'the natural number 2' is not invariant over all contexts. That is clearly part of what Benacerraf has in mind. But it is still reference to objects by a singular term. I find Benacerraf's train of thought at this

§18. A noneliminative structuralism 107

point difficult to follow, but that he means to deny this is suggested by the turn toward eliminative structuralism and perhaps more explicitly by the following statement:

> That a system of objects exhibits the structure of the integers implies that the elements of that system have some properties not dependent on structure. It must be possible to individuate these objects independently of the role they play in that structure. But this is precisely what cannot be done with the numbers.[39]

Whatever the interpretation of Benacerraf, what I'm most concerned to reject is the idea that we don't have genuine reference to objects if the "objects" are impoverished in the way in which elements of mathematical structures appear to be.

Another problem that may arise for our view is what to say about a number of properties of pure mathematical objects apparently shared by all of them: that they are abstract, that they are not in space and time, that they do not exist contingently. It might be objected that these properties go beyond being "places in a structure" but cannot be treated as we treated external relations. It then appears incompatible with structuralism to admit them.[40] I would regard these as what one might call metaproperties, consequences of the general grammar of the language of pure abstract objects. The scope for convention in identifying the natural numbers with given progressions implies that this grammar might sometimes be violated for special purposes. The structure of sets is too rich to be instantiated in space and time, but that is not true of the natural and real numbers. It may not be a good convention to use numerals to refer to objects located in space and time, but it is not logically absurd.

A further objection, apparently affecting any noneliminative structuralist view, was made briefly by John Burgess and more extensively by Jukka Keränen.[41] They see a problem concerning the individuation of objects on this view, if the structure that is their home has nontrivial automorphisms. To take some simple cases, for the complex numbers, there is such an automorphism interchanging i and $-i$. Thus it seems that the basic relations of the structure do not distinguish i from $-i$. Worse cases are Euclidean spaces of a fixed dimension (say the plane or

---

[39] "What Numbers Could Not Be," p. 291.
[40] That one might make such an objection was suggested to me by Øystein Linnebo.
[41] Burgess, Review of Shapiro; Keränen, "The Identity Problem."

3-space) or some simple graphs such as the four-element directed graph with edges so that $a \to b \to c \to d \to a$. In these cases, for every pair of elements of the structure there is an automorphism carrying one to the other.

Elsewhere I have replied to this objection by distinguishing basic from constructed structures.[42] Basic structures are those that are assumed in mathematics without the obligation to construct them within other structures. The prime examples would be the natural numbers, the real numbers, and structures of sets. Constructed structures, such as examples relevant to some general statement about structures of a certain type (such as fields) will naturally have properties inherited from the basic relations of the underlying structure, which can go beyond the relations of the structure that is constructed. The main claim of the reply is that the principal basic structures do not have nontrivial automorphisms. The complex numbers do not in my view have to be regarded as a basic structure. But the case of the Euclidean plane and 3-space, which of course do have nontrivial automorphisms, is more complicated.[43]

Although this reply meets the requirements of the structuralist view, a more dismissive attitude toward the objection is also possible. Given the conception of object presented in Chapter 1, why should we insist, for objects to be distinct, that there be *anything* that distinguishes them? This view might be implemented by treating identity as one of the basic relations of the structure. That is proposed by Hannes Leitgeb and James Ladyman in a paper building on a debate in the journal *Analysis* on the connection between structuralism and the Identity of Indiscernibles.[44] They defend this view by considering graph theory and give examples from graph theory of structures with elements that are otherwise

---

[42] "Structuralism and Metaphysics," Section IV.
[43] Ibid., pp. 70–71. In this case one element of what is called below the dismissive attitude toward the objection is required. In a prerelativistic view of the physical world, we have a structure consisting of points and physical bodies (and perhaps more) in which points are distinguished by their relations to bodies. Some points are for example literally on the surface of bodies. The claim is that this model is at least mathematically coherent. We can think of pure geometry as arrived at from a model like this by a Dedekind abstraction, in which only the geometrical structure is taken up. But then identity has to be taken as a basic relation of the structure for the points to be distinguished within the structure.
[44] Leitgeb and Ladyman, "Criteria of Identity and Structuralist Ontology." A more "metaphysical" point of view, which in general I resist, has been maintained by Fraser MacBride. See, for example, his "Structuralism Reconsidered." Some of the same issues arise with respect to particles in physics. See Saunders, "Are Quantum Particles Objects?" I am indebted to Saunders for calling the *Analysis* debate to my attention.

§18. A noneliminative structuralism

indiscernible.[45] A simple example is a two-element graph with no edges, which offers a mathematical model of some cases discussed in connection with the Identity of Indiscernibles. It appears from their discussion that graph theorists do not regard these graphs as constructed structures, but they are simple enough so that they could be if inquiry into the foundations of their subject required it. Leitgeb and Ladyman give to the dismissive attitude both intuitive force and some motivation from mathematical practice. But it seems likely that the debate will continue.

In this discussion, we have emphasized the point going back to Bernays that reference to mathematical objects is relative to a background structure. In some sense that is context-dependent, and it is also a matter of interpretation, particularly in the case of informal mathematical discourse. With respect to the former, we have emphasized cases in which context-dependence makes it not literally true to say that the same object is designated even by such straightforward terms as 'the natural number 2'. But that is not inevitable even when the background structure changes. In set theory, for example, the background structure might be the sets up to a certain rank; then if we consider the sets up to a higher rank, or even an absolute universe, we are not forced to say that terms designating sets of low rank change their reference. In fact, in the case of well-founded sets, it is not obvious that there are cases other than contrived ones where terms intuitively designating the "same" set have to change reference for the sort of reason we have given for numerals. A set like $\wp(\omega)$ will have to be different, for example, in the constructible sets and in a context where large cardinal axioms are assumed. But if one takes seriously the idea of the true power set, then one of these is nonstandard. A more likely analogue of the situation with numerals might arise in different category-theoretic constructions of sets. But I have to leave category theory out of account because of insufficient knowledge.

How are we to identify the background structure in a serious discourse in unformalized mathematics? Let us imagine it to be number-theoretic, so that the natural numbers will be especially salient. Still, we will have further objects of familiar kinds: functions, sets, numbers of other number systems, probably even structures as mathematical objects, such as groups, fields, and topological spaces.

---

[45] Considering this structure as a basic structure can be defended on the same grounds as are applied to the four-element graph mentioned in the previous paragraph. Any defense of the Euclidean plane as a basic structure will carry over to these cases.

Even before we consider ontological questions, two different ways suggest themselves of describing such a situation in structural terms. Each offers a formulation of the idea that in talking of mathematical objects, there is always a background structure in which the objects "reside." The first idea is that the objects referred to all fit into a single comprehensive structure, of which the most obvious example would be sets with the membership relation. The other is that we might have a number of different structures that provide "homes" for objects of different kinds. The second fits better with the language of informal mathematics, before logicians undertake to formalize it. It also provides a good setting for discussing the issues that arise for structuralist views.

Nonetheless, there is a reason from mathematical practice for adopting the first picture.[46] Work on an actual mathematical problem can often not be confined to the consideration of a very limited range of structures, such as those actually invoked in the statement of the problem itself. If the problem is difficult enough, it may be necessary to make connections with ostensibly remote parts of mathematics in order to arrive at a proof. It then becomes natural to represent the mathematician as working within a comprehensive theory, within which all the mathematical domains that turn out to be relevant can be constructed. Set theory is the theory that has been developed most fully for this purpose and remains the leading candidate for the role of framework for all of mathematics.[47]

This point of view leads to an objection to structuralism only if one takes it with excessive metaphysical seriousness. The framing of a part of mathematics in set theory involves representing as sets at least the numbers of various number systems, spaces from geometry, other structures described abstractly, and mappings from one structure to another. Of course, this can be done in well-known ways. But it is equally well

---

[46] This was urged on me by Yiannis Moschovakis.

[47] I use the term "framework" rather than the more common "foundation" because the latter has to a philosopher's ear misleading associations. Although she uses the latter term in this way, Maddy gives a very instructive discussion of the sense in which set theory is and is not a foundation for mathematics in *Naturalism in Mathematics*, Part I, Chapter 2. The terms "framework" and "framing" derive from Müller, "Framing Mathematics."

That there is a single comprehensive structure within which all of mathematics can be constructed can still be questioned on the ground that there is no single absolute universe of sets. This claim (with which I have elsewhere expressed sympathy) raises large and difficult issues. But nearly all actual mathematical work outside of higher set theory itself can be framed in a rather limited set theory. ZFC is more than sufficient.

§18. A noneliminative structuralism 111

known that the ways of doing it are not unique, and in many instances the choice of one rather than another is largely conventional and made for a particular purpose that is likely to lead to a different choice under other circumstances. It is an illusion that the framing of mathematics in set theory yields a determinate ontology for each individual part of mathematics.

The second picture perhaps captures how mathematics would proceed without making the choices involved in the construction of each structure by means of sets. According to it, natural numbers are given as natural numbers, rational numbers as rational numbers, real numbers as real numbers, and so on. Nothing forces questions of identity of elements of different structures among these to arise, and that is still the case if we consider functions mapping one of them into another. They might be prevented from arising by adopting a many-sorted or typed language, as we have suggested earlier.[48]

Another element of context-dependence concerns which relations and even objects count as the basic ones for what we would intuitively consider the "same" structure. Informal mathematical discourse would not choose among different interdefinable possibilities, for example, in Euclidean plane geometry between a formulation in which points and lines are individuals and incidence is a basic relation and one in which only points are individuals.[49]

Let us now turn to the concept of structure itself. I have resisted the interpretation of structuralism that would make it an interpretation of mathematical statements as *about* structures, thus giving rise to the question what manner of objects these are. On the contrary, a structuralist account of a particular kind of mathematical object does not view statements about that kind of object as about structures at all (except in the special case where structures are themselves mathematical objects, as in model theory). Still, a concept of structure is needed to state the view itself. The most fundamental notion of structure for this purpose is metalinguistic: The "domain" is given by a predicate, and the relations

---

[48] Such a framework would not allow the formation of mixed sets, whose elements are of different sorts or types. My conjecture is that most of mathematics gets along without such sets, but there is no principled reason for ruling them out. The framework envisaged would allow them to be coded in one way or another, for example as functions.

[49] This example was brought to my attention by Frank Veltman. One might offer it as an objection to eliminative structuralism; however, logical reconstructions in general tend to make choices where unformalized discourse usually does not.

and functions by further predicates and functors.[50] The most concrete way of giving a structure would be by predicates and functors that are antecedently understood, so that there is some independent verification of fundamental propositions about the structure (say, axioms for the theory of this kind of structure). Brouwer and Hilbert could be taken to have been attempting something like this in their descriptions of the intuitive basis of arithmetic. A less problematic, but also less philosophically interesting, case is where a type of structure is defined in the abstract and then examples are given from different branches of mathematics, but in this case, the "realizations" of one type of structure are simply found in another structure.

We can now see the difference between arithmetic and set theory in a different light: In the case of arithmetic, a realization of the structure can be described in the more concrete way by giving an intuitive model of which the most developed example is Hilbert's strings of strokes. In the latter case, the predicate 'x is a string of strokes' defines the domain, and the zero element and successor functor can also be directly explained.[51] Induction can be stated, at the cost of vagueness, in the metalinguistic way mentioned in the discussion of second-order logic in §13.[52]

We can then think of discourse about *the* natural numbers as introduced by Dedekind abstraction in the sense of Tait, although for reasons given above this is not the last word about reference to natural numbers.[53] The idea that the natural numbers are a unique structure, although it influences the particular form that a structuralist account of them will take, does not seem essential to the idea of a structuralist account of a kind of mathematical object. The corresponding uniqueness claim about the universe of sets is much more questionable, but that fact tends to make the application of the structuralist idea to set theory more rather

---

[50] As in §5, by a functor I simply mean a singular term with one or more argument places; hence the use of a functor does not of itself commit one to the existence of any function.
[51] See §28–29. In §29 and §41 we defend the claim that the elementary Peano axioms are intuitively known to be true when interpreted with respect to this model.
[52] Issues concerning induction will be pursued in Chapter 8.
[53] Because first-order arithmetic is not categorical, it might seem that impredicative second-order logic is needed for the proof of Dedekind's categoricity theorem. That is not true. To show that two predicates N_1 and N_2 define isomorphic domains, it is sufficient that for each we can apply induction to first-order predicates containing the other. See §49.

Tait, using an idea of F. W. Lawvere, points out that the categoricity theorem also holds in an intuitionistic setting. See "Against Intuitionism," p. 177 and n. 12.

§18. A noneliminative structuralism

than less plausible.[54] The absence of uniqueness would, however, give a kind of vagueness to quantification over all sets and possibly to other parts of set-theoretic language.

Many will not agree with this picture of number theory according to which the nonvacuity of the conception of natural numbers is made out by an example in which the objects are quasi-concrete. However that may be, the situation in set theory is different, and our understanding of set theory has to proceed by pulling ourselves up by our bootstraps. In its developed form in set theory, the concept of set draws on two elementary notions, neither of which is a purely structural one: that of a set as a "collection" or "totality" of its elements, and that of the extension of a predicate, that is, of an object associated with a predicate, with extensional identity conditions. A third notion of "plurality," motivated by plural expressions such as those discussed by Boolos (see §13) also might be distinguished.[55] Although these ideas would allow the construction of in some sense intuitive models, the idea of collection seems to me to depart from concrete intuition at least when it admits infinite sets, that of extension when it admits impredicatively defined sets. The result of these extensions, however, is that the elements of the original ideas that are unquestionably preserved in the theory have a purely formal character, as will be argued more fully in Chapter 4. For example, the priority of the elements of a set to the set, which is usually motivated by appealing to the first of these informal conceptions, is reflected in the theory itself by the fact that membership is a well-founded relation. It is important to our understanding of set theory and to the possibility of structures of sets, however, that for the bottom of the set-theoretic hierarchy we do have more intuitive models.

In this situation, what does the metalinguistic notion of structure mean? We can define "structures" to interpret the language of set theory by using our mathematical vocabulary, including the predicates '( ) is a set' and '( ) is an element of [ ]'. Without such vocabulary, or other

---

[54] On this point, earlier expositions of mine might have been misleading. I am indebted to a member of an audience at the University of Amsterdam for prompting clarification.

A paraphrase of set-theoretic language along eliminative structuralist lines will have "relativist" features like those of the first-order versions for number theory, as on no reasonable theory is the universe of sets even second-order definable. Because the intuition that the universe of sets is a unique structure is not so strong, this is a more acceptable result; even the consequence that the meaning of a set-theoretic statement can depend on a presupposed axiom system does not seem to me to be a fatal objection to an eliminative structuralist account of set theory.

[55] These conceptions are discussed more fully in §20.

mathematical vocabulary of similar abstraction, we will not be able to describe a structure satisfying the axioms of set theory; that is the "bootstrapping" aspect of the understanding of set theory. But otherwise, the metalinguistic conception of structure works in the same way as it does in the more elementary cases.

Why should I regard the metalinguistic conception of structure as more appropriate for my purpose than the set-theoretic? I have just denied an obvious reason, that it would enable one to describe structures for set theory independent of the concept of set. Indeed, the "bootstrapping" involved in the understanding of the concept of set weakens resistance to the use of the set-theoretic concept of structure here. What decides in favor of the metalinguistic conception in the present context is a feature of set theory itself: We want to talk of structures for set theory without supposing that their domains are sets. Also, the set-theoretic conception requires a decision about the ontology of set theory itself (say, in favor of ZF rather than a von Neumann-style formulation), which the most fundamental conception of structure should not make. These considerations are a symptom of the fact that giving a structure in the sense of introducing a (one-place) predicate, together with certain further predicates (in general not one-place) and functors that apply to the objects of which the first predicate is true, is, on a structuralist conception, the most elementary way of describing a kind of mathematical object; any method of generalizing about this, whether by taking structures to be set-theoretic objects or otherwise, will come later and will be subject to the limitations discussed already in §5.

At this point, we should revisit the difficulty about mathematical modality that arose at the end of §15. The upshot of the version of the structuralist view we have presented is that mathematical possibility is the most fundamental mathematical modal notion. We have just sketched ways of making out the mathematical possibility of the structure of natural numbers and of a structure of sets, possibly not a unique one. I would argue that it follows from what I would call the coherence of the theory of the structure. But much more would need to be said about coherence than I am prepared to say in this work.[56] It is by no means certain that this notion of possibility obeys classical modal logic. The assumption that it does underlies modalist eliminative structuralist constructions, and one can justify it by consistency proofs relative to standard mathematical

---

[56] This term has been used with a similar intention by other writers, for example Shapiro, *Philosophy of Mathematics*, esp. pp. 95, 132–136.

§18. A noneliminative structuralism

theories (for set theory, say, second-order ZF). But that justification may depend on the restricted context of the application.

But if mathematical existence is relative to a structure, it follows that the necessity of mathematical truths has a more internal character. We could express the necessity of the existence of sets of uncountable rank as the necessity that within the structure of sets as we normally conceive it, there must be sets of uncountable rank. The qualification "as we normally conceive it" is needed because we do have conceptions of more truncated such structures, for example, natural models of Zermelo set theory or of predicative theories. We could represent this necessity as conditional only by being more explicit in characterizing the structure than we normally wish to be, or than we could be without begging the question in favor of a negative answer to the disputed question whether we can take quantification over absolutely all sets at face value. Still, it resembles conditional necessity in having a presupposition.

Although I have presented what I consider to be a defensible version of the structuralist view, an outcome of our investigation is that structuralism is not the whole truth about mathematical objects. In the statement of the view using the metalinguistic conception of structure, we appeal to linguistic objects such as predicates and functors. These are quasi-concrete objects, and so long as they are viewed in this way the structuralist view will not hold for *them*. The relation of linguistic types to their tokens (and in general of quasi-concrete objects to their concrete "representations") is not an external relation in the sense of §14.

It will be objected that any mathematical theory, that of linguistic objects included, can be interpreted as talking about objects for which the structuralist view holds. With regard to mathematical structuralism based on the metalinguistic conception of structure, the proposal is that syntax be viewed in this way, with the notions of string and concatenation, and perhaps others, as basic relations. There will be vagueness about this, but not necessarily more so than on the construal I have proposed; with regard to the domain of objects over which predicates are interpreted, such vagueness is unavoidable.[57] I also don't think it can be objected to on the ground of circularity.

---

[57] In response to the difficulty I state in §9, Resnik considers a theory in which structures themselves are considered directly as "only structurally determined" objects ("Mathematics as a Science of Patterns," pp. 538–539). The interpretation of this theory will give rise to the same systematic ambiguities. My preference for the metalinguistic conception rests in part on the fact that on it such systematic ambiguity arises at a place, with the notion of truth, where it will arise anyway.

I am not prepared to argue that unifying in this way the concept of structure used in stating the structuralist view with other reference to objects in mathematics is wrong. The philosophical gain it achieves, however, is only apparent. In this case as well, it carries out the transition in the development of mathematics from dealing with domains of a more concrete nature to speaking of objects only in a purely structural way. But this transition leaves a residue. The more concrete domains, often of quasi-concrete objects, still play an ineliminable role in the explanation and motivation of mathematical concepts and theories. In particular, this is true of any mathematical treatment of formalized or natural languages. Thus, if the structuralist view of mathematical objects is taken to mean that *all* mathematical objects are only structurally determined, it has to rest on legislation about what counts as a mathematical object. The explanatory and justificatory role of more concrete models implies, in my view, that it is not the right legislation even for the interpretation of modern mathematics.

4 A Problem About Sets

§19. An objection

In the last section, we put aside the question whether sets should simply be an exception to a general structuralism about pure mathematical objects. Structuralism about sets is not immediately implied by what has been argued up to now, even the argument against eliminative structuralism in §17. Indeed we would save eliminative structuralism in application to other kinds of objects (such as those of pure geometry) by denying it here, although, to be sure, we would not thereby save eliminative structuralism as a program for eliminating all mathematical objects or even all pure mathematical objects. We still have to consider whether the universe of sets consists of more than a domain of "objects" related by relation called "membership" satisfying conditions that can then be stated in the language of set theory.

We can see the difficulty from our examination of how attempts to use the structuralist idea to eliminate reference to mathematical objects of various kinds founder when applied to higher set theory. Such attempts still require that one make out that a structure of the kind required is in some way *possible*. I argued in §17 that in the present state of knowledge, this possibility can be made out for higher set theory only by domains of pure abstract objects, of which pure mathematical objects are a paradigmatic instance.[1] That this possibility be made out is, however, still required by the noneliminative structuralism of §18. Does this imply that, to establish the possibility of the kind of structure described in axiomatic

---

[1] See also *Mathematics in Philosophy*, pp. 22, 191–192, where, however, I did not distinguish between a strictly structuralist conception of the universe of sets and the "ontological" view to be stated presently, where the latter view allows for ontological relativity and "incompleteness."

set theory, we need to appeal to a system consisting of *sets*? Such an admission would threaten a structuralist view of set theory with circularity and perhaps force us to embrace the horn of our dilemma that takes sets as an exception to a structuralist view of mathematical objects. The purpose of the present chapter is to explore this objection to a structuralist view of the objects of set theory and to argue that it is not in the end sustained.

In stating the objection, I said "threaten" and "perhaps" because what would defeat the structuralist view is not reliance on a system of sets *per se*, but such reliance where what we are appealing to is something more than a domain of "objects" related in a relation called "membership" satisfying conditions that can then be stated in the language of set theory. We can put aside a case of such reliance that is trivial from the point of view of the present problem. Where individuals (*Urelemente*) are involved, they might be identifiable independently and not be preserved in an isomorphic structure. Surely, however, that is just an element of generality in the notion of the "universe of sets": It is not a unique structure because set theory does not determine what is an individual, how many there are, or what structure they have other than belonging to the sets they do. A structuralist view of set theory with individuals should treat the specific domain of individuals as belonging to the realization of the structure rather than as fixed.[2]

There is, however, a more serious reason arising from this consideration for thinking a structuralist view of set theory to be false. It comes from intuitions about sets of a more general ontological character. For example, there is the conception of a set as a totality "constituted" by its elements, so that it stands in some kind of ontological dependence on its elements, but not vice versa. This would give to the membership relation some additional content, still very abstract but recognizably more than a pure structuralism would admit. The "iterative conception of set" (a conception, not primarily of what a set is, but of what the "universe of sets"

---

[2] The principle that there is a set of all individuals requires separate discussion, since it implies that there are not as many individuals as there are sets. On the point of view just stated, however, it would be true for most realizations that would be at all natural.

If we think in terms of the formation of sets from their elements, then it is natural to think of the individuals as "available" prior to the formation of any sets. Then that there is a set of them is a consequence of the principle that any available objects can constitute a set. On the usual understanding of an individual as an object other than the null set that has no elements, it might be thought a mathematical accident that this point of view is workable: Other objects that must be posterior to some sets, such as ordinals and functions, can as it turns out be represented as sets. However, if this were not so the remedy might be just to revise the definition of individual.

is, and therefore of what sets there are) is usually explained with appeal to such general ontological conceptions, although also with appeal to ideas that cannot in the end be taken literally. One might put the objection in the following way: The coherence or nonvacuity of the concept of set, which on the structuralist view needs to be made out, cannot be made out except by describing structures consisting of *sets* (or some functional equivalent), where a set may still be ontologically impoverished compared to a concrete object but is still more than structurally determined.[3] We might call this view the "ontological" conception of the objects of set theory, as opposed to the structuralist one.

It should be stressed at the outset that the objection to structuralism based on this ontological point rests on an epistemological one. The structuralist conception of set theory is rejected on the ground that without a richer conception of sets we cannot know that set theories describe coherent possibilities. Although it then becomes natural to interpret set theory by that richer conception, it does not follow that a structuralist interpretation of set theory is not possible at all and will not sometimes be found at work in mathematics, as it obviously will be in the study of models of set theory.

§20. Ontological conceptions of set

The claim made in the objection is that in order to convince ourselves of the possibility of rich enough structures to interpret set theory, we need to appeal to certain "ontological" marks of the concept of set. That in fact such ontological conceptions are appealed to in justifying the axioms of set theory can readily be seen by examining well-known expositions on the subject. The question is then whether this appeal in fact has the implications the objection claims.

One consideration that might make us doubtful is the following: Differing characterizations of what a set is have been advanced in the history of the subject, and it appears that at different points in the development of the subject different characterizations are appealed to. Characterizations have used three basic ideas: the idea of collection, the idea of plurality, and the idea of extension. By a collection, I mean an object that consists of or is constituted by its elements. It is typically thought of as in some way "formed" from them by some kind of activity, though that is not essential.

---

[3] An objection along these lines was sketched to me very briefly in conversation several years ago by Timothy McCarthy. Whether its further development here corresponds to what he had in mind I do not know.

If one wanted to give a genetic story about the concept of collection, one would start, as some writers on the subject do, with the activity of collecting. Because the formation of collections needs to be thought of in an abstract and general way for the notion even to begin to be adequate for set theory, it is not difficult to cast the notion into disrepute. But it has survived, perhaps because it motivates something central to the avoidance of paradox in axiomatic set theory, the priority of the elements of a set to the set: thought of as a collection, a set is formed from objects that are already "available" or "given."

Pluralities I think of as given by plural constructions: the cats in my house, the prime numbers, the critics who, in a famous sentence, are said to admire only one another.[4] Writers who talk of "multiplicities" or "multitudes" often have this idea in mind. In Cantor's informal explanations of the concept of set, one can discern both the idea of collection and the idea of plurality.[5] Where the notion of plurality occurs most clearly in the history of the foundations of set theory is in Russell's *Principles of Mathematics*. What he there calls a "class as many" is what I want to call a plurality. (A "class as one" is a set, roughly as explained by Cantor.) One may doubt that there is a clear distinction between the idea of collection and that of plurality; one might think of them as rather two different ways of arriving at the same conception. Or one might resist the identification only by purism about the plural: It is already regimentation to think of the cats in my house as *a* plurality, but if one resists such regimentation, plural descriptions like the preceding simple examples needn't be taken to designate objects. The fact that this question arises in a way it doesn't for the idea of collection shows that there is an initial distinction. George Boolos relies on purism about the plural in his use of what I have been calling the idea of plurality to interpret second-order logic and the notion of class in set theory.[6] Even if one does not agree with the claims Boolos makes in his writings on this subject, the fact that plural logic has in recent years come to be developed as a subject in its own right is a consideration in favor of Boolos's purism.[7]

---

[4] See (2.7) in §13.
[5] See "Genetic explanation," p. 283. Cantor thinks of a set as a plurality that is a unity; that could specify what a collection is. If pluralities derive from the plural, and the plural can be applied to any predicate, then this appears to yield comprehension principles without independent appeal to the idea of extension. But what it actually accomplishes is to restate the question what predicates have extensions as the question what pluralities are unities.
[6] "Nominalist Platonism."
[7] See §13 and references given there.

§20. Ontological conceptions of set

The last conception, that of extension, is possibly the clearest, because of its explanation by Frege: Roughly, extensions would be objects correlated with predicates, such that coextensive predicates have the same extension. It is an "extensionalized" version of the very traditional idea of a property or attribute, which is so entrenched in our ordinary thought that even well-trained philosophers take for granted that predicates designate properties, in spite of the vulnerability to Russell's paradox of an unrestricted version of that view.[8] In most discussions of the basic ideas of set theory, the idea of extension plays rather a background role. But for one important philosophical writer on set theory, W. V. Quine, it is the central concept.[9]

The view I would defend of the role of these conceptions of what a set is, is that they do play an indispensable role in the development of the concept but that none of them is adequate to set theory as it now stands (or even as it was constructed in Cantor's time). They certainly appear at important points in the history of set theory, especially, as one might expect, at the beginning, and it is entirely reasonable that they should be used to explain set theory to beginners. Here one has to consider beginners of two different types: both absolute beginners with the concept of set, who nowadays might be elementary school children, and beginners in *axiomatic* set theory, who have some familiarity with the set concept and its use in mathematics, but who are beginners in set theory as a systematic axiomatic theory. It is the latter audience that is addressed in the type of exposition commented on later.

I won't try to argue here that none of these conceptions by itself is adequate to set theory, even if they are taken to be clear enough by themselves, as has been questioned especially for the notion of collection.[10]

---

[8] On this point, see §5. What I, following Frege, call extensions some writers call logical collections, in effect using the term "collection" as a generic term for any entity that might do duty for sets or classes. An example is Penelope Maddy, *Realism in Mathematics*, pp. 102–103. That usage seems to leave us without a term for what I call a collection, at least if one distinguishes, as I argue in the text should be done, between the notion of collection and that of plurality. Maddy's term 'combinatorial collection' (ibid., p. 102) is close to my term 'collection', but her usage is more specifically tied to the iterative conception of the universe of sets.

[9] This is documented and discussed in *Mathematics and Philosophy*, pp. 197–205; cf. also Section III of "Genetic Explanation."

[10] Frege's criticisms of some of his contemporaries amounted to the claim that this notion could not be clearly distinguished from that of a mereological sum. Nelson Goodman's well-known rejection of classes is again in the first instance a rejection of collections. (Cf. "Genetic Explanation," p. 284.) See also Max Black, "The Elusiveness of Sets." In my terms, Black is severe with the idea of collection but friendly to that of plurality.

I have made large parts of the case in previous writings.[11] It is relevant to our present concern that a more structuralist conception of set theory also played a role in its earlier history in the work of Zermelo.[12] All this is a pretty good indication that an ontological conception of set theory can't be right. But it doesn't prove it. For all three of our rough and ready notions may be just crude forms of a more sophisticated and adequate notion of set, which is still richer than the structuralist view allows. In order to come to some sort of conclusion as to whether such a conception is needed, I shall examine some considerations offered in justification of axioms of set theory. These arguments, however, leave the question how they themselves should be taken. Where something more than a structuralist conception would allow is involved, is it intended literally? Does it possibly belong to a genetic account, something that belongs to the conception of set at a certain point in its development but not to that of mature set theory? And if the answer to this question is affirmative, where does that leave the matter of commitment to sets in some richer sense?

§21. The iterative conception of set

A common method of explaining the axioms of Zermelo-Fraenkel set theory (ZF) relies on what was called in §20 the notion of collection, more specifically on the idea that sets are "formed" from their elements, one could almost say "constructed," in stages. That this "formation" proceeds in stages is a consequence of the ontological priority of the elements to the set. It belongs to the picture that the stages are well-ordered. Thus, speaking of Russell's paradox, Joseph Shoenfield writes:

> The explanation is not really difficult. When we are forming a set z by choosing its members, we do not yet have the object z, and hence cannot use it as a member of z. . . . Putting the matter in a positive way, a set z can have as members only those sets which are formed before z. . . . Carrying the analysis a bit further, we arrive at the following: Sets are formed in *stages*. For each stage S, there are certain stages which are *before* S. At each stage S, each collection consisting of sets formed at stages before S is formed into a set.[13]

---

[11] In particular "Sets and Classes" and "What is the Iterative Conception of Set?" (Essays 8 and 10 of *Mathematics in Philosophy*) and "Genetic Explanation."
[12] "Genetic Explanation," pp. 285–6.
[13] "Axioms of Set Theory," p. 323. In a footnote Shoenfield states that 'before' is to be understood "in a logical rather than a temporal sense."

It is possible, given other assumptions about stages, to derive the statement that they are well ordered (which will yield the axiom of foundation) from the assumption that

§21. The iterative conception of set

Shoenfield proceeds to argue for the axioms of ZF using these ideas. Similar presentations can be found in other writings on the elements of set theory. These explanations make use of somewhat pictorial ideas, but the authors generally do not make clear to what extent they are to be taken literally. Thus, in the explanation quoted, Shoenfield talks of the "forming" of a set from its elements by *us*. It seems that this could be a practical possibility at most in the case of finite sets, and not too large ones at that. By analogy with other parts of the foundations of mathematics, one could claim that it is possible "in principle" to form any finite set of given objects. But already in the case of an infinite set defined by a clearly understood predicate, this way of speaking is metaphorical, at least if the set is to be formed from its elements. When we come to sets of sufficiently high rank, moreover, it is very difficult to take seriously the idea that all the intermediate sets that arise in the construction of this set from individuals can be formed by us. This "forming" would have to take place in time and could require many more stages than there are points in time. The idea that for any ordinal number time could have had a structure so rich that that ordinal could have been embedded into it is hardly plausible enough to lend intelligibility and plausibility to the axioms of set theory. It is incompatible with the idea that time is modeled by a structure whose underlying set has the cardinality of the continuum.

These remarks summarize a critical discussion I gave some years ago, where I also attempted to reformulate some of these justifications in such a way that this sort of objection will not apply.[14] In considering the relevance of "intuitive" justifications of the axioms to structuralism, however, we should allow for the possibility that no such explanations will be successful and that any attempt to justify the axioms directly will make use of metaphors or other notions that cannot be taken literally. Another relevant point is that the arguments that have actually been offered have rested on different ideas that don't clearly blend into a whole. Thus some writers have offered serious reasons for seeing these arguments as heuristic principles that have a certain suggestive role but that are not adequate

---

they are partially ordered. See *ibid*., p. 327. (This was first shown by Dana Scott; see "Axiomatizing Set Theory.") For fuller discussion see George Boolos, "Iteration Again." I doubt that this somewhat remarkable technical fact is of much philosophical significance, since it is hard to see how the existence of a partially ordered sequence of stages is more evident than the existence of a well-ordered sequence.

[14] Essay 10 of *Mathematics in Philosophy*.

for the justification of the whole system of axioms.[15] That parallels the way I treat conceptions of what a set is.

On these grounds, one might question the objectivity of set theory or at least whether our concept of set is sharp enough to determine objectively all questions of set theory or to give rise to a structure determined up to isomorphism. The latter position would fit well with a structuralist conception, since it would imply that the ontological conceptions concerning sets that cause difficulties for the structuralist view are part of an explanation by analogies and not necessarily part of the literal truth about sets. Conversely, looking at the structuralist view itself, one might ask whether it concedes to set theory the degree of objectivity that many set theorists are themselves inclined to claim, following the example of Gödel in "What is Cantor's Continuum Problem?" I am not undertaking here to defend a definite view on the still actively disputed question how far the objectivity of set theory extends. I do wish to argue that to the extent that they are persuasive, the considerations offered in favor of axioms of set theory are compatible with a structuralist view.

§22. "Intuitive" arguments for axioms of set theory

At this point one might observe that the set theorist has more argument for the axioms that he uses than the direct, intuitive, and perhaps metaphor-infused arguments that occur in expositions like Shoenfield's. Many of the analogies used are of a much more mathematical kind, especially in the case of large cardinal axioms where it is a matter of generalizing properties of ω and other smaller infinite cardinals. Another very important type of argument might be called a posteriori: It is argued that an axiom is likely to be true because it leads to a theory of some domain that unifies a number of phenomena in a satisfying way and settles questions that cannot be settled otherwise and in a way that is in harmony with the results based on axioms already accepted. As Kurt Gödel put the matter:

> There might exist axioms so abundant in their verifiable consequences, shedding so much light on a whole field, and yielding such powerful methods for solving problems . . . that, no matter whether or not they are intrinsically

---

[15] I take this to be the intention of the somewhat skeptical discussion of such justifications by Maddy, "Believing the Axioms." Similarly Boolos, "Iteration Again," sees some axioms as coming naturally from the "iterative conception of set" and others quite independently from the idea of limitation of size.

necessary, they would have to be accepted at least in the same sense as any well-established physical theory.[16]

For example, such arguments are used for the assumption of Projective Determinacy, which made it possible to extend through the whole hierarchy of projective sets the theory developed at the lowest levels of the hierarchy by the classical descriptive set theorists. An alternative in this case might have been the axiom of constructibility, which had also been shown to yield a theory, but the theory yielded by determinacy was regarded as more satisfactory and cohered with large cardinal axioms, inconsistent with constructibility, that set theorists were disposed to accept. A kind of holistic justification is possible in set theory, in which there is a certain amount of mutual reinforcement of axioms and their consequences. In this case, there is also reinforcement of axioms of different kinds: determinacy axioms and large cardinal axioms.[17] In this context, evidently the term "a posteriori" does not have its Kantian connotation of "empirical."

A posteriori arguments of this kind seem to me to have no definite bearing one way or the other on the question whether a structuralist view of sets is correct. The reason is simple: They undertake to justify the axioms by their logical consequences, and what these consequences are is independent of the issue between the structuralist and the ontological view of sets. It follows that one important structuralist, W. V. Quine, is immune to the objection I am considering. In his view, except perhaps for the "obvious but false" idea that is embodied in the universal comprehension schema, the justification of axioms of set theory is entirely a posteriori.[18]

---

[16] "What is Cantor's Continuum Problem?" (1964 version), p. 261.

[17] For a rather brief account of the theory of projective sets and of Projective Determinacy and its consequences, see Martin, "Descriptive Set Theory." A fuller account of the theory is Moschovakis, *Descriptive Set Theory*. The introduction to that work gives a lucid sketch of the historical development and methodological problems of the subject.

In the 1980s, after the just mentioned publications, it was shown by Martin, Steel, and Woodin that strong large cardinal axioms imply that determinacy holds in $L[R]$ (the sets constructible relative to arbitrary sets of integers), a much stronger principle. I don't think, however, that this reduction abolishes the mutual reinforcement of determinacy assumptions and large cardinal axioms, since its yielding given determinacy principles and their consequences in descriptive set theory would be offered as an a posteriori argument for a large cardinal axiom. On the variety of considerations used by set theorists in justifying their assumptions, see Maddy, "Believing the Axioms," and on the special case of determinacy Martin, "Mathematical Evidence."

[18] This view is the one most explicitly expressed by Quine and the dominant view before *The Roots of Reference*. The latter work makes some concessions to more usual ways of looking at set theory. See my "Genetic Explanation."

For this reason, although a posteriori arguments are of great importance for the larger question of the objectivity of set theory, I will largely leave them out of account in discussing the relevance to structuralism of arguments for the axioms. Also, the difficulties for eliminative structuralism already arise in application to ZF, which, moreover, codifies modes of reasoning in set theory that in their main lines had developed before the specific axioms were formulated. Moreover, some expositions seem to rest on the presupposition that ZF can be justified without a posteriori arguments;[19] that makes it a better testing ground for the significance of more intuitive justifications. I shall therefore consider only the axioms of ZF and not large cardinal, determinacy, or other axioms.[20]

We should begin with the idea that Shoenfield expresses with his talk of formation but which at least can be divorced from the idea that sets come into being by human *action* (or perhaps by Husserlian "acts"). A somewhat vague principle used by Shoenfield and others in explaining axioms of set theory is that a set can be formed from *any* given objects. A way of putting it is to say that any given objects can constitute a set. The latter formulation removes the implication of formation as an action; it still contains the unexplained term 'given', which is meant to express the priority of the elements of a set to the set. But in the modal formulation this term may be superfluous, because the word 'can' already expresses the fact that the set is potential relative to its elements. In other words, 'given' means roughly actual, and the principle says that there can be a set with these objects as elements. Russell's paradox will then show that in some cases there is not an *actual* such set.

The modality here would still give rise to a number of problems, but I would like now to point to another: the principle uses a plural quantifier 'any objects'. This is the point where Cantor used the terms *Vieles* and *Vielheit* and other writers have used such terms as 'multiplicity' and 'multitude'.[21] In our earlier terms, the notion of plurality is invoked. If, with Boolos (see §13), one regards such plural quantification as clear and ontologically noncommittal, even in such an abstract setting as this, one

---

[19] This will be questioned in §23.
[20] Although the axiom of choice is not considered later in spite of the fact that it is treated by most set theorists as part of basic set theory and thus comparable to (the rest of) ZF, I do not believe it would raise any new problems.
[21] Cantor, *Gesammelte Abhandungen*, pp. 204, 443, and elsewhere; Wang; *From Mathematics to Philosophy* pp. 281–284. The choice of language already seems to force a choice as to whether one will regiment the plural by the singular. The more conservative choice is not to do so, and I will try to avoid it.

§22. "Intuitive" arguments for axioms of set theory

will not see a problem. I argued in §13, however, that it is at least not ontologically noncommittal, and, moreover, other views of the semantics of such plural quantification in natural languages introduce reference to sets or classes. In particular cases, however, either the plural quantifier can be cashed in by first-order quantifers or our principle about sets takes the form of a comprehension principle. A case of the former kind is that of the axiom of pairing. Given objects x and y, our principle says that they can constitute a set, that is, a set whose elements are just x and y.

It should be noted that if we formalize the preceding instance by a modal quantificational logic of the usual kind, the variables will have objects assigned to them rigidly, so that (viewing the matter in terms of possible-worlds semantics) in a possible world in which the set $\{x, y\}$ exists, its elements will be the very x and y for which it is claimed that a pair set can exist.

In the general case, however, Boolos's way of reading second-order formulas is helpful. Our principle could be stated a little more explicitly as follows: For any objects there can be a set such that every object is an element of it if and only if it is one of them. This suggests something that is essential if the formulation is to escape Russell's paradox: The plural quantifier is outside the scope of the modal operator; what objects are referred to is settled before one considers alternate possibilities. If one thinks of the formula in the usual syntax of second-order logic, that is, as

(1) $\quad (\forall F) \diamond (\exists y)(\forall x)(x \in y \leftrightarrow Fx)$,

then with the interpretation of second-order variables usual in second-order modal logic one has to either have scope operators that put 'Fx' outside the scope of '$\diamond$' or, for arbitrary F, assume only that the possibility claim holds for an equivalent of F that is fully rigid.[22] The intuitive conception of pluralities or "multiplicities," however, we have taken to imply full rigidity.[23]

---

[22] *Mathematics in Philosophy*, pp. 315–318. The idea is that what F is true of does not change if one goes to an alternative possible world.

[23] Thus if we assume that the second-order variables range only over pluralities, we can use (1) without modification. Moreover, we can simplify the second-order logic. We would have for the second-order variables rigidity *axioms* (the RS, RS/, and BFS of *Mathematics in Philosophy*, p. 335), and the comprehension schema would be expressed just as in nonmodal second-order logic, i.e., as follows:

$\quad (\exists F)(\forall x_1 \ldots x_n)[F x_1 \ldots x_n \leftrightarrow A]$

(A is any formula; typically, but not necessarily, $x_1 \ldots x_n$ are just the free variables of A). But its force is not that of the usual comprehension axiom in second-order modal

Let us return to the very simple case of the axiom of pairing. Does our relating it to the general principle tell us anything about what the set is that either confirms or disconfirms the structuralist view? The principle surely is not informative by itself, without either explicating its content further or examining why it seems evident. One way of reading the term 'constitute' is so that in the case at hand x and y are constituents, that is to say analogous to *parts*, of the set $\{x, y\}$. This way of thinking about the set-element relation can be found in Cantor and, given the prominence of conceptions about wholes and parts in both ordinary thinking and in the metaphysical tradition, derives no doubt from the origins of the mathematical concept of set.[24] If it is regarded as a general mark of the element-set relation, it motivates the priority of element to set that is the keystone for the accepted strategy of avoiding the paradoxes.

Although this analogy of the set-element and whole-part relations may contribute to our understanding of the set-theoretic notions and even, therefore, to the evidence of principles such as the pairing axiom, the question is really whether it is essential that sets be thought of in that way. I would argue that it is not. For just the pairing axiom, this is even obvious, since it is certainly true on the conception of "sets" as extensions of predicates. But there are other intuitive representations of the set-element relation that do not rely on the whole-part concept. One frequently appealed to by set theorists is the conception of sets as trees, where the set is the root, its elements are immediately connected with it, their elements are immediately connected with them, and so on, so that the end points either represent the empty set or are "labeled" by individuals. To make this intuitive, one is likely to think of it spatially, although that brings in some extra structure (the typical plane representation in diagrams of trees would entail an ordering of the elements of a set) and, what is more important, it does not generalize to higher cardinalities. The intention behind this kind of representation seems to me really structuralist: It illustrates the idea that what matters among sets is

---

logic, but rather of extensional comprehension (*ibid.*, p. 336, from Gallin, *Intensional and Higher-Order Modal Logic*, p. 77). This simplification, unfortunately, is of limited use for the axiomatization of set theory in a modal language. To obtain the force of the usual axioms of separation and replacement, we need to instantiate for the second-order variables predicates that are not fully rigid. That situation is likely to obtain more widely than for the particular theories developed in Essay 11 of *Mathematics in Philosophy*.

[24] One can see this clearly in one of the earliest explanations of the concept of set as a mathematical concept, in Bernard Bolzano's *Wissenschaftslehre*, §82; see also *Paradoxien des Unendlichen*, §3.

§22. "Intuitive" arguments for axioms of set theory

which ones are connected with which others by the set-element relation. It seems to me that one might also think of the "formation" of sets from given objects as a kind of labeling of the objects, such that to each object (many) labels are attached with a proviso of extensionality: No two distinct labels can agree with respect to what objects they are attached to. This picture also has a kind of structuralist character: The principle "any objects can constitute a set" is taken to mean that objects ("labels") can be assigned to given objects in a many-to-one way, quite arbitrarily except for the proviso of extensionality. Thus it seems to me that quite generally, the principle that any given objects can constitute a set does not depend for its persuasiveness on a conception of what sets are that is incompatible with structuralism; rather, the existence of different ways of expressing it "intuitively" tend rather to favor the structuralist conception.

One can also see the relation between the set-element relation and the whole-part relation in a somewhat different way, possibly more satisfactory because in terms of it the connection of set theory with a formal theory of the whole-part relation, that is, mereology, has been worked out by David Lewis.[25] According to this conception the parts of a set are its non-empty subsets, so that one-element sets are mereological atoms, lacking proper parts. In a monadic second-order version of mereology, it is possible to develop set theory with a single set-theoretic primitive, the singleton relation $y = \{x\}$. On this account, the analogy to the whole-part relation of the relation of $\{x, y\}$ to x and y is condemned as misleading: $\{x\}$ and $\{y\}$ are literally parts of $\{x, y\}$, and x and y themselves can seem to be so only by confusing an object with its singleton. In particular, x is in no way a part of $\{x\}$. One could argue for this by the observation that if x were a part of $\{x\}$, it would have to be a proper part, since otherwise the distinction between x and $\{x\}$ would collapse and we would not have set theory but just mereology; but it is hard to see how x could be a proper part of $\{x\}$.

In his book, Lewis explains his conception of set theory in a thoroughly ontological way. One could not cite him as an authority in favor of structuralism. It is noteworthy, however, that the ontological conception from which he starts is quite different from those appealed to in the earlier history of the subject. The kind of constitution by elements peculiar to the conception of sets as collections could be assumed only for one-element sets, and that must, for reasons just given, be distinguished from the

---

[25] *Parts of Classes*; see also "Mathematics is Megethology."

part-whole relation.[26] The rest of the work is done by mereology. That is, however, a somewhat misleading statement because the mereology Lewis uses is a monadic second-order theory understood as a logic of plurality, along the lines proposed by Boolos, as discussed in §13. Even if (as Lewis does not) we take this logic to be committed to pluralities, a plurality is not a set. To assume that any objects can constitute a set is not only not part of Lewis's theory but would not be easy to reconcile with it.[27] A lot more should be said about Lewis's treatment of the foundations of set theory. But provisionally, we can find in it some support for our view, because he arrives at ZF on the basis of ontological intuitions that differ importantly from those usually appealed to. It should be pointed out, however, that Lewis does make use of the idea of "limitation of size," of which more shortly.[28]

§23. The replacement and power set axioms

I shall now consider two cases that are harder for the justification of the axioms, whatever one's point of view on the issue that immediately concerns us. These are the axiom of replacement and the power set axiom. In both cases, there are intuitive plausibility arguments for these axioms, but

---

[26] Lewis seems to me in fact not to rely on the idea of the priority of the elements of a set to the set. The paradoxes are avoided by two features of the theory: the fact that comprehension occurs in the setting of second-order logic (so that it does not yield a Fregean extension) and the use of the limitation of size idea, which is ingeniously adapted to the mereological setting.

Lewis does consider a view he calls structuralism; it is in effect eliminative structuralism applied to the specific case of the singleton relation. Because he allows himself only monadic second-order logic as a framework, this poses a technical problem of how to obtain dyadic second-order quantification. Solutions (dependent on mereological assumptions about how close "Reality" is to being atomic) are offered in the Appendix to *Parts of Classes* by Burgess, Hazen, and Lewis.

Should Lewis's construction lead me to reconsider my rejection of eliminative structuralism as an approach to set theory? Even if one grants him monadic second-order mereology as in effect a part of logic, and the atomicity assumptions needed to obtain dyadic second-order quantification, the answer is no. For the question still arises what domain of objects could possibly make out the possibility of the structure satisfying the formal properties of the singleton relation with all the mereological fusions the logic then yields. Lewis's own answer to this problem is in effect that such an hypothesis is needed for standard mathematics. I can accept this answer, but then the distinction between an eliminative and noneliminative structuralism about sets as Lewis conceives them virtually disappears.

[27] Our own reading of that principle would fit uneasily into Lewis's metaphysics, because of his reduction of modality to possible worlds.

[28] In "Mathematics is Megethology," Lewis takes a more structuralist view.

§23. The replacement and power set axioms

arguments of different kinds are to be found in the literature. There is not complete agreement about the "intuitive basis" of either, and that inclines to the conclusion that a posteriori considerations play an important role in the acceptance of both.

Informally, the axiom of replacement says that if we have a set a and a relation R that to an element x of a associates at most one object R_x, then there is a set whose elements are the R_x. A picture is that we can "replace" in a each element x by its counterpart R_x (deleting it if it has no counterpart), and the result will be a set. The axiom of replacement was proposed independently by Fraenkel and Skolem at about the same time. The precise formulation used today is due to Skolem, but this more intuitive conception of the axiom is due to Fraenkel.

In the iterative conception, the objects R_x will be objects that are available at certain stages; this will be the case if they are either individuals (assumed to be available before the "formation" of any sets) or sets.[29]

One type of argument for the axiom proceeds by considering, for each $x \in a$, the stage S_x at which R_x is "formed" (more literally, the first stage at which all the elements of R_x are available). Shoenfield, for example, claims there will be a stage S after all the S_x for $x \in a$.[30] Clearly all the R_x are available before S; hence a set consisting of all of them can be formed at S. Unfortunately, Shoenfield does not argue for the claim that there will be such a stage S; in his book *Mathematical Logic* he states the principle involved as a "cofinality principle" and argues that since the set a can be "visualized as a single object," the same is true of the stages S_x for all $x \in a$ (p. 240). But that visualizability as a single object is preserved by replacement of x by S_x is, given the point of view of Shoenfield in this work, just the principle of replacement itself.[31]

This argument might be interpreted in two ways: The claim that the S_x can be visualized as a single object may be taken to imply that there is a set of all of them, and then S is just the stage at which it is formed. Then Shoenfield is literally assuming a case of the principle of replacement. The other reading, however, could simply be to take the assumption as one about givenness or "availability" of objects: The elements of a are all "given"; and this possibility of being simultaneously available is what is

---

[29] Although this does not happen in the usual way of doing set theory, it would be theoretically possible that there could be objects other than sets that are generated in stages; for example, to avoid the Burali-Forti paradox it is natural to think of ordinals that way quite independently of the von Neumann treatment of them as sets. Cf. note 2.
[30] "Axioms of Set Theory," p. 326.
[31] As already argued in *Mathematics in Philosophy*, p. 280 n. 16.

preserved by "replacement." Then S is just the stage at which all the S_x are available. If we look at the argument this way, then we can proceed (as does Hao Wang[32]) without reference to stages: Since the elements of a are given objects, so also are the R_x, and therefore they can constitute a set. One way of thinking of the axiom of replacement suggested by these considerations is that what underlies it is a principle about when objects can be considered as given or "available" or, in Cantor's language, "existing together."[33] Then, although an ontological intuition of a kind is being appealed to, it is not one specifically about *sets* and thus not incompatible with a structuralist conception of sets.

The axiom of replacement arises naturally from the idea of "limitation of size," which has often been considered independent of the iterative conception. A crude way in which the idea is expressed is that a collection is a set if it is not too large, and that a collection fails to be a set if it *is* too large. This formulation uses the term "collection" in a generic way (cf. note 8), in particular, obviously, so as not to presuppose that collections are sets. But clearly other notions such as class or plurality can be substituted if one attributes to them the requisite generality. Applications of the idea are best understood if we think of it as a second-order principle, in the general sense that it involves generalization of predicate places in our language, whether or not we take the further step of formulating it in terms of second-order logic. Issues concerning such principles were already addressed in §5.

At least if we think of "too large" in terms of cardinality, the axiom of replacement falls rather directly out of the limitation-of-size idea: If a set a is given, then *it* is not too large, and there are not more R_x's for x in a, so that the "collection" of them cannot be too large either. (The second-order reasoning used here has minimal presuppositions.[34] A more serious

---

[32] *From Mathematics to Philosophy*, p. 186; cf. *Mathematics in Philosophy*, pp. 279–280.

[33] In one kind of modal theory, the "theories of potential sets" of Essay 11 of *Mathematics in Philosophy*, the role of the axiom of replacement is taken by a purely modal-logical reflection principle.

[34] In a second-order language, we can state a principle of limitation of size as an axiom: A plurality is a set if and only if it is not equinumerous with the universe. This axiom implies the axiom of replacement (and of, course, separation). A more anomalous consequence is that it implies that the universe is well ordered: For, by the Burali-Forti paradox, the ordinals are not a set; hence they must be equinumerous with the universe. That is, there is a one-to-one map of the ordinals onto the whole universe, which induces a well-ordering of the latter.

The axiom in question was put forth (in a variant formulation) as Axiom IV.2 in §3 of J. von Neumann, "Eine Axiomatisierung der Mengenlehre." Cf. the discussion in §2 of the paper.

§23. The replacement and power set axioms

question concerning replacement is what counts as an admissible relation; the standard first-order version is already quite strong, since it admits nesting of quantification over the whole universe of sets.)

Some care needs to be observed with the limitation-of-size idea. It has in the past been understood as a kind of device for separating sheep from goats among collections or pluralities, or as a criterion for a predicate to have an extension, either way arising from a diagnosis of the paradoxes. This also makes it more specific to sets than seems to me warranted. Behind this understanding of the principle lies an interpretation that represents the universe of sets as a determinate reality and quantification over it as not different from other forms of quantification.

On the contrary, "limitation of size" is more properly understood in terms of the idea of inexhaustibility, emphasized by Gödel as early as 1933[35] but in a way implicit in Cantor's distinction between the transfinite and the absolutely infinite. In a sense, there are no "collections" that are "too large" to be sets. If we try to form an idea of such a collection, as soon as it assumes a certain degree of definiteness we become able to conceive of a still larger collection. For example, we try to describe the universe of sets by its containing the natural numbers or some other basic infinite set and being closed under the operations of power set and replacement (taking for granted more elementary operations). But it is arbitrary to suppose that this gives us everything, we have, since 1930,[36] a clear conception of a *set* closed under these operations: the set of sets of rank less than α, for α a strongly inaccessible cardinal.

It is significant that the axiom of replacement follows from a reflection principle, that is, a principle saying roughly that something that holds in the universe also holds within some set. The principle required is the basic, first-order reflection principle, according to which for any formula A in the first-order language of set theory we have

(2) $A \to (\exists z)(z \text{ is transitive} \wedge A^z)$

Where A^z is the result of restricting all quantifiers of A (in primitive notation) to z.[37] A may contain parameters, but z may be chosen so that their values are all elements of z. This reflection principle and stronger

---

[35] "The Present Situation in the Foundations of Mathematics."
[36] Cf. Zermelo, "Über Grenzzahlen."
[37] In the usual formulation of the principle z is taken to be V_α for some α. Then the proof of the principle from the usual axioms of ZF requires Foundation. I do not know whether Foundation is required to prove (2) in ZF.

principles of the same kind that have been studied are meant to cash in one interpretation of inexhaustibility, that one cannot circumscribe the universe of sets by statements in the first-order language of set theory, or even certain higher-order statements.[38] How much this does to strengthen an "intuitive" case for the axiom might be questioned; for example, instances of the reflection principle quantify over all sets. However, the idea of taking reflection principles as basic does have a certain conceptual character. (See further Chapter 9, note 46.)

In practice the axiom schema of replacement seems to have been adopted partly from intuitions about inexhaustibility but largely because it had great technical convenience, particularly for developing von Neumann's theory of ordinals and for constructing the levels of the cumulative hierarchy within set theory. Some writers express discomfort about the fact that it plays this technical role but also yields rather high infinities, beyond those yielded by Zermelo set theory or theories developed to obtain the technical advantages of ZF without assuming higher infinities than were available in Zermelo set theory.[39]

Our way of looking at "limitation of size" illuminates the role of the power set axiom, which, as we shall see, is difficult to justify intuitively. We can't show, starting with the limitation of size conception and the other ideas I have canvassed, that the sets of integers would *not* be a too large totality or that they are not inexhaustible as are the ordinals or the sets themselves. As Michael Dummett observes, the concept "set of natural numbers" could be an indefinitely extensible concept, and the different forms of predicativism present a perfectly coherent conception according to which that is so. But the power set axiom does *cohere* with the limitation of size conception in a satisfying way, because it provides a way of going to larger and larger sets. Without it,

---

[38] Paul Bernays, "Zur Frage der Unendlichkeitsschemata," formulates a second-order reflection principle that implies the existence of inaccessible and Mahlo cardinals. The problem of formulating higher-order reflection principles faces difficulties where parameters of higher than second order are used, but there is a workable solution due to Tait; see "Constructing Cardinals from Below." However, Peter Koellner has shown that the resulting principles can all be satisfied in V_κ where κ is the Erdös cardinal $\kappa(\omega)$. This cardinal is well below a measurable, and its existence is consistent with $V = L$. For a brief summary, see Koellner, "The Question of Absolute Undecidability," §2. (For the definition of Erdös cardinal see footnote 19.) An elegant presentation of a theory based on Bernays's principle is in Burgess, *Fixing Frege*, §3.6, and in connection with plural logic, "*E pluribus unum.*"

[39] An example is the theory ZU of Potter, *Set Theory and its Philosophy*. Potter's very informative and acute discussion of the reflection principle and of Replacement is accordingly skeptical. (See §§13.2–13.6).

§23. The replacement and power set axioms

we could not prove it false that all sets are either finite or denumerably infinite.

The idea of limitation of size does not make any demand on the concept of set that is incompatible with structuralism. We might note that the inexhaustibility that underlies it applies as well to ordinals and cardinals as to sets. But the main point is that it is a maxim that tells us something about when sets exist, without requiring anything further of what they are to be; that is, that there will be a set *a* such that anything is an element of it if and only if a certain condition holds, and no more. One could thus view it as a principle about extensions (the most "structuralist" of the three conceptions mentioned in §20), but in its application it is combined with principles that the conception of extension would not give rise to. If we think in genetic terms, limitation of size tells us that we can go on, that there is more to the "universe of sets" than we have so far got hold of, but it doesn't tell us that this "more" is more than a richer structure, more objects, some of which stand in the membership relation to some others.

I now turn to the power set axiom, which is probably the most difficult case for the whole enterprise of justification of axioms of set theory by intuitive considerations. There is no agreed upon approach. The power set axiom does not arise directly from an intuitive conception in the way in which, for example, Pairing arises from the idea that any objects can constitute a set, or Replacement from the idea of limitation of size. It is, to be sure, a pillar of the iterative conception. But in practice that conception *assumes* the power set operation; according to it, sets are what is obtained by iterating the power set operation starting with an initial domain of individuals.

It is possible to understand the iterative conception in a weaker way, according to which sets are obtained by iterating the operation of *forming arbitrary subsets*. Then, however, suppose we have at a certain stage S a set a. Then any subset of a can be formed at the next stage. However, that allows the possibility that the formation of subsets of a will stretch out indefinitely through the stages. In that situation there is never a stage at which all the subsets of a are available, which would be necessary to form the power set. In other words, this version of the conception does not rule out the idea that the totality of subsets of a set is irremediably potential, like the totality of sets itself.

At this point, writers on the subject appeal to a principle of plenitude: If a is formed at a certain stage S, then its elements are available at S, and therefore any subset of a *could* have been formed at S. It is then concluded that all subsets of *a* are formed at S and are thus available at the next stage, at which $\mathcal{P}(a)$ can be formed.[40]

This principle makes a nice connection between the power set axiom and an idea from the metaphysical tradition, but it is not at all clear why we are obliged to accept it. It does, however, distinguish the power set from the totality of sets or ordinals. Because there is no stage at which any set or ordinal could have been formed, one cannot infer by plenitude that a set of all sets or ordinals can be formed at the next stage.

The situation is not essentially changed by Paul Bernays's classic description of reasoning about arbitrary sets and functions of natural numbers as "quasi-combinatorial": By analogy of the infinite to the finite, we "imagine functions engendered by an infinity of independent determinations which assign to each integer an integer, and we reason about the totality of these functions."[41] As Bernays implicitly recognizes, the analogy in question has two distinguishable aspects: an understanding of an arbitrary function on the natural numbers (or of an arbitrary set), and the further step of regarding the "totality of these functions" as completed, or as giving rise to a set. The weak iterative conception can appeal to the first aspect of the analogy, and then the question raised concerns the second. Both for their argumentative force and for their relevance to the structuralist view what is important is that what is offered are *analogies*, so that not everything in the picture is intended to be literally true. In particular, we cannot take Bernays's "independent determinations" literally as *choices* (which would certainly be incompatible with structuralism).

I think we have to admit that an essential part of the justification of the power set axiom is a posteriori. It enables us to formulate in terms of set theory the ideas about the continuum that underlie and come out of classical analysis, in particular as it was reformulated in rigorous form in the nineteenth century.[42] Once one accepts the continuum as a set, the power set axiom is a general principle to which the set-theoretic construction of the continuum can be reduced. Power sets of denumerable

---

[40] Such an appeal to plenitude is implicit in Shoenfield, "Axioms of Set Theory," p. 326. It is embodied in axiom IX of George Boolos, "The Iterative Conception of Set."

[41] "Sur le platonisme," p. 54, trans. p. 260.

[42] Shaughan Lavine argues persuasively that the power set was not assumed in the theory of sets and transfinite numbers as Cantor originally worked it out, and that reasoning that would require the power set axiom appears in his work only after 1890. In Lavine's view, Cantor's assuming at that point of an equivalent of the power set axiom led to the consequence that the real numbers are a set, something which Cantor had earlier hoped to prove by, in effect, proving the Continuum Hypothesis. See *Understanding the Infinite*, Chapter IV.

§23. The replacement and power set axioms

sets already present the conceptual problems that the idea of power set gives rise to. It would be arbitrary to accept the reals or the set of sets of natural numbers as a set and not to accept the power set operation in general. And as we have noted above, acceptance of the power set coheres with other principles motivating the axioms of set theory.

This discussion of the arguments that are actually in the literature should make plausible that there is not a set of persuasive, direct and "intuitive" considerations in favor of the axioms of ZF that are incompatible with a structuralist conception of what talk of sets is. If we add to this the neutrality of a posteriori considerations with respect to this issue, we can conclude that an ontologically richer conception of set is not needed for ZF. Although I have not argued this here, in my view, the considerations that are offered in favor of axioms going beyond ZF are not fundamentally different, although the role of a posteriori considerations is certainly greater when one comes to the stronger axioms of infinity postulating measurable and larger cardinals.

5 Intuition

§24. Intuition: Basic distinctions

The concept of intuition occupies an uneasy place among the different notions deployed by philosophers and others in order to describe knowledge and belief. Some such notions, such as that of knowledge itself, have been thought to be definite enough in prephilosophical usage, and of sufficient importance, so that considerable philosophical effort has been devoted to their "analysis," in the hope of giving, in clearer terms, necessary and sufficient conditions for knowledge, or, failing that, stating interesting general principles. Others have been from the beginning technical terms, so that no one has hoped to get from them a better understanding than can be derived from philosophers' explanations. Kant's terms 'analytic' and 'synthetic' would be examples.

The word "intuition" undoubtedly has a prephilosophical use from which its use in philosophy in some way derives. But there does not seem to be conviction on the part of philosophers that there is a fundamental notion behind that use, which philosophical analysis might make clear. By contrast, at least in comparatively recent discussion, there is also not a technical usage to which writers discussing intuition have been ready to defer, except in rather restricted contexts such as the interpretation of Kant. Thus, in his useful encyclopedia article on intuition, Richard Rorty goes so far as to say that "nothing can be said about intuition in general."[1] To be sure, one use may be an exception to this state of affairs, namely the

---

[1] Rorty, "Intuition," p. 204 (1st ed.), p. 722 (2nd ed.). Rorty does, however, proceed to distinguish four principal meanings of the term. Rorty's article and the two addenda by George Bealer and Bruce Russell in the second edition of the *Encyclopedia* illustrate well the lack of fixed anchors in the use of the word.

§24. Intuition: Basic distinctions

use of the term in linguistics, when the "intuitions" of native speakers of a language are discussed. But probably this generalizes at best by analogy.

In application to mathematics, which of course particularly concerns me, this general difficulty I mentioned is very much in evidence, and it is not at all clear that those who defend the idea of mathematical intuition, and those who attack it, have the same concept in mind. In our time, there has not been a developed positive conception of mathematical intuition that is sufficiently salient either as a model to be developed and defended or as a target to be attacked. One might contrast this state of affairs with what prevailed in the late nineteenth and early twentieth centuries, when Kant offered a kind of paradigm of a philosophical conception of intuition applied to mathematics (whether or not he was interpreted correctly or even consistently). That is no longer the case.

In this chapter, I shall be concerned to develop a conception of mathematical intuition that is in a general way Kantian, although I will not claim that it was Kant's, and it has a more immediate historical antecedent in the work of Hilbert and Brouwer. But before doing so, I shall rehearse some elementary distinctions about intuition, which apply quite generally.

In the philosophical tradition, intuition is spoken of both in relation to objects and in relation to propositions, one might say as a propositional attitude.[2] I have used the terms 'intuition *of*' and 'intuition *that*' to mark this distinction. Probably the latter usage has dominated in the philosophical tradition, and it also occurs in ordinary speech. The distinction is simply an observation about how the term (and others that would be translated as 'intuition'[3]) have been used, and it should be obvious and noncontroversial. In particular, making it should be compatible with the views of those who reject the whole idea of mathematical intuition or are cool to claims of intuition in general.[4]

What gives intuition *of* an important place in philosophy is probably the fact that Kant's *Anschauung* is intuition of objects. However, Kant certainly allows for intuitive knowledge or evidence that would be a species of intuition *that*. I think it is quite clear that Kant has such a conception,

---

[2] I try to use the term "proposition" as neutrally as possible.

[3] In particular *intuitio* in Latin and *Anschauung* in German.

[4] Hale and Wright may miss this point when they preface a critical discussion of my own conception of what is called in §28 Hilbertian intuition with "Parsons' explication of *intuition of*..." ("Benacerraf's dilemma revisited," p. 105). Neither in the present work nor in earlier writings have I said that Hilbertian intuition (even beyond the simple version that I consider in detail) is the only, or even the only viable form of intuition of objects, even of abstract objects.

but he doesn't designate it by the term *Anschauung* or even use such a phrase as *anschauliche Erkenntnis*. It is not so clear to what extent this priority of intuition *of* is preserved in the rather loosely Kantian talk about intuition in late-nineteenth- and early-twentieth-century writing about the foundations of mathematics; but it is preserved in the most important developments of broadly Kantian conceptions of intuition in this period, in the work of Brouwer and Hilbert.

It would be natural to talk of intuition of objects and intuition of truths, but the latter usage adopts one resolution of a further ambiguity. In propositional attitude uses, 'intuition' is not always used for a mode of *knowledge*. In particular, it is not necessarily factive. If one has the intuition that p, it by no means always follows that p. When a philosopher talks of his or others' intuitions, that usually means what the person concerned is inclined to believe at the outset of an inquiry, or as a matter of common sense; intuitions in this sense not only need not be true: They can be very fallible guides to the truth. To take another example, the intuitions of a native speaker about when a sentence is grammatical are, again, not necessarily correct, although in this case they are, in contemporary grammatical theory, taken as very important guides to truth. It is not easy to find a contemporary propositional attitude use of 'intuition' that *is* factive. But what Descartes in the *Rules for the Direction of the Mind* called *intuitio* was not genuine unless it was knowledge.[5] Rorty, in the encyclopedia article mentioned earlier, distinguishes four meanings of the term. The first two are instances of intuition *that*; the last two of intuition *of*. But all but the first[6] are by definition knowledge. His sense (2), which he justifiably regards as philosophically most important, is worth quoting:

> Intuition as immediate knowledge of the truth of a proposition, where "immediate" means "not preceded by inference." (*ibid.*)

This is a good first approximation to a characterization covering the important uses of intuition *that*. But use of "intuition" with the

---

[5] I take this to be implied by his characterization in Rule Three of intuition as "the conception of a clear and attentive mind, which is so easy and distinct that there can be no room for doubt about what we are understanding" (*Philosophical Writings*, I, 14).

Concerning Descartes' mature notion of clear and distinct perception, however, the possibility that one has clearly and distinctly perceived that p and yet p is false is a live one until his argument for the truth of clear and distinct perception, appealing to the veracity of God, has been carried out.

[6] "Intuition as unjustified true belief not preceded by inference; in this (the commonest) sense 'an intuition' means 'a hunch'" (ibid.).

§24. Intuition: Basic distinctions

connotation of knowledge, and therefore truth, is likely to cause misunderstanding in the circumstances of today; it may even lead a reader to think one has in mind something like intuitions in the less strict senses we have mentioned with the extra property of being infallible. When one wants to make clear one is speaking of knowledge, it is probably best to use a term like 'intuitive knowledge' rather than simply 'intuition'. Here, however, what would count as intuitive knowledge would depend on the underlying conception of intuition, as is shown by the fact that in the philosophy of mathematics 'intuitive knowledge' has had a more special sense, connected with Hilbert's version of a Kantian notion of intuition.[7] That is the sense in which I will be using the term in §28 and later.

Thus I will use the term 'intuition', as in the above examples, so that it is not *ipso facto* knowledge, and one's having the intuition that p doesn't imply the truth of p. Within this usage, differences are possible as to how much is or ought to be claimed for intuition as a source of knowledge or as a guide to the truth. Philosophers' intuitions are not claimed to be an autonomous source of knowledge, and their reliability will vary greatly. But, as we noted, the intuitions of native speakers of a language do create a presumption of truth. Moreover, it could be that in some domain, intuition, if carefully enough cultivated, is a source of knowledge and a quite reliable guide to the truth, without actually constituting knowledge in the sense (again) that an agent's having the intuition that p implies p. It can, I believe, be shown quite convincingly that that is the way Kurt Gödel, the twentieth-century philosopher who claims the most for mathematical intuition, uses the term.[8] It should be clear that the issue between this usage and that of Descartes is first of all terminological: Suppose one has what is to all appearances a convincing intuition that p, but after a time something comes to light that causes one to withdraw assent to p. (Perhaps p, together with other equally convincing assumptions, leads to a contradiction.) On Descartes' usage, one will then say that one did not really have intuition that p; on the usage I am attributing to Gödel, it can be that one had an intuition that turned out to be wrong. Someone who believes that, under the right circumstances and by exercising sufficient care, one can get into a state which *guarantees* the truth of what one is

---

[7] Consider, for example, the use of the terms 'anschauliche Evidenz' and 'anschauliche Erkenntnis' in Gödel, "Über eine bisher noch nicht benützte Erweiterung des finiten Standpunktes," pp. 240, 242.
[8] See my "Platonism and Mathematical Intuition in Kurt Gödel's Thought," Sections IV and V.

claiming, is more likely to adopt the Cartesian usage, but such a belief is not a necessary condition for it.

Descartes contrasted intuition with deduction; on his usage, the conclusion of an inference would not be an intuition. In describing possible sources of knowledge, not only would intuition be distinguishable from the results of arguments involving inferences, but such results could not be intuitions, although possibly the same proposition could be, or could have been, known by intuition.[9] Descartes' usage agrees with the most common present-day usage; what is most distinctive about "intuitions" is that they are *not* the conclusions of arguments. More properly, the status of a proposition as an intuition cannot result from its being the conclusion of an argument, although in philosophy and other fields, even to some extent in mathematics, an argument may be defended on the ground that its conclusions agree with intuition.

In this usage, what is most characteristic of intuition is that it is belief, or inclination toward belief, independent of any articulation of its grounds, possibly coupled with expression of doubt as to whether it *could* be reinforced by grounds of another kind, that is, by argument (taking this word in its most general sense, which would incorporate deductive arguments, including mathematical proof, empirical or "inductive" arguments, and arguments of a less rigorous structure such as are characteristic of philosophy).

Evidently Rorty, in the definition I have cited, takes this as what the "immediacy" of intuition *that* consists in. That is no doubt true for many uses of the term. But it cannot be the whole story for the uses that most interest us. For if it were, the simplest and most evident *logical* truths would be intuitively known, and the immediacy of the simplest logical inferences would also be of the same type as what we would call intuitive. But the intuitive is contrasted not just with what is inferred but also with the conceptual. This is the usage of Kant, which has been followed by philosophers who are not particularly Kantian in other respects. For example, the "knowledge without observation" that we have of our own intentional bodily movements is not usually described as intuition or intuitive knowledge, although it satisfies Rorty's definition. On this point, the philosophical tradition divides. In insisting that only sensibility gives rise to intuitions, Kant is breaking both terminologically and substantively

---

[9] This is not to say that the individual inferences of a deduction might not be intuitions or rest on intuitions in some important sense, as was surely the view of Descartes and, for a more Kantian sense of 'intuition', traditional intuitionism and the Hilbert school.

from the earlier rationalistic tradition, for which "intuitive knowledge" can represent a high development of what we would call conceptual. This earlier tradition has not lost its influence, and, moreover, there are uses of 'intuition' that are neutral on the issues it would raise. Although my own development of the concept will be in the Kantian tradition, the other tradition will not completely disappear from what follows.

Another ambiguity is also worth mentioning: The distinction between intuition *of* and intuition *that* does not quite coincide with that between intuition with nonpropositional and intuition with propositional objects. For if one takes propositions seriously as objects, in principle one might grant that one could intuit a proposition, in such a way that one could conclude its existence and perhaps something about its structure, while emphasizing that such intuition of the proposition as an object is not apprehension of its *truth*. In other words, intuition *of* the proposition that p as object is not the same as intuition *that p*.[10]

§25. Intuition and perception

At this point we need to turn to intuition *of*. Although Kant, at least in his official explanations, did think of intuition as *Erkenntnis* (see for example A 320/B 376–377), it is cognition or knowledge *of objects* and not in the first instance propositional knowledge.[11] Kant's *Anschauung* is a case of

---

[10] A philosopher to whom this observation is relevant is Husserl. The basic notion for Husserl is intuition of; nonetheless intuition is what distinguishes different ways of entertaining a proposition from actively knowing it. But it seems to me that Husserl reduces intuition that to a form of intuition of, where the object is not what we would call a proposition but rather a state of affairs (*Sachverhalt*).

[11] It does not follow that there is essentially nonpropositional knowledge according to Kant. His view that intuitions without concepts are blind, and that the use the understanding makes of concepts is to judge by means of them (A 68/B 93), implies that knowledge of objects is manifested by propositional knowledge about them. On the other hand intellectual intuition, which he denies that we have, would presumably be nonpropositional.

It is a somewhat vexed question how far Kant's *Erkenntnis* corresponds to our term 'knowledge'. That was the rendering in the long-standard translation of Norman Kemp Smith, and the substitute 'cognition' adopted in much recent writing on Kant and in the translation of Guyer and Wood is rather transparently a technical term chosen for the purpose. 'Knowledge' conveys more immediately what Kant is talking about, and it is inevitable that one will paraphrase much of what Kant says in terms of knowledge. It has the disadvantage that it is not a count noun in English, while Kant uses *Erkenntnis* as a count noun, which led Kemp Smith to translate the plural *Erkenntnisse* awkwardly as 'modes of knowledge'. ('Items of knowledge' might have been better.) More substantively, it is not evident that where an *Erkenntnis* is propositional (that is, is a judgment), it must be true. Furthermore, Kant distinguishes *Erkenntnis* from *Wissen*, and it has been argued that this distinction is important for understanding some major claims of Kant,

intuition *of*. This is forced by what we might anachronistically call its "logical form": An intuition is a singular representation, contrasted with concepts, which are general.[12] Intuitions are also said to be "immediate"; the meaning of this has been a matter of controversy.[13] One aspect that is not controversial is that empirical intuition arises from perception. A situation in which a subject has an empirical intuition of an object, however it is described further, is one in which he perceives the object; in particular, he is affected by the object. Kant's notion of intuition in this sense generalizes the concept of a representation arising from perception. One element of the controversy about Kant's notion of intuition is how much of the presence to the mind of the object in perception carries over to other cases of intuition.

It is hard to see what could make a cognitive relation to objects count as intuition if not some analogy with perception. And, indeed, we find such an analogy claimed or appealed to by writers advancing very different conceptions. Thus, in our own time, Gödel famously claimed that "we do have something like a perception of the objects of set theory."[14] In the paper "Is mathematics syntax of language?" that he worked on in the 1950s but never published, he also stresses this analogy, going so far as to speak of an analogy between reason and an "additional sense."[15] But Gödel is not alone in discerning such an analogy.

The Latin *intuitio* is derived from *intueri*, meaning 'to look at, to gaze at'. Descartes' explanation of *intuitio* in the *Rules* relies on the analogy with perception only in this choice of terminology, on which no doubt nothing in particular turns. It is clearly intuition *that*; the examples that he gives in Rule Three are all propositional. The analogy with perception is, however, used in Descartes' later definition of clear and distinct perception:

> I call a perception "clear" when it is present and accessible to the attentive mind – just as we say that we see something clearly when it is present to the eye's gaze and stimulates it with a sufficient degree of strength and accessibility. I call a perception "distinct" if, as well as being clear, it is so sharply separated from all other perceptions that it contains within itself only what is clear.[16]

for example concerning things in themselves. But this issue is not especially relevant to the present inquiry.

[12] A 320/B 376–7, also *Logik, Ak.* IX 91.
[13] See *Mathematics in Philosophy*, pp. 111–115, 142–145, and the writings of Hintikka, Howell, and Thompson cited there, also my "The Transcendental Aesthetic," section I.
[14] "What is Cantor's Continuum Problem?" (1964 version), p. 268.
[15] For example pp. 353 n. 43, 354 (version III).
[16] *Principles*, I, 45 (*Philosophical Writings*, I, 207–208).

§25. Intuition and perception

By contrast, Leibniz does not use such analogies in his explanations of clear and distinct knowledge in "Meditations on knowledge, truth, and ideas" (1684). There he makes a contrast between intuitive and "blind or symbolic" knowledge; knowledge of a notion is intuitive when we can "consider all of its component notions at the same time."[17]

Edmund Husserl is a philosopher for whom the notion of intuition takes on a very general significance. He makes a sustained attempt to develop a theory both of knowledge of abstract objects and of rational evidence based on an analogy with perception. Basic to his theory of meaning is a distinction between meaning intentions and their fulfillment. The acts or intentional experiences that constitute our consciousness have intentionality, relation to an object. Such a relation is realized or fulfilled if the object is present in intuition (or at least represented in imagination);[18] in the case of actual intuition, where there is a certain kind of unity of the intended sense and the sense of the act fulfilling it, one has knowledge of the object. Intuitions are in one place described as "the acts that in knowledge are called to the fulfillment of other intentions."[19] The examples by which Husserl explains these ideas tend to be perceptual. Acts of outer perception have the characteristic that they contain both fulfilled and unfulfilled intentions; for example, a perception of a cup sitting on a table will represent it as having a bottom, but, since the bottom is not visible, that intention is not fulfilled.

In the *Logical Investigations*, the idea of intuition where the object may be abstract is explicitly represented as an extension of the concepts of intuition and perception. The intention/fulfillment schema invites generalizing the notion of intuition in a way that parallels the generalization of the notion of object:

> Thus, also in generally customary speech, aggregates, indeterminate multitudes, totalities, numbers, disjunctives, predicates ("being just"), and states of affairs become "objects"; the acts, through which they appear as given, become "perceptions." (VI §45, A 615/B$_2$ 143)

---

[17] *Philosophical Essays*, p. 25. I would conjecture that this essay had an influence on Gödel, but it is unlikely that Gödel derived from it the idea of thinking of reason as an "additional sense." As regards the influence of Husserl, the various versions of "Is Mathematics Syntax of Language?" were in all probability written before Gödel began his serious study of his writings in 1959.

[18] *Logische Untersuchungen* (hereafter LU), I §9, §14. The distinction between intention and fulfillment and the relations between them are explored more thoroughly in Investigation VI.

[19] LU VI §10 (A 511/B$_2$ 39).

Something common to Kant, Husserl, and Gödel is a close connection between what I am calling intuition *that* and intuition *of*. According to Kant, intuition (which as I have remarked is intuition *of*) in mathematics confers evidence that is immediate. Thus axioms are said to be "immediately certain" (A732/B760):

> Mathematics ... can have axioms, since by means of the construction of concepts in the intuition of the object it can combine the predicates of the object both *a priori* and immediately, as, for instance, in the proposition that three points always lie in a plane. (ibid.)

Evidently, the immediacy of the *judgment* derives from a construction "in the intuition of the *object*."[20]

Husserl seems to regard intuition *that* as a species of intuition *of*: Evidence of a judgment is a situation in which the state of affairs that obtains if it is true is "itself given." Because, typically, a proposition involves reference to objects, evidence will involve intuition of those objects, but they play the role of constituents of a state of affairs that is also intuitively present, at least in the ideal case.[21]

For Gödel, that we have "something like a perception of the objects of set theory" is supposed to be "seen from the fact that the axioms force themselves upon us as being true."[22] The latter appears to be a matter of intuition *that*; it is clearly a matter of the evident character of certain *statements*. Even if we grant this, why should it follow that there is *intuition of* "the objects of set theory"? Gödel's particular reason for thinking this was very probably that in talking of the objects of set theory, he had in mind not only sets but also concepts, and for him rational evidence of a

---

[20] Kant does not explicitly claim immediate certainty for all mathematical propositions; the question naturally arises whether it is limited to axioms. In the Hilbert school and later, the idea of intuitive evidence has been applied also to the results of proof, with, however, strong restrictions on the admissible methods. It is doubtful that Kant would make such an extension; for example, he glosses "propositions that are synthetic and immediately certain" as "*indemonstrabilia*" (A164/B204). Here, where he is talking about arithmetic, he clearly has in mind numerical formulae such as $7 + 5 = 12$. In these cases, however, one clearly needs to carry out the operations (which Kant certainly thought of as constructions). Although apparently a construction is sufficient to make such propositions evident, the construction can be quite complex, and in the case of large numbers evidently will be. For this reason one should not read "immediately certain" as "self-evident," as Frege did (*Grundlagen*, §5).

[21] In his discussion of truth, Husserl talks of the "ideal of final fulfillment" (LU VI §§37–39). Later, he concedes that this is in interesting cases not achieved or even achievable, so that final fulfillment is a kind of Kantian Idea.

[22] "What is Cantor's Continuum Problem?" (1964 version), p. 271.

§25. Intuition and perception

proposition involved "perception" of the concepts occurring in it.[23] What Gödel calls perception would in our terminology be intuition, since it is clearly perception only in an extended sense.[24]

That intuition *that* should in some way rest on, or at least be intimately connected with, intuition *of* is what one would expect if intuition *that* is analogous to perception, since one of the central elements of perception is the presence of the object perceived; one knows by perception *that* my bicycle is blue by seeing my bicycle. Someone who has never seen my bicycle might know this, but he would not know it by perception in the most straightforward sense.

It is this that makes Gödel's inference natural, even apart from the particular context of his views about concepts and perception of concepts. I propose to use the term 'intuition' so that a mode of evidence does not count as intuition unless it is analogous to perception in a definite way. In the case of some proposed kind of intuition *that*, one way in which the analogy can be made out is that it involves intuition of certain objects. Unlike Gödel, I will not argue that all rational evidence of principles that are not the conclusions of deductive or empirical arguments is a case of intuition, or even that intuition extends very far into the conceptual domain. The following inquiry will be in the tradition of Kant, for whom intuition and Reason are of a different nature, rather than in the tradition of Spinoza and Leibniz, for whom intuitive knowledge is possible at high levels of abstraction and rational integration. Taking the analogy with perception as what distinguishes intuition *that* from other forms of "intrinsic plausibility" that statements might have does not force one to be in Kant's rather than the other camp on this matter (as the example of Gödel shows), but the adherent of the other tradition is required to stretch the analogy much further, as one can see by comparing Gödel's remarks in "Is mathematics syntax of language?" with the conception of intuition presented in §26 and later.

Beginning in §27, I will argue that a form of intuition in which the objects can be described as mathematical exists. A conception of intuitive knowledge will be associated with this conception of intuition. But the very idea of such an intuition seems at first sight outrageous, and the defense of intuition in the tradition of Kant may seem to deny the

---

[23] This claim is documented in "Platonism and Mathematical Intuition in Kurt Gödel's Thought."

[24] When Gödel uses the term 'intuition', he usually means intuition that, but this is not always the case, in particular not when he is talking about a Kantian notion of intuition.

accomplishments in the foundations of mathematics since the late nineteenth century, so much of which was directed against either Kant's own views or views deriving from Kant. In the next section, I will consider some objections.

§26. Objections to the very idea of mathematical intuition

Many critics of theories of mathematical intuition such as Gödel's accuse these theories of postulating a special faculty of intuition. Even if this criticism can be answered in the particular case where it is advanced, behind it lies a point that has to be taken seriously. By its very nature, intuition is not the sort of thing that should be "postulated." If mathematical objects are given to us in a way similar to that in which physical objects are given to our senses, should it not be *obvious* that this is so? But the history of philosophical discussion about mathematics seems to show that it is not. Whatever mysteries and philosophical puzzles there may be about perception, it works to a large extent as a straightforward empirical concept. We can make a lot of assured judgments about when we perceive something, and confidence about this description of our experience can often survive doubt about what it is an experience of. Thus the proposition that I now *see* before me a computer with text on its screen is one that I expect that no other philosopher, were he now in the room where I am writing, would dispute except on the basis of skeptical arguments, and many of these would not touch weaker statements such as that it *looks* to me as if I see these things. There is a phenomenological datum here that is as close to being undisputed as anything is in philosophy.

It is hard to maintain that the case is the same for mathematical objects. Is it *obvious* that there is an experience of intuiting the number 7, or a triangle, or at least of its "looking" as if I were intuiting 7 or a triangle? But if it isn't obvious, how could it be true, or how could our intuiting these objects have a relevance to mathematical knowledge comparable to that my seeing my computer screen has to my knowledge that, for example, some sentence I have just typed contains a typographical error? One can put the question in another way by asking whether there are any experiences we can appeal to in the mathematical cases that are anywhere nearly as indisputed as my present experience of seeing the computer screen. If we don't know what to point to, that already appears to be a serious disanalogy between sense-perception and whatever consciousness we have of mathematical objects.

§26. Objections to the very idea of mathematical intuition 149

This embarrassment is connected with an obvious disanalogy. In normal cases of perception, there is a physical action of the object perceived on our sense-organs. Our perception is in some way founded on this action, and there are serious reasons for holding that such a causal relation is a necessary condition for perceiving an object. But it would be implausible to hold that in *mathematical* intuition there is a causal action of a mathematical object on the mind. In fact, this is no part of the view of the upholders of mathematical intuition that I have mentioned, though it is sometimes included in popular conceptions of Platonism, or attributed to philosophers like Gödel by their critics.[25] Though in the case of Gödel it is not justified by the texts, it is not an unnatural error, since one might expect that if mathematical intuition is like perception it should share with it this central feature.

In the first instance, these difficulties apply to intuition of mathematical objects, to which both Husserl and Gödel have to be interpreted as being committed. Kant does not say this straightforwardly, and indeed mathematical objects do not play an explicit role in his philosophy of mathematics. In my own view, he simply does not face the questions that in our own time surround the notion of mathematical object.[26] Although it does not follow that he is not in any way committed to them, other interpreters have held that Kant's conception of mathematics has no room for mathematical objects.[27]

---

[25] Thus in "Mathematical Truth" Paul Benacerraf writes:

> He [Gödel] sees, I think, that something must be said to bridge the chasm created by his realistic and platonistic interpretation of mathematical propositions, between the entities that form the subject matter of mathematics and the human knower. Instead of tinkering with the logical form of mathematical propositions or with the nature of the objects known, he postulates a special faculty through which we "interact" with these objects. (p. 416 of reprint)

The quotes around "interact" indicate that Benacerraf thinks that even for Gödel such talk is not to be taken literally, but it is not clear to me what he thinks Gödel wishes to substitute for literal action of mathematical objects on this "special faculty."

My own reading of Gödel's conception of mathematical intuition does not give place to anything like interaction between ourselves and either mathematical objects or the concepts that are more prominent in Gödel's discussion of these matters; see "Platonism and Mathematical Intuition," Sections IV and V. I agree with the earlier criticism of Benacerraf's remark by Tait, "Truth and Proof," footnote 3, although I do not entirely agree with the interpretation Tait wishes to put in its place; see pp. 66–67 and n. 44 of my paper. (That interpretation does not correspond entirely to Tait's present views.)

[26] See "Arithmetic and the Categories," section I, which expands *Mathematics in Philosophy*, pp. 147–149.

[27] Friedman, *Kant and the Exact Sciences*.

Still, Kant expresses puzzlement about how intuition can be a priori, related to the difficulty about causality. In §8 of the *Prolegomena*, after introducing the notion of pure intuition, he writes:

> An intuition is such a representation as would immediately depend on the presence of the object. Hence it seems impossible to intuit spontaneously (*ursprünglich*) *a priori* because intuition would in that event have to take place without either a former or a present object to refer to, and in consequence could not be an intuition.... But how can the intuition of an object precede the object itself? (*Ak.* IV, 281–82, Beck trans.)

It is clear from the context that by 'object' Kant means here *real* object, in practice physical object. The problem does not directly concern intuition of mathematical objects, except insofar as such intuition is to yield mathematics that is applicable to the real world. The question is how it is possible for a priori intuition to be "of" objects that are not given a priori.

Kant's own solution to the puzzle, given in §9, appeals to the idea that a priori intuition contains only the form of our sensibility. This evidently removes the causal dependence of intuition on the object. It is a nice question what is left of the characterization of intuition that gives rise to the puzzle.[28] Kant's solution seems to allow the *phenomenological* presence of an object to be preserved, but it is a further question whether what one has is a representation of a physical object, not individually identified and not really present, or a representation of a mathematical object. The former is not ruled out by the a priori character of pure intuition, as the "presence" might be that characteristic of *imagination* rather than sense. In fact, a number of passages in Kant indicate that just that is his position.

Kant's puzzle may have force for us, but we are not likely to accept the view that pure intuition contains only the form of our sensibility, a central part of Kant's transcendental idealism, at least not as Kant understood it. In what follows, I will not claim that mathematical intuition is a priori, but this concession does not fully remove the force of the puzzle. For suppose that intuition of certain objects underlies our knowledge of a mathematical truth, say $7 + 5 = 12$. We have this intuition at a certain place and time, and yet the proposition is general in its implications; it is applicable to the "first three minutes" after the Big Bang and also to the world long after we are dead. Moreover, we generally understand such

---

[28] In fact, the characterization does not comport fully with Kant's conception of intuition as explained and deployed elsewhere.

§26. Objections to the very idea of mathematical intuition 151

statements to be necessary, and I myself have defended that view.[29] I shall return to the problem in §30.

Another difficulty for the idea of intuition of mathematical objects is posed by the structuralist view of such objects, already discussed at length in Chapters 2 and 3. According to it, the properties and relations of mathematical objects that matter for mathematics are those determined by the basic relations of some system or structure to which all the objects involved belong, or perhaps several such structures and mappings between them. For applications, what in §14 are called external relations also matter, but they are independent of the choice of a system of objects to realize the structure.

On the structuralist view, mathematical objects are in a way not individually identifiable at all. But then how is it possible that they should be objects of intuition? For example, unless one is presupposing a structure including numbers and sets, it seems indeterminate whether the number 2 is identical to the one-element set $\{\{\varnothing\}\}$, the two-element set $\{\varnothing\}, \{\varnothing\}\}\}$, or neither. How can this be if numbers and sets are objects of mathematical intuition? Can such intuition be a significant source of mathematical knowledge if it does not determine the answers to such simple questions?

It will turn out in Chapter 6 that numbers are not objects of the intuition that will be affirmed in this chapter; this is generally true of the pure mathematical objects for which the structuralist view is most persuasive. The outcome of our examination of the structuralist view, however, was a qualification on it; we still had to admit quasi-concrete objects that are properly mathematical, and yet the structuralist view is not true for them (§18).

However, that does not quite dispose of the objection in this case. It appears that the quasi-concrete objects that one might admit in mathematics, such as expression-types and geometrical figures, are still incomplete in Leibniz's sense. That they have intrinsic concrete instantiations is not enough to endow them with the range of properties and relations a concrete object would have, and it is not clear that they do not suffer from the same indeterminacy of identity that pure abstract objects appear to suffer from. This is shown by the fact that in the development of mathematics nothing that the mathematician normally attends to is lost by thinking of them in a structural way, so that expressions, for example, are simply built up from arbitrary objects called "symbols" by a relation called

---

[29] See §15 and especially Essay 7 of *Mathematics in Philosophy*.

"concatenation" that is specified in a purely structural way. On that way of looking at them, their relation to their tokens is an external relation.

At present I will not concern myself with whether these claims about quasi-concrete objects are true. They are plausible enough so that a view according to which there is intuition of such objects will have to deal with them. The matter will be addressed again when I present my own conception of intuition.

The objections put forth in this section have all been to the idea of intuition of mathematical objects. Intuition as a propositional attitude is not at the outset as questionable an idea. First of all, there is no doubt that principles in mathematics have "intrinsic plausibility,"[30] and reductionist programs in mathematical epistemology that would remove any necessity to rely on this plausibility have not fared very well. But then, as I have structured the question, whether we should talk of intuition depends on whether we can make out a significant analogy with perceptual knowledge. If, as I have suggested is likely, part of this analogy will be dependence on intuition *of*, the above objections become relevant. The "obviousness" objection does have some independent force. One might put it this way: If on reflection we do not find a certain mathematical principle (such as an axiom of set theory) obvious without appealing to a proof, then it is doubtful that we can regard it as a deliverance of intuition.

Because I prefer to use the term 'intuition' in a way related to Kant's and to view questions of rational evidence as for the most part not involving intuition, the idea of intuition *that* as more or less independent of intuition *of* will not be pursued further. However, some of the issues that have been discussed historically under this heading will come up in Chapter 9.

§27. Toward a viable conception of intuition: Perception and the abstract

Thus, in undertaking to present a defensible conception of intuition, my focus will be on intuition *of*. I propose to show that there is at least a limited application of this notion that will be able to meet the objections.

First, let us review briefly the reasons why one might introduce the concept. Intuition *that* becomes a persuasive idea when one reflects on the obviousness of elementary truths of mathematics. Two alternative views have had influential advocates in this century: conventionalism,

---

[30] On this notion, see Chapter 9.

§27. Toward a viable conception of intuition

the view that at least some mathematical propositions are true by convention, and a form of empiricism according to which mathematics is continuous with science, and the axioms of mathematics have a status similar to that of high-level theoretical hypotheses. Both these views have unattractive features. Conventionalism has been much criticized, and I need not repeat the criticisms here.[31]

The empiricist view, even in the subtle and complex form it takes in the work of Quine, seems subject to the objection that it leaves unaccounted for precisely the *obviousness* of elementary mathematics (and perhaps also of logic). It seeks to meet the difficulties of early empiricist views of mathematics by assimilating mathematics to the theoretical part of science. But there are great differences: first, the "topic neutrality" of logic, which receives considerable recognition in Quine's writings, although he insists that it depends on a specification of the logical constants that is at bottom arbitrary; second, the very close connection of mathematics and logic. The potential field of application of mathematics is as wide as that of logic, in spite of the fact that the existence of mathematical objects makes mathematics not strictly topic-neutral. Connected with this is a third difference: In mathematics there are very general principles that are universally regarded as obvious, where on an empiricist view one would expect them to be bold hypotheses, about which a prudent scientist would maintain reserve, keeping in mind that experience might not bear them out. A fourth is that differences about logic and elementary mathematics, such as the issues raised by intuitionism, are naturally explained as differences about *meaning*. Quine recognizes this by the role that logic plays in his theory of translation, but the obviousness of logic is a premise of that theory.

For the moment I will bracket the question of the obviousness of logic. The most elementary logical evidences are of logical truths and inferences that do not involve any distinctly logical or mathematical reference, at least on the view taken in §5 that the understanding of predicates and sentences is prior to any apprehension of or commitment to such objects as propositions and attributes. For this reason, the idea I want to pursue

---

[31] It has been recently argued that the view of Carnap, at least in *Logical Syntax*, is not what is usually criticized under the name of conventionalism; see in particular Goldfarb and Ricketts, "Carnap and the Philosophy of Mathematics." Carnap's view still rests on assumptions that few would accept today.

now, of intuition of mathematical objects, is not of any help in dealing with it.

In the case of arithmetic, the situation is different because, unlike logic, it has ontological commitments. It may be that the idea of logic as true by virtue of meaning can still be defended. But it would be another matter to extend it to arithmetic. That a structure such as the natural numbers should exist, or at least should be *possible* in some mathematically relevant way, is hard to make out as true by virtue of the meanings of arithmetical or other expressions.[32] One might reasonably regard a structure as not the natural numbers if it is not infinite, and let us grant that one might express this by saying that it is contained in the concept "system of natural numbers" that it should be infinite. But that doesn't make it by virtue of the concept, or of our meaning what we do by talk of natural numbers, that there is or could possibly be such a system. Gödel was prepared to assume that the axioms of type theory or set theory are analytic in something like this sense, but, appreciating this difficulty, said that it did not contradict his view that mathematics is "based on axioms with a real content."[33]

Just at this point, the idea of intuition *of* suggests itself. We are taking as a gross fact about arithmetic, that a considerable body of arithmetical truths is known to us in some more direct way than is the case for the knowledge we acquire by empirical reasoning. And this knowledge takes the form of truths about certain objects – the natural numbers. What is more natural than the hypothesis that we have direct knowledge of these truths because the objects they are about are given to us in some direct way? The model we offer of this givenness is the manner in which a physical body is given to us in perception.[34]

---

[32] Gödel, in his remarks in "Russell's Mathematical Logic" about the analyticity of the axioms of *Principia Mathematica* and in other writings about axioms of set theory, attempts to do just that. Although I don't regard Gödel's attempt as successful, for the present it should suffice to remark that it is based on the premise that understanding the propositions of higher-order logic or set theory involves apprehension of concepts, so that some ontology *is*, so to speak, already given with meaning. Just for that reason Gödel's affirmations of analyticity go with his realistic point of view. In particular, he admits that the question of the existence of a structure like the natural numbers is pushed back to the question of the "existence" of the concept of natural number.

[33] "Russell's Mathematical Logic," note 47. Cf. Gibbs lecture, pp. 320–321, and "Is Mathematics Syntax of Language?"

[34] Just what this is, is of course a very large question. Precisely by talking of physical bodies as given, the sense is different from that in the notorious "myth of the given."

§27. Toward a viable conception of intuition

We will see in Chapter 6 that this is an oversimplified picture of our knowledge of natural numbers. As we remarked in the last section, that is already suggested by the structuralist view of numbers. For working out a positive conception of intuition, the case of quasi-concrete objects (§7) has the advantage of being simpler and closer to perception. Discussion of mathematical intuition has suffered from aiming too high (e.g., at set theory) and not looking closely at simple cases. Moreover, in §18 we saw the usefulness of quasi-concrete objects for the structuralist view in making out the possibility of an instance of the structure of natural numbers.

Even in this case, however, one should not demand too much in the way of closeness to perception. Here we can be guided by Kant and Husserl. Kant viewed pure intuition as giving the *form* of empirical intuition; our own conception will echo that, although probably not in the sense that Kant intended. Although Husserl is even prepared to call categorial intuition "perception" (*Wahrnehmung*), he contrasts sense-perception as *schlicht*,[35] in which the object is "immediately given," with categorial intuition which is *founded* in other "acts" such as ordinary perceptions and imaginings.[36] Without adopting Husserl's whole apparatus of acts, we will still view the intuition that concerns us as founded on perception and imagination.

The observation from which I will start is that consciousness of objects that in the usual classification count as abstract is pervasive and commonplace and closely intertwined with perception. In my view, it is closely enough intertwined so that in many cases it is appropriate to call the consciousness itself perception or intuition. An example with which I might

---

[35] LU VI §46. Findlay translates *schlicht* as "straightforward." It also might be rendered as "simple"; very likely, Findlay thought this would have misleading associations. I will follow his translation.

[36] *Ibid.* The sense in which an act of ordinary perception is straightforward, and its object immediately given, seems to amount just to the fact that it is not founded on other acts. "Categorial" intuitions, by contrast, are founded on perceptions and imaginings. (See in particular A 618–619/B_2 145–146.)

Gödel also says that "mathematical intuition need not be conceived of as a faculty giving an *immediate* knowledge of the objects concerned" ("What is Cantor's Continuum Problem?" p. 268). Although this remark was written at a time when Gödel had recently been studying Husserl quite seriously, it is doubtful that he means by "immediate" something close to what Husserl means, since the "something besides the sensations" that he asserts to be "immediately given" are evidently conceptual elements of knowledge, which according to Husserl would be the objects of *founded* acts. Nonetheless, it can hardly be doubted that Gödel found congenial Husserl's view that there are objects of intuition that are abstract and conceptual.

begin, although it is not mathematical, is what in the empiricist tradition are called sense-qualities. We all perceive colors, and a natural way of describing such experiences is as seeing this or that color, say red or blue. To see red is not the same as to see that some object is red, although of course they are related. Looking to the left of my computer, I see bright blue. It is also true that I see a folder that *is* bright blue, but I could see bright blue without seeing that the folder is blue, either because I fail to identify it, or because there is something in the setting that misleads me about the color of the folder and in fact it isn't bright blue although it looks so. It is the latter possibility that is of greatest importance in the philosophical discussion of color, but I would like to dwell for a moment on the former. If in fact there is a bright blue folder before me, and looking in that direction I see bright blue, then in the absence of some grossly abnormal conditions I do see a bright blue folder. But seeing the color is different from seeing that the folder has that color, or even from seeing, of some demonstratively identified object, that it has that color. One thing that marks the difference is individuation. If, in similar lighting, I look at another folder just like this one, I will see the same color but not the same folder. In semidarkness, I may be able to see the same folder but not see its color at all. This difference is also enough to distinguish seeing bright blue from seeing a bright blue object, even though under normal conditions they go together (at least if one counts something like a flash of light as an object).

There are two kinds of conditions in which they appear not to go together. The first is where lighting or other conditions might cause an object to look some color other than its real color. Then it seems that I might look at a green object and see bright blue, because the object looks bright blue to me. That seems to me to be the natural way to describe this sort of case, particularly since if my color vision is normal the blue that I see is a real physical phenomenon. The second type of case is an afterimage or hallucination, where I might "see" a color but there is no object out there that I am seeing at all. In this sort of case, someone might deny that I really see the color, because the experience is not a perception of a real color phenomenon in my immediate environment. For the purpose of the present discussion it is not of great moment whether such cases are, or are not, accepted as genuine cases of seeing colors.

It appears that the color we see is a universal, in the sense that it can be instantiated at a great variety of places and times. Although I see bright blue in a certain location, that is not to say that the color itself is located. What is located is not the color but its manifestations, typically objects

§27. Toward a viable conception of intuition

that have that color or at least look that color to an observer.[37] It may be that what we see is still something more particular than a color as identified by a color-word. For example, a blue computer disk on my table is not quite the same color as the folder. In each case, I can properly be described as seeing blue, but one might say that this is only by virtue of my seeing a color that is a variety of blue, so that my seeing blue while looking at the disk and my seeing blue while looking at the folder are not cases of seeing the same color. This does not alter the fact that the individuation of what is seen is different from that of located spatiotemporal objects.

Color and color perception have many complications that have nothing to do with our main theme, and so I will not pursue the subject further. *Language* offers an example closer to mathematics to reinforce the claim that perception of objects that are abstract is a commonplace experience and acknowledged as such in our perceptual vocabulary.

The vocabulary of seeing and hearing is often used with objects that are linguistic and neither particular events nor physical objects. Thus, one often talks of seeing letters, words, and sentences, and of hearing words and sentences.[38] Often when what is heard or seen is a particular event, it is still naturally described in terms referring to linguistic objects, for example as someone's uttering certain words. The word 'hear' is also used with what would be called propositional objects; that is, one hears what someone said and could naturally report this using a that-clause. That-clauses also occur in descriptions where the object of hearing is legitimately construed as an event: One can say that one heard someone say that p.[39] One would take in the same information visually by reading it, but we usually talk of reading what someone says or writes rather than seeing it.

---

[37] Whether that should count as a "manifestation" of the color is a question that would have to be addressed in a fuller account of color.
   Some philosophers have maintained that there are "abstract particulars," in our example what one might call the particular blue of my folder. On such a view that might properly be called the manifestation of the color. This would allow one to hold that seeing bright blue is a case of seeing that something is bright blue; namely one sees the color-particular, and sees that it is bright blue. Husserl, for example, would distinguish between the color-moment of the folder, which is just such an abstract particular, and the color blue, which is a universal. The former is quite properly located. Presumably even the folder's looking bright blue now (even if, perhaps, it isn't really) has a particular moment of the folder corresponding to it.

[38] Hearing letters probably does occur, but only when a particular letter sound is salient.

[39] If I hear someone say that p, then I might sensibly say that I have heard that p. If I see someone write that p, something further is needed for me to see that p.

The cases of reports by that-clauses obviously introduce a serious complication, as is indicated by the continuing disagreement as to what, if any, the objects of indirect discourse and propositional attitudes are. I want to leave that at one side, as there is no doubt that we do talk of seeing and hearing words and sentences, and at least seeing letters, and there is not the same disagreement as to what these objects are. Here, however, it may be objected that in talk of syntactic objects such as words and sentences, the objects might be types or tokens. If we understand written tokens as physical inscriptions, and spoken tokens as particular events,[40] then perception of tokens is no counterexample to the view that the objects of perception are always concrete. But it seems to me evident that the vocabulary of perception is used where the natural reading is to take the objects as being types, as for example when I report what I heard someone say by giving the exact words. The question whether another hearer heard the same will be answered in the affirmative if he reports the same words, and not otherwise, where it is the words spoken that are at issue; that is, an indirect-discourse report that would involve paraphrase or translation is not what is called for. And I may also report that I heard a certain word or sentence in a situation where I could not identify the speaker or the location from which the sound came. In such a case I, in a sense, identify what I heard, namely, this or that linguistic expression, but I do not identify a particular object or event. This kind of case is less significant for my claim, however, because in cases where abstract objects are not at issue one's identification of what one perceives can be very partial, and this kind of case can be described in that way. I also think one can talk of perception of linguistic expressions in situations where the perception of an actual object or event is illusory or mistaken, and there was not actually a speaker in the neighborhood who uttered those words; but, even if that is agreed, its significance can be read in different ways, and I will not insist on it.

Just what is involved in the perception of words and sentences will be as complex as other questions about perception, perhaps more so just because the objects are linguistic, and specifically linguistic abilities are exercised. None of this is evidently incompatible with the possibility of describing cases of perception of linguistic types in ways that remove

---

[40] I don't find the concept of a particular event as clear as some do. But however that may be, it seems evident that events are concrete rather than abstract, according to the rough criterion of §1, as they are located in space and time and stand in causal relations. One would have to take a position on controversial questions about what events are in order to use perception of events to buttress the idea that perception where the object is abstract is a common phenomenon.

the commitment to types.[41] All I have tried to do in this discussion is to make the case that talk of perception of types is something normal and everyday. A convinced nominalist might still respond to this by saying that there must be a reduction in which such talk is shown not really to be committed to types.

§28. Hilbertian intuition

I want now to describe a situation that resembles that of perception of linguistic types but is much simplified. Following Hilbert we begin by considering the "syntax" of a "language" with a single basic symbol '|' (stroke), whose well-formed expressions are just arbitrary strings containing just this symbol, i.e., |, ||, |||, ...[42] This sequence of strings is isomorphic to the natural numbers, if one takes '|' as 0 and the operation of adding one more '|' on the right as the successor operation.[43] It gives rise to an interpretation of arithmetic as a kind of geometry of strings of strokes. For the purpose of making out a conception of intuition, we could equally have taken a more complex formal language, with more than one primitive symbol. The one-symbol case already contains what is essential to our discussion, as the fact that Hilbert's language is a model of arithmetic indicates.

I think it is clear that we stand in a perceptual relation to the expressions of this simple formal language. The same reasons for talking of perception of expression-types with reference to natural language arise also with

---

[41] Bromberger and Halle, in "The Ontology of Phonology," argue that expression-types are not fundamental objects in phonology, that phonology need not be committed to them, that they are merely a *façon de parler* (p. 226). I am not entirely convinced by their case. However their claims do not contradict mine directly: I am claiming that the straightforward reading of certain ordinary locutions about perception of language will be in terms of perception of types. Whether reference to types can be eliminated in a certain linguistic theory, phonology, is another matter. We will, however, return to the question of the possibility of eliminating reference to types.

[42] David Hilbert, "Über das Unendliche," p. 171 (trans. p. 377), and a number of other writings. In the exposition of finitary arithmetic in volume I of *Grundlagen der Mathematik*, Hilbert and Bernays treat the symbols of this sequence as the objects of the theory (p. 21).

Both in Hilbert's papers and in Hilbert and Bernays, the symbol '1' is used instead of '|'. For our purposes it seems desirable to have the greatest possible typographical simplicity; moreover, the term 'stroke' is convenient.

[43] It is perhaps more natural to think of '|' as 1, thus obtaining a model of the positive integers. So long as one has only an initial element and a successor function or relation, the two structures are isomorphic; that, of course, ceases to be so as soon as one has addition.

reference to this language. I shall, however, prefer to use the term 'intuition' for reasons that will emerge. Although more needs to be said about it, we know when we are confronted with the same symbol, or the same string of symbols. The objects count as abstract by the rough criterion of §1, because two tokens that instantiate the same symbol or string can be separated in space, without one having arisen from the motion of the other. Moreover, we don't think of strings as causal agents; what acts on our sense organs are, roughly, their tokens. Still, they are quasi-concrete; for example, the basic situations in which we have intuition of types are typically those in which we perceive tokens of that type.

We can think of the strings of strokes as forms, in that describing a token as a token of a certain string is characterizing it as of a certain form. A string is like a geometric figure: It is spatial in that it is a form instantiated by objects in space, but it does not itself have a particular location in space. Letters of a written language can differ considerably in shape and still be occurrences of the same letter, and can resemble each other in shape and be occurrences of different letters.[44] Strokes are individuated more strictly and more simply. What counts as a stroke is a symbol of the same shape as the samples that have been presented, and the same will be true of the strings that arise by concatenation.

The nature of the objects with which we are concerned can be made clearer by at first attempting to understand what I am calling intuition of types in this symbolism in a nominalist way. Let us call an inscription of the form '|' a *stroke-inscription*. (I will sometimes call such an inscription simply a stroke.) A stroke is a *string-inscription*; if to a string-inscription *a* we add another stroke on the right, then the result *b* is again a string-inscription, which we can call a successor of *a*.[45] These rules suffice to generate all string-inscriptions; we might think of them as the

---

[44] This is pointed out by Linda Wetzel in "Expressions vs. Numbers." She makes the strong claim that two tokens of different types can be more similar to one another than either is to more "distant" tokens of its own type (p. 187). This is probably correct. I mean to avoid this complication by the design of the symbolism in the text. I don't think Wetzel's considerations show, even for natural language, that identification of types cannot be perceptual, but it does tell against a nominalist or modal nominalist understanding of expression-types of natural language, where the relation "same type" is understood in terms of physical or visual or auditory similarity. A deeper consideration of what understanding of the relation *would* be satisfactory (as in Bromberger, "Types and Tokens in Linguistics") introduces ideas (such as speakers' intentions) that hardly yield a nominalist construal of talk about types.

[45] Although they do not attempt a nominalist explanation, Hilbert and Bernays (p. 21) appeal to the same inductive generation in explaining the *Ziffern* that are to be the objects of intuitive, finitist arithmetic.

§28. Hilbertian intuition

introduction clauses of an inductive definition. Now we could define, for string-inscriptions, being "of the same type" in the same inductive way: Any two stroke-inscriptions are of the same type; a stroke-inscription is not of the same type as a string-inscription resulting from adding a stroke to such an inscription; if a and b result from c and d each by adding a stroke, then a is of the same type as b if and only if c is of the same type as d. We might explain the notion of being of the same type in other ways, for example, we might say that two strings are of the same type if they can be placed side by side so that strokes correspond one-to-one to strokes.[46]

For the moment we can put aside an obvious difficulty of this purely nominalist way of describing this language, namely, that it is not evident that for any string-inscription, there is a string-inscription that extends it by an additional stroke. This is an instance of a general problem facing nominalistic syntax, which has already been discussed in §12. For the present discussion, the possibility that (in non-nominalist terms) our language admits only finitely many types does not have to be ruled out. Moreover, when it does, that is, when we need the claim that this language is a model of arithmetic, a retreat to some form of modal nominalism seems an obvious way of dealing with the difficulty.

A more immediate problem with the nominalist formulation is that it does not accurately render our perceptual consciousness of strokes. It would make what I want to call intuition of a string an instance of seeing a certain inscription *as* of the type '...', where '...' is a certain string. But in actual cases, the identification of the string will be firmer and more explicit than the identification of any physical inscription that is an instance of the type. That the inscriptions are real physical objects with definite physical properties plays no role in the mathematical treatment of the language, which is what concerns us. An illusory presentation of a string, provided it is sufficiently clear, will do as well to illustrate a mathematical notion as a real one.

We can indeed talk, after Husserl, of intuition of a type as *founded* on perception of a token. It might under some circumstances be imagination rather than perception. In ordinary cases, what we will have will be full-blooded perception, involving the physical presence of an inscription and its action on our senses. Such is normally implied by ordinary talk of hearing words; although the token may be pushed into the background by

---

[46] Cf. *Mathematics in Philosophy*, p. 44.

the type or a meaning assigned to it, it does not follow that the token is not an object of perception. However, even in normal cases, the background and further experience that are necessary to the perception's being of something physically *real* are irrelevant to its being of the *form* given by the type. In most cases, physical reality is important not for taking in the type, but for further considerations: What is likely to be of interest about the words is that they were spoken by a speaker at a certain time, or stand written in a certain book or other piece of writing.

What is distinctive of intuitions of types is that the perceptions and imaginings that found them play a paradigmatic role. It is through this that intuition of a type can give rise to propositional knowledge about the type, an instance of intuition *that*. I will in these cases use the term 'intuitive knowledge'. A simple case is singular propositions about types, such as that ||| is the successor of ||. We see this to be true on the basis of a single intuition, but of course in its implications for tokens it is a general proposition. Let b be the token of ||| two sentences back;[47] let a be the token of || at the end of that sentence. Our statement implies that if c and d are respectively of the same type as a and b, then d consists of a part of the same type as a, and one additional stroke on the right. We can of course buttress the statement that ||| is the successor of || by considering arbitrary tokens of the relevant types and verifying the above consequence. But we have to verify it in the same way, by instances that we take as paradigmatic. This situation is not peculiar to our artificial framework. The same is true of calculations done on paper and of formal proofs, such as the deductions done in elementary logic courses. (I mean what the teacher might write on the blackboard or a student on a homework paper.) This paradigmatic role of instances also appears in Kant's examples of construction of concepts in pure intuition.

The nominalist or modal nominalist formulation makes this paradigmatic role of instances difficult to understand by importing the notion of physical object with its standards of individuation and actual existence into the content of a mathematical theory. But what is fatal to it is that, understood in this way, the notions of sameness of type and of successor are *vague*. Although its mathematical use would require it to be an equivalence relation, it is not clear that the notion of being of the same type, as explained above, is transitive. For example, it might be vague with regard

---

[47] The reader is to take it as the particular token in whatever copy of this work he is reading.

§28. Hilbertian intuition

to strings of strokes what the distance between them might be. Consider the examples

At what point does the inscription cease to be of the same type as the initial one? They all satisfy an obvious criterion in that they each consist of four strokes. So one might say at *no* point. But that is not at all plausible; no one would say, for example, that the first of the above strings continued by '|' one thousand miles to the right constitute a single string of five strokes.

We have explained the strings as generated inductively, beginning with a simple stroke and successively attaching an additional stroke on the right of a given string. The above example illustrates the potential vagueness of the relation 'y results from x by attaching a stroke to the right of x'. Vagueness could equally infect the notion of being a stroke (as a predicate of inscriptions). This would be clear if one relies on the usual ostensive explanation, given by writing or printing a sample.[48] Regarding the transitivity of the same-type relation, we can define it inductively parallel to the proposed inductive definition of string. There is no reason to believe that this explanation will remove the vagueness, given the vagueness of the notions of stroke-inscription and succession that underlie it. Although we can prove by induction according to the definition that sameness of type is transitive, we should distrust this proof for the well-known reason that it relies on application of induction to vague predicates.

The nominalist may not be troubled by these matters; he will point out that one might characterize in an exact way, with the help of physics, just what inscriptions count as strokes and what the operation of adding a stroke on the right is. That reply would have force if what were at stake were the ontological commitment of a high-level theory. But it is hard to see how its use could square with the idea that consciousness of types (on this construal, consciousness of tokens together with an equivalence relation

---

[48] Concerning the official explanation of Hilbert and Bernays, the question arises whether anything that we would recognize as a numeral '1' counts as an initial *Ziffer* for them, or only something whose size and shape are the same (or sufficiently close to) that of '1' as actually printed in their text. Their remark on the subject (p. 21) does allow some deviation but probably not so much that anything that would be a '1' in any German typeface would count.

of "being the same type") is essentially *perceptual*. Such an exact concept of a stroke inscription will make a finer discrimination between strokes and non-strokes than perception can make.[49] For example, a stroke might be required to be exactly 0.3 cm high, but one that is 0.30000000001 cm high would not be distinguishable from it.[50]

Because it seems to me evident that there is perceptual consciousness that one naturally describes in language ostensibly referring to types, for this description we have to abandon the nominalist reduction strategy. What I have already called attention to, the ubiquity in our everyday experience of linguistic expressions and other "quasi-concrete" abstract objects such as sense-qualities and shapes, should lead us to question why a reduction should be necessary or desirable. Behind the technical difficulties arising from vagueness is the rather basic consideration that one may identify such a form, as it were in the flux of experience, without identifying an object *of* which it is the form. A spoken expression is a type of which the tokens are presumably events of some kind, but to identify just what event a given token is presumably implies locating it in space and time (and possibly in a causal nexus), although the expression itself does not have a location. I hold that neither the identification of the token nor that of the type is a necessary condition of the other.

If I read a sentence, and take in what sentence it is, of course I do see a certain physical inscription. In that sense, my perception of the sentence is founded on perception of the inscription. But I can identify what sentence it is without identifying what inscription it is, and certain kinds of illusion about the latter would not disturb the correctness of my perception of the former. And it is not an evidently necessary condition of such perception that there be a criterion, stated in terms of physical characteristics of inscriptions, for an inscription to be an inscription of a particular sentence; similarly, it is not evident that there need be a criterion, stated somehow in terms of sensible appearances, for a given perception to be *of* the sentence. We do not demand such criteria in the case of everyday objects.

Thus, as a first answer to this difficulty, the phenomenological claim that our identification and reidentification of expression types is perceptual is not undermined by the fact that we cannot give a sharp equivalence relation of more "basic" objects that serves as a criterion of identity. There is not such a relation for physical objects either.

---

[49] Compare Dummett, "Wang's Paradox."
[50] Of course, handwriting and even printing fail to make such fine discriminations.

§28. Hilbertian intuition

At this point, however, I can make clear why, in the case of our artificial symbolism, I do not speak of *perception* of types, but rather of intuition. If I see a written string from this language, then I see a certain inscription even if for some reason I fail to identify it or misidentify it (say, as a string of four strokes instead of three). This is an instance of a general feature of the use of 'see', shared also by 'hear' and other perceptual verbs, that one can see a certain object even if one does not identify it as such, and a description of what is seen can use concepts that the subject does not possess. This feature certainly extends to events: Suppose I have never heard of the game of lacrosse and have no conception of it. If I come upon a field where a game of lacrosse is being played, I see a game of lacrosse even though I have no idea that that is what it is. I believe this extends also to expressions of natural language. Suppose I say to someone who has never before heard English, "Where is the American Embassy?"[51] We would say that that person has heard that sentence of English, even though he certainly does not recognize it as such and would not be able to recognize an utterance of the same sentence by someone with a different accent.

In this case, perception of a certain utterance counts as hearing a sentence of a language of which the hearer knows nothing, presumably because that is what the speaker produced, and the hearer has at least the general linguistic capacity that humans are born with. I do not want to say, however, that seeing a stroke-inscription necessarily counts as intuition of the type it instantiates. One has to approach it with the *concept* of the type; first of all to have the capacity to recognize other situations either as presenting the same type, or a different one, or as not presenting a string of this language at all. But for intuiting a type something more than mere capacity is involved, which, at least in the case of a real inscription, could be described as seeing something *as* the type. This is not an aspect that is absent in normal perception, but, as I have said for seeing or hearing an F, one does not have to identify it as an F although one at least recognizes it under some description or other.

Something more is involved in speaking of intuition in the case at hand. What makes intuition *mathematical* intuition is that it gives objects that instantiate concepts that have a sharp, precise character. At least for statements in the mathematical vocabulary, there is no vagueness in their application to strings of strokes. There may indeed be vagueness as to whether what is before us is, or is not, a token of a given string, but not

---

[51] This example is from "Mathematical Intuition," note 24, written in late 1979. Today one might wonder whether a normal adult can be found who has never heard English.

about the question whether one string, say, consists of two more strokes than another. When I spoke above of perception or intuition of types, was I in a position to rule out vagueness or puzzle cases about their identity, such as there might be in the case of everyday objects? Pure intuition as Kant understood it was evidently supposed somehow to get us across the divide between the fuzzy *Lebenswelt* with its everyday objects and the sharp, precise realm of the mathematical, in terms of which mathematical conceptions of the physical world are developed.[52] Does intuition of types as I have described it present objects that are conceived in a sharp way?

Let us consider first a somewhat different case from that of types, where we can see very clearly where a problem about vagueness will arise: the idea, developed in some detail in early writings of Husserl, of intuition of finite sets. According to this, a certain way of taking a surveyable plurality of objects of perception, say the eggs in a carton,[53] is intuition of a set whose elements are the eggs in the carton. Eggs belong to what I have called the *Lebenswelt*, but on the face of it sets of eggs are mathematical objects. Given such a set a, one would expect that it would be quite determinate, for a given x, whether or not $x \in a$. But suppose x is one of the eggs, and let x' be an egg that we encounter some time later. If $x' = x$, then of course $x' \in a$. But suppose (as it seems that we cannot rule out) that x' is an egg such that our criteria of identity for eggs are not able to settle whether or not it is x, although they determine that it is not one of the other eggs in a (perhaps they were consumed before we encountered x').[54] Then it

---

[52] One might call the former the "phenomenal world," but that would not be exactly in a Kantian sense.

[53] The example is from Maddy, *Realism in Mathematics*, p. 58. Maddy defends the idea of perception of sets. She holds, for reasons not relevant to the present discussion, that a case like that in the text is one of perception, in a sense distinguishable from intuition.

The use here of the Husserlian idea of intuition of finite sets for purposes of illustration and comparison is not meant to express commitment to it. It will be considered critically in §34.

[54] If we take the letters x and x' to be true variables, then it appears we are admitting x and x' as "vague objects" such as were famously rejected by Gareth Evans (see Evans, "Can there be Vague Objects?"). Clearly, however, we can admit vague identity statements without admitting vague objects. To avoid the commitment to vague objects, we should understand our letters as schematic letters for singular terms.

Since the question whether vague objects can be admitted is controversial, the formulation that allows them is still of interest. The discussion in the text would then argue that if x and x' are vague objects, such that it is not determinately true or false that $x = x'$, then sets containing them are also vague objects; in particular, it is not determinately true or false that $\{x, y, z\} = \{x', y, z\}$. The situation is analogous to the rigidity of membership when set theory is combined with modal logic. (On the latter, see for example *Mathematics in Philosophy*, pp. 298–308.)

§28. Hilbertian intuition

seems that it is, after all, indeterminate whether $x \in a$. Cantor thought of sets as consisting of "well-distinguished objects"; and the practice of set theory ever since has been to assume that problems of individuation of the basic elements will not arise. This is a case of "idealization": The application of a theory of sets of eggs, even small finite sets, might break down if it is made to cases where the eggs are not "well-distinguished," that is, if there are questions of the identity of "two" potential elements of a set that cannot be resolved.[55] For example, there might be an unresolvable question about the cardinality of the set. In the present case, however, it seems clear that one can see that there are three eggs in the carton.[56] Let x and x' be as before. Then if $x = x'$, x' does not add another egg, and if $x' \neq x$, then our hypotheses imply that x' was not in the carton.[57]

It seems from these considerations that the idea of perception or intuition of sets is ambiguous. If we perceive the set of eggs in the carton in the above-described situation, then we do not get from that the knowledge that it consists of well-distinguished objects. Thus, on this view, we perceive the set a; perhaps when we have encountered x' we can perceive or intuit the set $a' = \{x, x', y, z\}$. But of a' it seems to be undetermined whether it has three or four elements; likewise, whether or not $a = a'$. The conclusion to draw is that individual sets of concrete objects can inherit from their elements any troubles about the individuation of the latter.

It could be a matter of dispute whether the concept of set that allows vagueness of this kind is really the mathematical one. Set theory would not lose its applicability to everyday situations by ruling out such sets, as they will in any case be rare, and one will often get correct results by ignoring such vagueness; thus in the present case, one would get the right cardinality for the set of eggs in the carton.

The idea that sets must consist of well-distinguished objects may be interpreted simply to exclude vague objects as elements of sets, or it may exclude from the language of set theory singular terms that can enter into vague identity statements. In the text I am taking it in the latter, stronger sense.

[55] This does not mean that we would have to be able to *know* whether $x = x'$.

[56] For the sake of argument, I am assuming, with the early Maddy, that seeing that there are three eggs in the carton involves perceiving or intuiting the set of eggs in the carton. This will be questioned in §32.

[57] One might argue that we cannot see at a time t, before we encounter x', that there are three eggs in the carton, on the grounds that to know that there are three eggs in the carton we have to know that x' is in the carton at t only if $x' = x$, and we can't know that at t because, by hypothesis, we are totally unaware of x'. But we don't have to know it; it is implied by 'there are three eggs in the carton' only together with certain facts about x'.

What, now, is the impact, if any, of the vagueness we have called attention to on concepts of expression-types and a mathematics that deals with such types? One can see that a similar vagueness will arise on an attractive conception of strings, namely, that they are sequences of symbols. Some have wanted to think of strings in this way, conceiving the notion of sequence in some abstract, possibly set-theoretic way, while thinking of the individual symbols as sets of their tokens or in some other way where the concrete tokens play a constitutive role.[58] Such an approach will leave the identity of strings no more precise than that of individual symbol-types; for example, given the two-symbol strings st and uv, $st = uv$ if and only if $s = u$ and $t = v$. But then any indeterminacy left by our criteria of identity for individual symbol types will infect the identity of strings.

If we are not antecedently committed to this construal, the case of types is different from those of sets and sequences in that it is possible to shunt off any threat of vagueness in their individuation onto the relation to their tokens or possibly the individuation of the tokens. Thus it may seem to be vague whether the type instantiated by

(a) ||||

is the same as the type instantiated by

(b) ||| |.

But that conclusion is forced on us only by the unwarranted idea that "the type" instantiated by each of these inscriptions must be uniquely determined. If we allow ourselves to talk of the type instantiated by (a), presumably because we can reidentify it and relate it to other types we have admitted, and then find it vague whether (b) is *of that type*, then our difficulty concerns a relation between a type, the type of (a), and the *token* (b).[59] That relation can be vague without the relations between types becoming vague. The vagueness may generate ambiguity as to what type 'the type instantiated by (b)' designates (if we do not take it to fail

---

[58] This should be distinguished from a more purely structuralist way of construing strings as sequences, in which the alphabet is just some arbitrary finite set of objects, about which nothing is assumed except the number of its elements. Such an approach is perfectly appropriate to the mathematical study of formal languages, but it simply does not engage the question how the relations of symbol- and string-types are related to those of concrete objects or are manifested in perception.

[59] The reader is again asked to take that token to be the particular inscription marked (b) in the particular copy of this work that he has at hand.

§28. Hilbertian intuition

of reference because of failure of uniqueness). But if we admit that it does designate a type, then the identity statement the type instantiated by (a) = the type instantiated by (b) cannot be vague, although it may be ambiguous.[60]

Considering strings of strokes as types, does someone who has the concept of such a string and (we will suppose) intuits a string when he perceives (a) also intuit a string when he perceives (b)? The answer is not altogether clear. Let us suppose that he approaches the situation with the concept of a string of the language that concerns us, and not with the concept of any other kind of string. Then, it seems, the only string he could intuit would be (a); that is, he would intuit a string only if (a) and (b) are of the same type. Let us assume he will read (b) as consisting of four tokens of the unique symbol of our alphabet. So then the question is whether he will recognize the attachment of the last symbol as an instance of the concatenation relation. But obviously he may or may not. We must then ask whether he would be *right* in making one or the other choice. But this is just where the vagueness comes in: If (b) is really a borderline case of being of the same type as (a), then our agent's situation must be a borderline case of intuition of the type that we are assuming to be instantiated by (a). This is only true, however, if we assume that his intuition must be founded on his actual perception. Perception of (b) might prompt imagination that would found intuition of the type, which our agent would then presumably recognize to be only imperfectly instantiated by the perceived token (b).[61]

These considerations show that the fact that the notions of stroke-inscription and sameness of type of stroke-inscription are vague does not imply that the notions of stroke, string of strokes, or identity applied to such strings are vague. But we have not shown that the notion of stroke, say, is not vague. Vagueness of that notion would infect the notion of string of strokes, however we proceed from one to the other. One might ask whether the stroke of the present work and the stroke of my paper "Mathematical Intuition" are identical. Differences in printing make it natural

---

[60] Even if we do not reject vague objects on general grounds (see note 52), types are not vague objects, and singular terms designating them can be indeterminate as to their reference only when they involve nonmathematical relations such as relations to tokens.

[61] If we ask, in the above case of sets, whether someone intuits $\{x', y, z\}$, we again have a borderline case, but one that rests on a borderline case of an identity statement concerning what is intuited. And there is no opening, in the case where he has x, y, and z before him, for intuition of $\{x', y, z\}$ founded on imagination.

to say that they are not of the same shape.⁶² This result is counterintuitive, but it is a consequence of the constraint we imposed on the concept of stroke. But let us suppose that we are unable to decide the question. Will this show the concept of stroke presented here to be vague? No, what we are unable to decide is whether it applies to what was printed in "Mathematical Intuition," once again no longer a straightforward question about the symbols themselves but about their relation to the empirical world.

It is instructive to ask whether there is vagueness about the phonemes of a natural language or about the letters of the alphabet. Consider first the Roman alphabet. Once we have made certain decisions, such as that letters with the diacritical marks that occur in some languages are not distinct letters, and that capital and small letters are variants of the same letters, then it is hard to see what room there is for vagueness in the notion of letter-type: A letter is a letter of the Roman alphabet just in case it is one of the twenty-six letters a, b, c, . . . z. (A "letter" such as the German ü or the French ç counts either as a variant version of a letter or a combination of a letter (u and c, respectively) with something else that is not a letter.) Once something is identified as a letter-type of a written language, it is hard to see how it could be vague whether or not it is, say, a. If it is not a letter of the Roman alphabet, it is not.⁶³ What makes this work is that there is a definite finite number of alternatives, and the identity of the types goes with a system (the Roman alphabet) to which they belong. I believe that the same holds for the phonemes of a natural language.

The same considerations apply to the present case. There is only one stroke, and each string has only one successor. Where the relations of tokens cannot be too vague is for their role in making the relations of the types intuitive, not for the precision of the relations of the types themselves. A reader trying to determine whether a letter in difficult handwriting is an a, is trying to judge whether it was *intended* to be an a. Such a judgment may well be necessary even in the case of high quality printing in a single font and type size, although in that case a reader can make it immediately and spontaneously, and it can probably be made once and

---

[62] Compare the immediately preceding pages with pp. 153 and 155 of the 1979–1980 volume of the *Proceedings of the Aristotelian Society*.
[63] It does follow, for example, that the Greek capital alpha, although printed A, is not the same letter as the Roman A.

for all for each letter, since subsequent occurrences will be pretty nearly exactly similar in appearance to the first one. To sidestep the necessity of such judgments of intention, I wish to describe a symbolism where variation in size, shape, and arrangement of the symbols is at a minimum. That makes the determination of sameness of type for stroke-tokens and short strings purely visual. Similarly, such a matter as the distance from a stroke to the next is, although determined ostensively, not thought of as varying. That means that the stroke and the strings in the present work and in "Mathematical Intuition" are different and belong to different symbolisms, although of course both illustrate the same conception. We can of course in practice identify them, but once we do so we are using a more complicated notion of type, closer to that of written natural language.[64]

§29. Intuitive knowledge: A step toward infinity

We have already encountered simple examples of propositions about our stroke-language that are known intuitively. At first approximation, what this means is that they are known by intuiting the expressions that they are about. This parallels perceptual knowledge of the objects we perceive. The concept of perceptual knowledge does not have at all precise boundaries. We should expect that the same would be true of intuitive knowledge. However, our general discussion of intuition would suggest that what we would call intuitive knowledge would be intuition *that* in the epistemically loaded sense, which in §24 we said was a potentially misleading way to use the term 'intuition' but would be appropriate enough for intuitive knowledge. Then an item of intuitive knowledge would be something that can be "seen" to be true on the basis of intuiting objects that it is about.

The actual conception of intuitive knowledge to be deployed here will generalize this conception in a respect that is forced on one if it is to have any useful application to mathematics at all. Evidently, at least some simple, general propositions about strings can be seen to be true. I will argue that in at least some important cases of this kind, the correct description involves imagining *arbitrary* strings. Thus, that will be included in "intuiting objects that a proposition is about." It is clear that Hilbert and Bernays, whose guidance I am following here, meant to include this when,

---

[64] We shall briefly consider generalized notions of type in Chapter 6.

in their description of the intuitive, finitary method, they wrote:

> What is characteristic of this methodological standpoint is that considerations are put forth in the form of *thought experiments* on objects that are assumed to be *concretely present (konkret vorliegend)*.[65]

This matter will be explored shortly.

There is another, more essential, respect in which intuitive knowledge as here conceived goes beyond what our general discussion of intuition might lead us to expect. Any conception of intuitive knowledge applicable to mathematics will have to embody a decision as to whether knowledge acquired by a rather complex proof, and presumably not acquirable otherwise, is to count as intuitive. This is just the question of the boundaries of intuition that arose earlier in our remarks about Descartes. We shall have to consider whether rules of inference, either logical or belonging to a particular domain of mathematics, preserve intuitive knowledge. Now, one might want to say that only the starting points of proofs, that is the axioms, are properly described as intuitively known, if even they are. It is they that one can "see" to be true. There may be other such propositions that for reasons of simplicity or convenience are treated as conclusions of proofs rather than as axioms, but most of what one establishes by proof in a mathematical theory will not be knowable without a deductive argument, possibly of quite a number of steps, building on intermediate conclusions. The suggestion is that no such proposition should be regarded as intuitively known, even if the conceptual resources used in its proof are very restricted, for example, if (in arithmetic) the proof is finitist.

I do not wish to argue that the term 'intuitive knowledge' should not be used in that way. Our sense, following that of the Hilbert school, is a more extended one that allows that certain inferences preserve intuitive knowledge, so that there can actually be a developed body of mathematics that counts as intuitively known. This seems to me a more interesting conception, in addition to its historical significance.[66] Once one has adopted this conception, one has to consider case by case what inferences preserve intuitive knowledge. This will be considered with respect to the theory

---

[65] *Grundlagen der Mathematik* I, p. 20, emphases in the text.
[66] That finitary mathematics in the sense intended by Hilbert is characterized as intuitively evident mathematics is most explicitly stated by Gödel, "Über eine bisher noch nicht benützte Erweiterung," p. 240. See §40, where evidence that the Hilbert school accepted this characterization is given.

§29. Intuitive knowledge: A step toward infinity 173

of the strings of strokes in Chapter 7. If we admit (as I will argue that we should) that certain logical steps preserve intuitive knowledge, then we will be confronted with degenerate cases of intuitive knowledge, in which the role of intuition is limited to that of an element of understanding, or securing reference for, components of the statements involved. Self-identities such as $|| = ||$ would be an example; a simple tautology such as $|| = ||| \to || = |||$ would be another. We shall consider the role of logic further in §42.

I now turn to the matter of intuitive knowledge of general propositions about types, which have in their scope indefinitely many *different* types. This would be admitted also by the more restricted conception that would not allow logical or other inferences to preserve intuitive knowledge. It is this kind of case that prompts us to follow Husserl in saying that sometimes *imagination* of the token can found intuition of the type. Consider, for example, the assertion that each string of strokes *can* be extended by one more. This is an essential element of the idea that our "language" is potentially infinite. It is essential for using the strings to interpret the language of arithmetic. Something similar is needed to use strings from a larger alphabet to interpret elementary syntax more directly.

However, we cannot convince ourselves of this by perception or by the kind of mathematical intuition we have talked about so far, founded on actual perception. But if we imagine any string of strokes, it is immediately apparent that a new stroke can be added. One might imagine the string as a *Gestalt*, present all at once: Then, since it is a figure with a surrounding ground, there is space for an additional stroke. However, this leaves out an important aspect of the matter, since the imagination of an *arbitrary* string in this way will have to leave inexplicit its articulation into single strokes. Alternatively, we can think of the string as constructed step by step, so that the essential element is now succession in time, and what is then evident is that at any stage one can take another step.

Either way, one has to imagine *an arbitrary string of strokes*. We have a problem akin to that of Locke's general triangle. If one imagines a string in a specific way, one will imagine a string with a specific number of strokes, and therefore not a perfectly arbitrary string. The first way of imagining an arbitrary string involves imagining a string of strokes without imagining its internal structure clearly enough to imagine a string of n strokes for some particular n. Such imagining is common enough. I might for example imagine the crowd at a baseball game, without imagining a crowd consisting of 34,793 spectators.[67] To return to the case of the string, that there is no *n* such that I imagine it to have *n* strokes does not imply that I imagine it as not having a definite number of strokes. We might speak of the imagination as vague,[68] but that can be misleading since it suggests that *what I imagine* is indeterminate in some way.

We naturally think of perception as at least sometimes uncorrupted by thinking, in that without conscious thinking one can take in some aspect of the environment and respond to it, and one can take a stance toward one's perceptions that is largely noncommittal with respect to the judgments we would ordinarily be prepared to make. However that may be, it is clear that the kind of thought experiments I have been describing can be taken as intuitive verifications of such statements as that any string of strokes can be extended only if one carries them out on the basis of specific concepts, such as that of a string of strokes. If that were not so, they would not confer any generality.

Brouwer may have been trying to meet this difficulty, in a special case of this sort, with his concept of two-one-ness, according to which the activity of consciousness brings about "the falling apart of a life-moment into two qualitatively distinct things," of which the moment then present retains the structure of the original, so that the resulting "temporal two-ity" can be taken as a term of a new two-ity, giving rise to temporal three-ity.[69] Thus the process can always give rise to a new moment, which for Brouwer is the foundation for the infinity of the natural numbers.[70] What plays this role in the description above is the figure-ground structure of perception,

---

[67] Even if I am imagining the crowd at an actual game, where the attendance was in fact 34,793, it does not follow that it belongs to the content of my imagining that the crowd consists of that number of spectators.

[68] As I did in "Mathematical Intuition," p. 156. There I offered another suggestion, that one might imagine a single string and take it as a paradigm. I don't think this option differs substantively from what is proposed in the text. Otherwise Hale and Wright's criticism of this suggestion ("Benacerraf's Dilemma Revisited," p. 107) would be well taken. They remark concerning this option and the one in the text that they differ from that concerning a geometrical diagram in that in the latter case, "seeing the irrelevance of the specific values of the other angles, and the specific lengths of the sides, consists in the fact that none of the steps in our accompanying *reasoning* relies on any assumptions about them" (ibid.). However, it is equally true that reasoning from an axiom saying or presupposing that any string can be extended does not rely on its specific length, unless that is introduced at some other point in the argument.

[69] L. E. J. Brouwer, "Mathematik, Wissenschaft, und Sprache," p. 153, my translation. The same conception is expressed in many places in Brouwer's writings, going back to *Over de grondslagen der wiskunde*, p. 81 (of original).

[70] It is of course natural also to view the generation of strings temporally. I believe that the structure that results, and the issues concerning it, are the same as in Brouwer's case.

§29. Intuitive knowledge: A step toward infinity 175

so that what is imagined after the addition of a stroke is still a bounded geometric configuration, perceptually a figure that has its ground as we had before. Thus, after the step of "adding one more" one has essentially the same structure.

Both in our own case and in Brouwer's, that is the reason why we think of whatever step it is as one that *can be iterated indefinitely*. But a concept such as that of a *string* of strokes involves the notion of such iteration. To spell that out, we are led into the circle of ideas surrounding mathematical induction. Although the view has been attributed to Brouwer that "iteration" is the fundamental intuition of mathematics, my view is that the particular concept of intuition I am explicating runs out at this point.[71] There is no intuition of an object, or of objects, fitting this conception that would yield induction. But there is much more to be said on the relation of induction to intuition; a little will be said shortly, somewhat more in Chapter 7.

Although the concept of a string of strokes involves iteration, the proposition that every such string can be extended is not an inductive conclusion. A proof of it by induction would be circular. Such a proof would be called for only if we really needed the fact that every string of strokes can be obtained by iterated application of the operation of adding one more. Only a proto-conception of string is needed to see that every string can be extended. I think the matter stands thus: We have a structure of perception, a "form of intuition" if you will, which has the essential feature of Brouwer's two-one-ness, that however the idea of "adding one more" is interpreted, we still have an instance of the same structure. But to see the *possibility* of adding one more, it is only the general structure that we use, and not the specific fact that what we have before us was obtained by iterated additions of one more. This is shown by the fact that, in the same sense in which a new stroke can be added to any string of strokes, a new stroke can be added to any bounded geometric configuration.

We do not acquire in this way any reason to believe it *physically possible* to extend any string of strokes, embodied in a physical token. What is at stake here is at most the structure of space and time. Physical possibility requires something more, whatever makes the difference between the space of pure geometry and the physical universe, consisting at least of space containing matter. Actually, we require less than the space of pure

---

[71] Brouwer is not as clear as he might be about the distinction between intuition of and intuition that. Writers about Brouwer tend to be even less so.

geometry, even if we hold to the spatiality of the strokes. Only very crude properties of space are appealed to, in particular not its precise metric properties.[72]

It is sufficient for the purpose to take the possibility in question as an instance of *mathematical possibility* (see §15). This expresses the fact that we are not thinking of the capabilities of the human organism, and it may even be extraneous to think of this "construction" of adding another stroke as an act of the *mind*. Brouwer would see situations of this kind in that way, and it seems reasonable to attribute the same viewpoint to Kant. It will be very tempting if we want to say that any string of strokes *can* be intuited or that a token of it *can* be perceived or imagined. The idea is that no matter how many times the operation of constructing one more stroke in imagination has been repeated, "I" or "we" can still construct one more. However, there is really a hidden assumption that the only constraints on what "we" can perceive are the open temporality of these experiences and some very gross aspects of spatial structure. Kant and Brouwer thought these were contributions of our minds to the way we experience the world. Kant of course thought that we could not know these things a priori unless our minds had contributed them. I am not persuaded by this, and in any case I do not want my argument to rest on the notion of a priori knowledge. If we express the *content* of the proposition as independently of the description of the insight as possible, then it is just that for an arbitrary string of strokes, it is possible that there should be one that extends it by one stroke.

Even if we have the stronger proposition that for every string given in intuition, one can intuit one that extends it, the route to the conclusion that every string can be intuited is not altogether clear. It seems that we might reach the conclusion by induction. I am postponing discussion of issues concerning induction, so that I will only in §§44–45 revisit this issue.

Although intuition yields one essential element of the idea that there are, at least potentially, infinitely many strings, we have already indicated that more is involved in that idea, in particular that the operation of adding an additional stroke can be indefinitely *iterated*. The sense, if any, in which intuition tells us that is not obvious. Cashing in this idea again brings us into the circle of concepts surrounding induction and recursion. It will be

---

[72] Compare the remark of Gödel: "What Hilbert means by 'Anschauung' is substantially Kant's space-time intuition confined, however, to configurations of a finite number of discrete objects" ("On an Extension of Finitary Mathematics," note b).

§29. Intuitive knowledge: A step toward infinity 177

argued in §44 that one can make a beginning with induction and recursion and stay within what is intuitively known. In a sense, that the operation of adding an additional stroke can be iterated is intuitively known. But it is part of our usual notion of natural number that *any* operation on numbers can be iterated,[73] and as it seems obvious that the strings are a model of arithmetic, the same should hold of operations on strings. Such a principle, because of its higher-order character, cannot be intuitively known in the sense that concerns us now, where the underlying notion of intuition is that of §28. The discussion of §§44–45 will raise doubts as to how far one can go in obtaining intuitive knowledge of particular cases of the principle, but will not close off the possibility that the operations that can be seen intuitively to be well defined are closed under primitive recursion.

Let us return to the proposition that any string can be extended. The idea that this rests on a capability of the mind is a very natural one and is in certain respects acceptable. I have proposed two different ways of seeing this, one resting on the figure-ground structure of perception and one (Brouwer's) resting on temporal experience. The former is a matter, among other things, of the way space occurs in perception. To be sure it involves the world: What one perceives is perceived as being in a surrounding space. But it is a fact about perception that one can shift one's attention so that what was previously ground is now figure, and the result still has the same figure-ground structure. In the other case, we experience the world as temporal, and have the conviction that we can continue into a proximate future, in which the immediate past is retained.

In both situations, given the structure of our spatial and temporal experience as it now is, it is not unreasonable to say that it can be continued in the way that the extension of a string requires, where the "can" in question is quite practical. However, it is only a quite abstract aspect of these situations that we actually use, that space extends beyond our present figure, and that time continues beyond the present. It is for this reason that I say that the actual capabilities of the human organism are not in question, even though an actual capacity obtains in the kind of situation that concerns me. Something, however, of the structure of the mind survives this abstraction, as we still have the figure-ground structure in the first case and the present-retained past-anticipated-future structure in

---

[73] This amounts to saying that such operations are closed under primitive recursion.

the second. But it is still on a more abstract plane than what is usual in talk about the mind.[74]

It is clear that the practical capacity need no longer obtain when we consider the possibility of indefinite iteration of the addition of a stroke, or of the transition from figure to ground-become-figure or from present to future-become-present. I may be, without knowing it, at the point of death. Maybe it is *now* possible for my experience to continue in time, but that possibility will not become actual, and I think we have to deny the practical possibility of even modest iteration. And starting from any point during my life, if there is some minimum time required to add another stroke, then my mortality places a limit on how many I can add, so something of what is contained in saying that such addition can be indefinitely iterated does not obtain. It is thus on this abstract structural plane that the iterability does obtain, and it is for this reason that the possibility of continuation that is relevant is a mathematical possibility; it is not properly speaking the possibility of an action. By the same token, the *necessity* that every string can be extended is mathematical necessity, dual to mathematical possibility. Our insight into it is insight into experienced space and time.

Once one has seen that every string can be extended, it is still another question whether the string resulting by adding another symbol is a different string from the original one. For this it must be of different *type*, and it is not obvious why this must be the case. For example, why can there not be a one-one correspondence of the strokes in the new string with those in the previous string? Thinking of the matter intuitively, it seems that in order to see this we have to appeal to the step-by-step construction of strings. The thought experiment to show that every string can be extended does not rely on the fact that every string is constructed from a one-symbol string in finitely many steps. Here again, although it will follow from considerations advanced in Chapter 7 that it is intuitively known that every string can be extended by one of a different type, ideas connected with induction are needed to see it. This fits with the fact that induction is needed to prove $Sx \neq x$ in elementary arithmetic.[75]

---

[74] It is imaginable that it might be given a functionalist interpretation, although a transcendental one comes more readily to hand.

[75] That it is unprovable in Robinson arithmetic Q (arithmetic with addition and multiplication but without induction) is shown in Tarski, Mostowski, and Robinson, *Undecidable Theories*, p. 55. (Thanks to George Boolos for this reference.) To see the independence of $Sx \neq x$, it is sufficient to add to a model a single nonstandard element a and stipulate $Sa = a$. However, a model with two nonstandard elements will show all at once the

§30. The objections revisited

The first of the objections raised in §26 was that if there is intuition of mathematical objects, it should be obvious that this is so, and it is not. My reply to this objection should be clear from §27. There I pointed to a number of phenomena that surely are obvious and are naturally glossed as something like perception where the object is abstract according to the criterion I have used. That this gloss is correct is admittedly less obvious. To that extent I do not in the end agree with the premise of the objection. The closeness between our awareness of shapes, colors, and linguistic expressions and undisputed cases of perception is something that has to be pointed out and about which there is room for argument. One might compare my thesis with the thesis that there is perception of *physical objects*. In principle, one might deny that even this thesis meets the obviousness condition, since, although that appears to be the way we ordinarily talk (and also talk in reporting scientific experiments), a lot of room is still left for argument, in view of the old representationalist view that the only objects of our "direct" perceptual consciousness are ideas or sense-impressions, and the possibility of a phenomenalist reduction of talk about bodies.[76] An analogue of the latter in the case of types has been discussed, namely, a possible nominalist reduction. There is no so direct analogue of the former, but the idea that types have to be thought of as equivalence classes or properties of tokens might be developed so as to have the analogous feature, that types are a theoretical construction based on a prior concept of token. This view was disputed in §§27–28.

Some arguments about these matters take the form of disputes about how close a cognitive phenomenon is to perception. A disanalogy already mentioned between ordinary perception and any intuition of objects where the objects are abstract is that we can't say in the latter case that the object itself causes either the intuition or something underlying it (as physiological processes underlie perception), because that would violate the acausality of abstract objects. Although this is a genuine disanalogy, I wish to say that it is not so great as might appear at first sight. One has to recall two things. One is the point made in §28 that intuition of an abstract object requires a certain conceptualization brought to the situation

---

unprovability in Q of a number of elementary arithmetic statements. See Boolos and Jeffrey, *Computability and Logic*, exercise 14.2 (any edition).

[76] Although the possibility of such a reduction is not taken very seriously today, it was so taken not so long ago, at least up through the 1950s.

by the subject. The other is that although it is in our usage a criterion for being abstract that an object not stand in causal relations, this is in a way a grammatical point. Abstract objects can be referred to in descriptions of events that do stand in causal relations, and these descriptions can figure in causal explanations. Intuition of a string of strokes is typically founded on perception of one of its tokens. That a token *of that type* acted on the sense-organs of the perceiver could perfectly well be part of a causal explanation of the perceiver's perceiving a token of that type and thus of his having intuition of the type. But we do not gloss this in such a way that causal efficacy is attributed to the type. The underlying reason is no doubt that the type does not emit light or otherwise transmit energy to us.

Some locutions might seem to attribute causal efficacy to linguistic expressions or shapes or colors. Suppose, for example, someone were to say to me, "You have no right to call yourself a philosopher." I might comment on this by saying, "His words made me furious." But if we are thinking of the words as types, surely what made me furious was his uttering them. That was what made them *his*, and his uttering them in that context made them addressed to me. Once again we have a situation where the cause is an event that we describe in a way involving reference to an abstract object (the sentence type) that is instantiated in it.[77]

---

[77] Although that isn't exactly what the imagined comment intends, it might be more accurate to say that what caused my anger was his saying *that* I had no right to call myself a philosopher. Then the event is described using indirect discourse. Some accounts of indirect discourse have it involving reference to a sentence, others to a proposition. Although more nominalist accounts do exist, it is not clear that any can be successful. The influential account of Davidson, "On Saying That," appears to introduce only reference to utterances, but carrying it through may require modifying it to introduce reference to sentences and their structure; see Higginbotham, "Linguistic Theory and Davidson's Program in Semantics."

In response to Paul Benacerraf's problem of reconciling reference to mathematical objects with a causal theory of knowledge, Mark Steiner pointed out that reference to such objects is typical of many causal *explanations* (*Mathematical Knowledge*, Chapter 4). The remarks in the text are clearly on the same general lines as Steiner's. Jaegwon Kim criticizes the whole idea that mathematical objects, at least the most elementary ones, are causally inert. However, on his view of causality it is properties, relations, and states of affairs that stand in causal relations. His point is much the same as Steiner's, but in the context of a rejection of the distinction between causal relations between events and causal explanation that relies on descriptions of these events. See "The Role of Perception in A Priori Knowledge," pp. 346–347.

Kim would evidently question the criterion of abstractness that we have relied on since §1. (See note 1 of Chapter 1.) I am inclined to uphold the Davidsonian view about the logical form of causal statements, and when the distinction between causal relations and causal explanation is taken into account, the criterion does not have the drastic

§30. The objections revisited

Compare now the case where someone, A, sees a token of '|||' but does not have the concept of strings of this language, and so does not think of it as a string of our language, still less intuit such a string, with the case where another person, B, intuits '|||', perhaps on the basis of a perception of the same token very similar in its other outward aspects such as orientation and lighting. The fact that the token is of the type it is, will very likely enter into the explanation of why B's experience is as it is, but it is not clear that its role will be causal, since the difference between B's situation and A's, is that B is exercising a conceptual apparatus that includes the capacity to recognize tokens as being of a certain type. Moreover, if a token of a certain type is present to a perceiver, then it follows *necessarily* that the type is instantiated there. Thus, one can't distinguish the two cases by saying that in the outer world the type is present in B's case but not in A's.[78] It is hard to say that the *type* is responsible for the difference.

The disanalogy between intuition and perception is reduced by these considerations but does not disappear altogether. A further disanalogy arises from the fact that intuition may be founded on imagination rather than on actual perception; in that case, the causal role of objects in the outer world is much more indirect. Furthermore, there is no phenomenon in perception corresponding to that of imagining an arbitrary instance of an intuitive concept, which we have seen in §29 to play a central role in basic cases of intuitive knowledge.

Let us now turn to Kant's puzzle. Are we, in order to accept the view that single intuitions can have general implications, forced to accept the conclusion Kant drew from this, particularly in the radical form that the spatiotemporal form of our experience is contributed by our own minds? A problem we face, already in interpreting Kant, is to see what it is specifically about intuition that should drive us to this conclusion. We know that if the earth still exists two thousand years from now, and the earth is (then) a satellite of the sun, then at least one satellite of the sun will

---

consequences for the relation of abstract objects to the causal order and the possibility of knowing about them that Kim fears.

[78] As noted above (note 53), in *Realism in Mathematics* Penelope Maddy holds that there is perception of sets whose elements are perceptually present, for example of the set of three inscriptions of '|' in the text. (See ch. 2 §2.) She attributes a causal role in this perception to the *set*. It is noteworthy that in order to perceive a set according to Maddy, it is not necessary to perceive it *as* a set or even have an articulate concept of set (*ibid.*, p. 63 n. 70). The kind of distinction we make between the cases of A and B would apparently not be possible on her account of perception of sets, although it is not entirely clear to me whether according to her *any* situation of, say, seeing three eggs together would be one of perceiving the set of the eggs. Maddy's views will be discussed further in §34.

exist two thousand years from now. This is a truth of logic and presumably not in any way a deliverance of intuition. Yet it talks of the world two thousand years from now, and obtains its evidence without the use of whatever procedures might be used in physics or astronomy to extrapolate so far and come to a reasonable hypothesis as to whether the earth will exist two thousand years from now. We don't, however, think of its truth as contributed by our own minds in a way that other truths, such as that the earth existed one thousand years ago, are not.[79]

With respect to Kant, a difficulty in understanding the Aesthetic, and the parallel parts of the *Prolegomena*, is why it should be exactly a priori *intuition* that requires that its content should be due to our own cognitive faculties and not to how things are in themselves. Kant's statements on the matter are often dogmatic (e.g., B41), and he does not seem to me to make a convincing case. Once we take account of the fact that the object of the intuition that concerns us is abstract (as the formulation of the puzzle in the *Prolegomena* does not), then the difference of intuition and concepts is much less evident. On the Hilbertian interpretation (thinking of numbers as strings) it is, I will argue in Chapter 7, intuitively evident that $7 + 5 = 12$. That statement will be equally true, whatever tokens of the relevant strings are produced two thousand years from now. But on the interpretation at issue it concerns strings as types. Is the relation of type to token so different from that of concept to instance so that, in the first case, our knowledge of a truth concerning types should have the consequence that it is only possible if the ground of the truth is a factor in knowledge contributed by ourselves, while this need not be true in the latter case?[80]

---

[79] The reader, thinking of conventionalist views of logical truth, will naturally object: Yes, but we do consider that, though all truth depends both on the world (outside us) and language (a human creation), some philosophers have thought that for some truths, in particular logical truths, the contribution of the former is zero. But if they thus depend only on language, then they are contributed only by us. Kant's view should probably be seen in a similar light. I am not concerned to show that no *general* view of this kind can be right. They have the feature that there is some factor contributed by "us" that *all* knowledge depends on. That seems to me to be the primary point, not whether there is some knowledge that depends only on this factor. As indicated below in the text with reference to Kant, I do have difficulty with the latter conclusion (for Quinean reasons). But my main concern is to argue that the kind of intuition I am concerned with does not force the particularly Kantian version of this dependence.

[80] In the end, Kant holds that it is true in the latter case whenever the proposition involved expresses genuine knowledge of objects. In view of the role of the forms of intuition in the deduction of the categories, it is very doubtful that he intends an argument for this conclusion that would be independent of the claims concerning intuition, although the

§30. The objections revisited

Moreover, there is a difficulty of principle that Kant's position suffers from: Given our inability to know things as they are in themselves, how can it be possible to identify a particular part of our knowledge that is due *entirely* to ourselves, or, in general, to separate our own contribution to our knowledge from that of the external world? Strictly, we may not have to make such a separation, at least in the sense of making an independent identification of the role other than that of ourselves in our knowledge. Generally, our knowledge results from our own cognitive faculties and from the world. What Kant's position requires is that one be able to identify some knowledge in which the role of the "world" factor is zero, even though there is no knowledge in which the role of the other factor is zero. Without being able to get at this other factor ("things in themselves"), how are we to identify the cases where it plays no role?

The last objection put forth in §26 derived from the structuralist view of mathematical objects. Although that view, as explained in Chapters 2 and 3, does not hold for quasi-concrete objects, the fact that they have intrinsic concrete instantiations may not be enough to endow them with the range of properties that a concrete object would have; moreover, it is not clear that they do not suffer from the indeterminacy of identity. We cannot rule out the possibility that they are incomplete in the sense that not every predicate is determinately true or false of them.

Before we turn to our artificial symbolism and its strings, let us consider sentences of a natural language. Ordinary thought seems to me to treat them as of a different category from physical bodies or other ordinary concrete objects, so if we ask whether a sentence has a property that such objects might have, such as being red or round, the answer will be "no" straight away; a slightly more sophisticated version of the answer will be that it makes no sense to say of a sentence that it is red or round. Linguistic expressions may be limited in their properties and relations to linguistic ones (broadly construed) and certain relations having to do with their instantiations, such as being uttered on certain occasions. If that is so, then certainly the general tendency of ordinary thought about them would be to hold other predicates false or nonsense in application to them. Some ordinary properties such as being located at a particular place and time are applied to them but not in quite the same way as to bodies. To say that an expression is at a certain place means that a token

---

transcendental unity of apperception is given in some passages a legislative role that parallels the "forming" role of the forms of intuition.

of it is there, and that of course does not exclude its being elsewhere at the same time. These facts about expressions don't present any particular obstacle to intuition or perception of them beyond those that arise from their being abstract, and those have already been discussed.

I don't think that is changed by the suggestion of Richard Wollheim that types can have secondary qualities such as color.[81] Wollheim offers as an example the Union Jack, which would have as tokens individual British flags and presumably also the many printings of it as a logo. It is, indeed, natural to speak here of a type that is red, white, and blue; it's hard to see how we could speak of a type at all without identifying it by its pattern of color. But even such a type would lack many of the kinds of properties of physical objects, in particular those going with location in space and time. But it doesn't follow that sentences of a natural language have properties of this kind. Consider a printed sentence that happens to be printed in red. One might identify a type that would be instantiated only by red inscriptions of that sentence. But that type wouldn't be the sentence or even the written sentence, since there would be tokens of the sentence that would fail to instantiate it. About spoken language, however, the question arises whether types have properties having to do with sound. Present views in phonology would imply that the properties it has have to do with the manner in which the sound is produced and are not acoustic properties or secondary qualities in the usual sense.

Similarly, the strings of our symbolism will initially have only the properties and relations that either come with their being instantiated in the way they are or are mathematical properties and relations internal to the system of strings. Strings will be, in effect, a logical type. As we have explained the symbolism, types will have spatial properties but not secondary qualities such as color.[82] Wollheim's observation implies that we might talk about strings of the sort we have been discussing that are, say, red. If we made this stipulation, however, the claim that every string can be extended would be much more doubtful.

---

[81] *Art and its Objects*, pp. 75–78. I am indebted to Frode Kjosavik for pointing out the relevance of Wollheim's discussion to my own views on intuition of types.

[82] Michael Resnik may be suggesting the contrary when he takes me to hold that "it is part of our conception of the type that it 'looks' or 'sounds' a certain way – the way its tokens look or sound" ("Parsons on Mathematical Intuition and Obviousness," p. 224). Although I find this interpretation misleading where secondary qualities are involved I am grateful to him for raising the question.

§30. The objections revisited

There is a difference with expressions of natural language:[83] In the further development of a mathematics whose objects include strings, we may cease to treat strings of this language as a logical type. One way this could happen is by generalization of the notion of string-type, so that some tokens are admitted as being of the same type as the present string tokens that were not previously admitted. But a more relevant one is that strings might come to be talked about in a first-order theory that also talks about mathematical objects of other kinds. It appears that stipulations have to be made about the identity of the strings and the other objects; the initial position, according to which no string is identical to anything described in another way (except in the development of the theory of strings, where for example terms introduced by recursions might arise), is not legislative. Hilbert and Bernays, in their practice of finitist arithmetic, identified natural numbers with strings. To consider another more contrived case, if one began to talk of a language in which '|' was just one of the symbols of the alphabet, it would be extremely natural (indeed, almost forced by that description of the situation), to identify our previous strings with the strings of the new language that contained only '|'.

One might see a difficulty in this for intuition of strings. To simplify things, let L be our original language and let L' be the language with three symbols, '|', '\', and '/'. Consider now a person A to whom L has been explained and who thus acquires the ability to intuit strings of L. It seems perfectly conceivable that he might not have the ability to intuit strings of L' containing the additional symbols. But if our own conception makes L a part of L', then, according to it, A is intuiting strings of L', even though he may not be able to distinguish them, and might even, for example, take '| \' to be just a badly written version of '||'.

What this example seems to me to show is that the requirement that one have the concept of a string in order to intuit a string implies that 'intuit' generates an intensional context. Thus the more accurate description of the situation is that A is intuiting strings that are in fact strings of L'. How the behavior of 'intuit' resembles and differs from other referential attitudes is a question that may be worth some investigation. I will not pursue it here.

---

[83] It is conceivable that the difference could disappear or be mitigated when expressions of a natural language are considered in theoretical linguistics.

6 Numbers as Objects

§31. What are the natural numbers?

The discussion of intuition in the last chapter naturally leads to two further inquiries. Perhaps the most obvious one is about natural numbers: How are natural numbers given to us, and, in particular, are they objects of an intuition of the kind described in §28 or something similar to it? The second concerns intuitive knowledge: We gave, particularly in §29, some examples of propositions about the strings of our little language that are intuitively known. Since strings are evidently a model of arithmetic, the question arises how far intuitive knowledge in arithmetic extends, when we understand arithmetic by reference to this model. That formulation of the question sidesteps the first question, whether numbers properly speaking are objects of intuition.

Because our first question is probably the more urgent for most readers, I will take it up first, in this chapter. The other question will be the subject of Chapter 7. The question: "What are the natural numbers?" is motivated independently of questions about intuition, and we will examine it in general as well as answering the question about intuition of numbers. But both chapters will explore the limits of intuition as understood in Chapter 5, first by considering in this chapter whether intuition extends beyond quasi-concrete objects to some pure abstract objects such as numbers, and second by inquiring in Chapter 7 how far intuitive knowledge extends, referring to a domain that we have taken to be intuitive.

The natural numbers are what one obtains by beginning with 0 and iterating the successor operation. They can be crudely described as 0, 1, 2, 3, ... These simple descriptions conceal many problems. To begin with, what is meant by *iterating* the successor operation (or, in the second formulation, by ' . . . ')? This question leads us into the circle of ideas

§31. What are the natural numbers?

surrounding the principle of mathematical induction. Issues concerning induction will be largely put aside in this chapter but will be considered in Chapter 7 and especially Chapter 8. An ostensibly more elementary question is what we are referring to when we speak of 0 and successor, or of 0, 1, 2 and other small numbers, reference to which we can apparently understand without appealing to induction or any equivalent idea. But any account we give of these particular cases should be generalizable to arbitrary natural numbers.

One part of an answer to the question what the natural numbers are is quite uncontroversial. They are a structure satisfying the Dedekind-Peano axioms. Some of the content of these axioms can be stated (for future reference) in the form of natural deduction rules. The first two axioms in Peano's formulation, that 0 is a number and that the successor of a number is a number, are introduction rules for the predicate N meaning "is a natural number":

(R1) $N0$

(R2) $\dfrac{Nx}{N(Sx)}$

These rules will be referred to simply as the introduction rules, or if necessary to avoid ambiguity, the introduction rules for N. The rule of induction then takes on the character of an elimination rule for N:

$$
\text{(R5)} \quad \dfrac{A(0) \quad \begin{array}{c}[A(a)]\\ \vdots \\ A(Sa)\end{array} \quad Nt}{A(t)}
$$

A theory in the context of natural deduction with just these rules would be the theory of a domain consisting of an object and the results of iterated application of a unary operation to it. The successor operation on numbers, as we usually understand it, has the additional property that for any n, Sn is always something new; it is never among $0 \ldots n$. This is insured by the third and fourth of Peano's axioms, that Sn is never 0 and that S is one-to-one. The fourth can also be put as a rule:[1]

(R3) $\quad Sx \neq 0$

(R4) $\quad \dfrac{Sx = Sy}{x = y}$

In the context of classical first-order logic, most conveniently a natural deduction version, (R1)–(R5) together with the usual recursion equations for addition and multiplication yield the usual first-order arithmetic PA. With intuitionistic logic instead, we have first-order intuitionistic arithmetic HA. Second-order arithmetics also can be straightforwardly formulated in this way.

We have described the natural numbers as "a structure satisfying the Dedekind-Peano axioms," but the deductive way in which we have presented the axioms suggests the interpretation that they are a structure such that interpreting the axioms or rules with reference to that structure yields a sound theory. For actual formal logics, even second-order logic, that condition does not characterize a unique structure, since formal theories of arithmetic have nonstandard models. The very idea of a nonstandard model presupposes thinking semantically about the relation of the axioms to the structure. The natural numbers are a standard model, in the sense of second-order logic, of the axioms. That is essentially what Dedekind called a simply infinite system (see §10); terms used more recently for this type of structure are progression (Russell, Benacerraf) and ω-sequence. The former of these terms goes with taking the structure to be $\langle N, 0, S \rangle$, as Dedekind implicitly did and as is most usual in philosophical literature. I will use the term 'progression' with this specific meaning. The term 'ω-sequence' may suggest to some taking the structure as an ordered set $\langle N, < \rangle$. For that, however, I will use the term 'ω-ordering', thus avoiding the term 'ω-sequence' because of its ambiguity. Dedekind's isomorphism theorem implies that there is, up to isomorphism, only one progression (see §10). The second-order character of the characterization, and the existence of nonstandard models of arithmetic, raises the question whether our conception of the natural numbers really is a conception of a unique structure. This question will be taken up in Chapter 8. Much of the discussion in the present chapter is independent of that question, but where it is not, we will assume that Dedekind's theorem can be taken at face value and

---

[1] If one has a constant $\bot$ for an absurd proposition, then of course (R3) can be rendered as: From $Sx = 0$, infer $\bot$.

§31. What are the natural numbers?

that we are talking of a unique structure when we talk of the natural numbers.

We naturally think of the natural numbers as having additional structure beyond that of a progression, and the reason why they are generally characterized in this more limited way is that in a progression addition, multiplication, and order are second-order definable, as Dedekind already showed. But of course for a model of first-order arithmetic, an interpretation of addition and multiplication must be given. In an ω-ordering, '$x = 0$' and '$y = Sx$' are first-order definable,[2] so that an ω-ordering contains a progression. Order, however, is not first-order definable in terms of 0 and S, although if we have addition we can define '$x < y$' as '$(\exists z)(z \neq 0 \wedge x + z = y)$'.

Now a partial answer to our question "What are the natural numbers?" is that they are a progression, or a progression with some additional structure. We may find that this semantical way of putting things begs too many questions, but then at least we can say that discourse about the natural numbers will be a theory in which we have (R1)–(R5) and perhaps some additional apparatus such as addition and multiplication, in a suitable logical setting. Even so, the question will arise whether we can say more: In the semantical way of speaking, whether we might distinguish some one progression as being *the* natural numbers, or at least uncover constraints such that some progressions are eligible and others are not.

The structuralist view of numbers explored in Chapters 2 and 3 offers an answer to our question. If we are prepared to help ourselves to second-order logic, then the objection made in §17 to the eliminative version of structuralism is not fatal in the particular case of natural numbers, and the modal version has been worked out thoroughly in Hellman's *Mathematics without Numbers*. The noneliminative version proposed in §18, however, has the advantage of not helping itself to second-order logic, and it seems to me to do better justice to the actual role of second-order principles in mathematics. Could we simply end our inquiry at this point and say that we have said what reference to the natural numbers is?

In the earlier discussion, we did not say much about something central to a full-blooded conception of the natural numbers: their role as cardinals and ordinals. The structuralist view holds that this role does not

---

[2] We can characterize an ω-ordering as a well-ordered set with an initial element in the ordering, such that every other element y is a successor, that is, $(\exists z)[z < y \wedge \neg(\exists w)(w < z < y)]$. Of course this characterization too is second-order.

constrain what objects the natural numbers are, so long as they instantiate the right type of structure. We defended this claim in §14. But our discussion of the natural numbers will be incomplete so long as we have not gone into the concepts of cardinal and ordinal. This will give rise to some other approaches to the question what are the natural numbers, as well as to the question whether they are objects of intuition in the sense of §28. This will lead us to a more final view about the place of the version of structuralism from §18. In §37, we will arrive at a negative answer to the question about intuition.

§32. Cardinality and the genesis of numbers as objects

I want first to approach the concept of number in a somewhat phenomenological way, beginning with the elements of its application. A first point to observe is that the most elementary cardinal and ordinal applications of numerical language do not obviously require numbers as objects at all. I will take as a canonical form for a statement of cardinal number 'there are (exactly) **n** *F*s', where '*F*' replaces a predicate or general term, and **n** is a numeral. I don't think this needs be taken as making a commitment one way or the other as to whether **n** is a singular term. I do choose this in preference to a form making reference to a set. The most basic statements of cardinality do not have to be interpreted as making reference to sets. However, in §34, I will consider arguments for taking the most basic statement of cardinality to be of the form 'there are **n** elements of **a**', where **a** names a set.[3]

In determining by counting that there are n *F*s, one sets up a one-to-one correspondence between the *F*s and a sequence of n "counters." Something like a canonical way of verifying that there are n *F*s is achieved if we have the *F*s successively before our minds, by naming or perceiving them, while marking them with the successive counters. This gives to the mth counter the approximate sense 'the mth', that is, in the context, the mth F.[4] Of course, we must insure that no object is counted twice,

---

[3] Of course, the form given in the text does subsume one obvious one involving sets, since '*F*' might be 'is an element of **a**', where **a** is a name of a set. (**m**, **n**, ... are used as (metamathematical) variables for numerals.)

[4] An alternative would be to attribute to the agent a use of a demonstrative such as 'this', where, however, the uses on each occasion are different; a linguist might index them as 'this$_i$', $i = 1 \ldots n$. Then the counters serve simply to keep track of the different uses. The proposal in the text has the advantage that the "demonstrative" is something that in a case of counting out loud is actually uttered.

§32. Cardinality and the genesis of numbers as objects

that is, that at each stage the object marked was not marked before, and we must continue the process until every F has been marked. But these two conditions amount to the condition that the marking should be a one-to-one correspondence of the counters used onto the Fs.

The "counters" will typically be numerals of some standard system, perhaps either the number-words of a natural language, or Arabic numerals. I will suppose the system fixed for the present. Let us consider what is required for it to be verified by perception that there are n Fs. It is not sufficient that we be able to perceive each individual F; we must be able to survey all the Fs one by one and identify them as Fs and as constituting all of them. But this obtains in many everyday situations, where what is counted are objects that either are before one's eyes all at once or can be brought so successively. Consider such cases as counting the plates, glasses, and pieces of cutlery to be put on the table for dinner, or counting the coins in one's pocket.

I have not been very precise about the meaning of the statement 'there are **n** Fs', but it should be observed that the claim that in such simple cases it can be verified by perception would run into difficulties on certain conceptions of what it means. Let us take one construal:

(1)   There is a 1–1 correspondence between the Fs and the numerals from 1 to **n**.

This is, on the face of it, a second-order statement, and it involves reference to numerals, which for constancy of meaning in different occasions of use will have to be taken as types. Assumptions about the intuitability of finite sets would plausibly imply that, in the kind of favorable case we are considering, (1) might be intuitively known, but even on that assumption it will involve mathematical intuition.

The issues about intuition of sets that these assumptions would raise are of interest in their own right and will be discussed in §34.[5] But I don't think that, in order to understand the most elementary applications of the notion of cardinality, we need to give a statement of number so much

---

[5] The strongest such assumption would be that any finite set of intuitable objects is intuitable. If we think of the correspondence as a set of ordered pairs, then it follows that the correspondence is intuitable if we either understand the ordered pairs by the standard Kuratowski definition or assume (independently and, in my opinion, with as much plausibility) that ordered pairs of intuitable objects are intuitable. Grounds for questioning these assumptions will be considered in §35.

ontological baggage.[6] Consider the following equivalences, which underlie the standard paraphrase of statements of number into first-order logic with identity:

(2a) $(\exists_0 x) Fx \leftrightarrow \neg(\exists x) Fx$

(2b) $(\exists_{n+1} x) Fx \leftrightarrow (\exists y)[Fy \wedge (\exists_n x)(Fx \wedge x \neq y)]$.

One way of looking at counting is to suppose that each numeral **m** used in the count has the force of a demonstrative, designating in the context the object with which it is correlated, so that it has the force of 'the **m**th'. To avoid confusion, I will write it in that way. I have been imagining a very favorable case, where at each point the object can be observed to be F and to be distinct from those counted previously, so that we would have:

(3a) $F(\text{the \textbf{m}th})$

(3b)   the **k**th $\neq$ the **m**th (for each k, $0 < k < m$).

Moreover, at the nth stage the subject can observe that there are no further Fs. That is, he is also able to observe

(3c) $(\forall x)(Fx \rightarrow x = \text{the 1st} \vee \ldots \vee x = \text{the \textbf{n}th})$.

From (3a–c), the equivalences (2a–b) are all that is needed to infer '$(\exists_\mathbf{n} x) Fx$'.

That the subject knows (2a–b), or principles from which they follow, is a reasonable enough assumption. For a case of the sort we have in mind, we do not even have to assume that the subject knows them with the generality that someone who understands arithmetic as we do does; it is sufficient that he should know the instances that actually come into play

---

[6] I might remark that the second-order aspect of (1) would not have caused any concern to Frege, because he would have taken it to be a logical consequence of a statement of the form 'H is a 1-1 correspondence of the Fs and the numerals from 1 to **n**', for a specific predicate 'H'.

Of course, Frege would not have interpreted statements of number by way of reference to numerals. The statement 'the number of F's is **n**' would on the reading of the *Grundlagen* be a second-order consequence of the usual translation of 'there are **n** Fs' into first-order logic and instances of the Fregean criterion of identity for numbers:

the number of Fs = the number of Gs iff there is a 1–1 correspondence of the Fs and the Gs,

(which in turn is, as is well known, provable with the help of Frege's axiom V from his definition of 'the number of Fs' using extensions). In the kind of maximally favorable case we are considering, by Frege's lights 'the number of Fs is **n**' would thus be a logical consequence of facts verifiable by perception.

§32. Cardinality and the genesis of numbers as objects

in the counting. But more will be demanded shortly, when we come to numerical quantification.

How shall we view this knowledge? First of all, for the present we should regard them as rules in which not only 'F' but also 'n' is schematic, in the latter case of course for a numeral. But if it is replaced by a numeral, (2a–b) yield a procedure for eliminating the numerical quantifier in terms of first-order logic with identity. This might suggest that 'there are **n** Fs' is simply an abbreviation for its expansion according to (2a–b), so that it is simply a first-order consequence of (3a–c). Another view, perhaps representing the next step in the development of the concept of cardinality, reads '$(\exists_n x)$' as a kind of generalized quantifier, with some of the basic inferences involving it still treated like logical inference. This would provide an entry for the idea that 'n' occurs in a generalizable place. It is natural because the first-order expansion becomes very unperspicuous even before n becomes very large. But it also should be possible to see that there are n Fs in a more step-by-step way than the way just described. One such way would be to verify at each stage 'there are m Fs that are G', where 'G' is some predicate known to be coextensive with 'counted so far'. What 'G' is would depend on the circumstances, but presumably it would contain an indexical whose reference changes with m. In the case where the objects involved are before one's eyes, one might demonstrate a place, so that 'G' would mean 'in this place'.[7] In any case, a step-by-step procedure would consist of verifying '$(\exists_m x)(Fx \wedge G_m x)$' at the mth stage (where 'G_m' fixes the above-mentioned indexical in the appropriate way). To pass to the $(m+1)$st stage, one would observe that for any x

(4) $\quad Fx \to [G_{m+1}x \leftrightarrow (G_m x \vee x = \text{the } (m+1)\text{st})]$.

and then apply (2b).

One will clearly pretty quickly come to the point of using numerical quantifiers with variable n, and thus in ways that are no longer reducible to first-order logic. This should remind us that it is only by confining ourselves to a very simple kind of case that we avoid even the appearance of introducing numbers as objects. Given our starting point, the obvious way in which reference to numbers comes to seem forced on us is that we allow the numeral place in the quantifier to be generalized, that is, that we introduce quantifiers "over numbers."

---

[7] One could of course say that what is demonstrated is the *set* of so-far-counted Fs, so that 'G' means 'in this set'. This is clearly of more general application and thus represents one entry point for the set concept. But I shall argue in §34 that, in the elementary cases that concern us now, this is not forced on us.

We would also quickly enlarge the language so that different expressions come to be used to "refer to the same number." Here we might distinguish three kinds of cases:

(i) Terms arising from the introduction of computable operations, beginning with the most elementary ones such as addition. In these cases there will be procedures for reducing terms composed from numerals with these operators to numerals. This reduction must be understood as applicable in contexts such as 'there are n Fs'; otherwise it would not follow, for example, that 'there are 4 Fs' is equivalent to 'there are 2 + 2 Fs'.

(ii) Expressions such as 'the number of Fs' and other definite descriptions designating numbers. These presuppose some form of generalization of numeral places.

(iii) Alternative systems of numerals. These one might see as giving rise to questions of translation (or paraphrase if they are in our own language). There is, however, a constraint on a correct translation or paraphrase that is independent of the idea that the numerals refer to numbers. If we consider two systems, call them the N- and the M-numerals, then we will render the (perhaps foreign) M-numeral **n*** by (our own) N-numeral **n** just in case there are n M-numerals less than **n*** (assuming that the initial numeral is **0**; otherwise 'less than' is replaced by 'up to and including'[8]). Another way of looking at the matter that gives the same result and that may be more appropriate if the M-numerals belong to our own language, is to regard them as terms introduced by recursion.

Our remarks about (i) and (iii) suggest that, at least before we have constructions comparable to variables and quantifiers for numbers, we have a conception of what it is for expressions to "designate the same number" that does not presuppose an antecedent conception of the numbers as objects. And by an expression that "designates a number" we need at present mean no more than simply a numeral of our initial system or an expression that according to the conception just mentioned "designates the same number" as such a numeral. This encourages us to think of the introduction of variables and quantifiers "ranging over numbers" as in the first instance substitutional. And, in fact, substitutional interpretations of arithmetic have been constructed and defended on philosophical grounds.[9]

---

[8] Note that these relations are decidable by a computational procedure, in some cases a trivial one.

[9] See especially Gottlieb, *Ontological Economy*.

§32. Cardinality and the genesis of numbers as objects 195

Suppose that we have introduced "numerical" quantifiers in something like this way. Let us consider whether this understanding would have the consequence that numbers are objects of intuition. A substitutional interpretation of quantifiers in the language of arithmetic might be viewed either as an explanation of reference to numbers or as an elimination of it.[10] The second of these views clearly does not make numbers objects of intuition, as it amounts to a denial that numbers are objects at all.

Consider then the first view. On this view, numbers would be constituted by the use of language, and in particular by the expressions that refer to them. It is the understanding and use of certain linguistic constructions that counts as consciousness of numbers. This does involve ordinary perception, for example of the objects counted, and intuition of linguistic expressions, in particular numerals. Could we still describe the understanding of a numeral as intuition of the number? If so, how would this be related to perception? Not in the same way as in the conception of intuition that has concerned us. Unlike an expression-type, the number is not a *form* instantiated by the numeral; we don't explain what the number is by saying that its nature is to be instantiated in just that way. The numeral is here playing the role of a linguistic expression; one thing that shows this is that the number would be just as "present" if it were represented by a corresponding numeral in a different system, perhaps totally different perceptually. Thus if we were to describe this situation as intuition of a number, the analogy with perception will be less close.

The understanding of reference to numbers that we are considering is a special case of a general conception of object that has been applied in the foundations of constructive mathematics, according to which an object is given by a canonical expression for it, in the case at hand a numeral. In general, a canonical expression shows how it is constructed from the basic constructions for that domain of object. Other rules would give rise to noncanonical expressions, which should reduce computationally to canonical expressions. Such rules arise naturally from the constructions manifested in the canonical expressions; for example, if canonical numerals are constructed from (say) 0 by means of successor, the predecessor

---

[10] Gottlieb defends the latter view. The former view would be suggested by the general view of Essay 2 of *Mathematics in Philosophy*. With more specific attention to arithmetic, I examine the question further in my review article on Gottlieb, "Substitutional Quantification and Mathematics."

function δ (introduced by the trivial recursion $\delta(0) = 0, \delta(Sn) = n$) simply undoes the last step of the construction of a non-initial number.[11]

The conception of mathematical objects as given by canonical expressions for them does in certain respects model not the conception of intuition with which I have been concerned but rather that of Husserl. In general, terms for mathematical objects would express intentions that are to count as fulfilled by their reduction to canonical expressions. Having a canonical expression then counts as the presence of the "object itself"; the idea is that there is not more that can be demanded. In the constructive setting in which this conception arose, one can understand a mathematical proposition as an intention that is fulfilled by the construction that proves it.[12]

It is easy to see, however, that a purely substitutional theory of arithmetic along the lines we have intimated cannot be adequate. There is no difficulty in giving a substitutional interpretation of first-order arithmetic as a stand-alone theory,[13] and this can be extended to a ramified second-order theory, at least within the limits of predicativity. Moreover, it can be further developed, along the lines sketched above, to include a treatment of numbers as cardinals.[14] The most serious limitation of this treatment, however, is that it presupposes that the objects numbered come from an antecedent domain. But, of course, in mathematical practice we apply the notion of cardinality to *numbers* and to mathematical objects of other kinds that might depend on natural numbers. Moreover, we do not restrict it to the finite.

This rather commonplace observation could be offered as an argument in favor of the view that our explanation of quantification over numbers really is an explanation of a domain of objects. The "cash value" of this view in this context, however, is the same as that of what, on another

---

[11] The conception of mathematical objects as given by canonical expressions for them underlies Martin-Löf's intuitionistic theory of types, a powerful constructive theory that is susceptible of still further extensions. See especially Martin-Löf, *Intuitionistic Type Theory*. I discuss this ontological conception (mostly with reference to quite simple cases) in "Intuition in Constructive Mathematics." The rather obvious considerations given later to show the inadequacy of the straightforward substitutional interpretation do not apply to a theory like Martin-Löf's. In fact, the treatment of higher types in his theory rules out taking the conception as an elimination of ontology.

[12] Cf. Tieszen, *Mathematical Intuition*, esp. ch. 4 §5. Husserl exercised at least an indirect influence on the origins of intuitionistic logic in the early work of Heyting.

[13] See *Mathematics in Philosophy*, p. 63 n. 1, or for a fuller treatment Kripke, "Is There a Problem about Substitutional Quantification?" §§8–9.

[14] See Gottlieb, *Ontological Economy*, and my "Substitutional Quantification and Mathematics."

§32. Cardinality and the genesis of numbers as objects

view, would be the further conceptual leap of regarding the substitutional quantification as objectual. The point is that in mathematical practice and to some extent already in ordinary life numbers are treated as on a par with other objects, so that they themselves can not only be numbered (which implies that number variables mark an argument place of a *predicate*[15]) but, further, also can be elements of sets and sequences and arguments and values of functions.

But now in what would this further conceptual leap consist? It would consist of steps taken after we have substitutional quantification of numerals, perhaps segregated from other quantification. Perhaps after noting the formal analogy between the behavior of these substitutional quantifiers and that of object quantifiers, and the fact that the former could be generalized to numerical quantifiers in the same way that the latter were, we might then come to treat the number quantifiers as just restricted quantifiers with an underlying predicate 'is a number'. But the equality predicate that underlies both computable operations and the use of numerical definite descriptions would also have to be treated as identity; talking of sets of and functions on natural numbers would involve still further steps. We should note that it is only at a rather late point in this development that the questions to which the structuralist view responds can even arise. Without allowing numbers into a common domain of quantification with other objects, the question whether any identities hold between numbers and objects given in some quite other way can of course not be formulated.

At this point, some comments are in order about the famous Fregean criterion of identity of numbers, which in the *Grundlagen* played such a central role in his argument for the thesis that numbers are objects and thus in his explanation of what reference to numbers is.[16] A Fregean approach would interpret 'there are n Fs' as *saying* that there is a one-to-one correspondence of the Fs and the Gs, where 'G' is some canonical predicate such that there are n Gs; in Frege's own analysis, 'G' would in effect be 'x is a natural number less than n'. We said that verifying by

---

[15] Assuming the form of the basic statement of number we have assumed. But taking number to attach to sets would have the same implication.

[16] As Frege's references show, the criterion itself was not original with him or even with Cantor (whom he cites, *Grundlagen*, p. 73 n.), although it was Cantor who saw and exploited its applicability to the infinite. It is not easy to say who first stated it in a mathematically usable form. I don't think that could be claimed for Hume. But in this Frege was clearly anticipated at least by Cantor. What is original with Frege is his expressing the criterion in terms of second-order logic. (Cf. *Mathematics in Philosophy*, p. 164.)

counting that there are *n F*s involves setting up a one-to-one correspondence of the *F*s and, say, the numerals **1 . . . n**. It does not follow that 'there are n Fs' has to be interpreted as asserting the existence of such a correspondence. To give the right result, it is sufficient that the numerical quantifier obey the rules (2a–b). Therefore we do not have to give it such an ontologically committal interpretation in order to understand the kind of basic statements of cardinality that concern us. I think it could be argued further that such an interpretation would not be needed for a theory of the meaning of statements of number in natural language, where the users do not have in view uses in sophisticated mathematics.

This remark is not really an objection to Frege. The project of describing the genesis of discourse about numbers as a sequence of stages was quite foreign to him, and that of an account of the meaning of a class of expressions in unscientific natural language was not central. He would still argue that in fully developed arithmetic, 'there are *n F*s' at least has logically to imply the existence of such a correspondence. Where one might object to Frege's own use of his criterion is on structuralist grounds, because he treats natural numbers as first of all *cardinals.* But of course his criterion of cardinal equivalence is undoubtedly extensionally correct, and therefore cardinal numbers as objects have to satisfy it. Moreover, there seems to be no escape from taking this criterion as the most basic one in generalizing the notion of cardinal number to the infinite case. This fits with the fact that the historical situation in which the criterion came to be seen as the basic one was the creation by Cantor of the theory of transfinite numbers. In that context, it is the unique criterion that yields a coherent theory with anything that can reasonably be called an arithmetic of infinite cardinals. But then in standard set theory it does not play the role it played for Frege, as a criterion of identity that was part of an explanation of reference to a certain kind of object.

The concept of *ordinal* number also describes a quite basic aspect of the concept of number already in ordinary life, and, like the concept of cardinal, it plays a central role in set theory. We might ask whether a similar exercise to the one carried out above with the cardinal concept of number could be carried out viewing numbers ordinally. In our discussion of counting, it occurred naturally that a given counter came to have the force of 'the **m**th', for some numeral **m**. But if the context is that of determining how many *F*s there are, no ordering has been specified, and if one asks in what order a given object is the *m*th, the answer can only be that it is the order in which the *F*s are being counted. This will be the temporal order, but one also can say that the correspondence of the counters and the *F*s induces an ordering on the latter.[17] We have a properly ordinal concept of number when the counting is supposed to track a given ordering, so that one begins with the initial object in the ordering, and at a given stage the next counter is to be applied to the next object in the ordering. These conditions impose a strong constraint on the ordering, that it have an initial element and that each element (at least up to the point where the counting ends) have an element immediately following it in the ordering, that is, a successor. This is just to say that the ordering is an initial segment of an ω-ordering. But what is required of the counters is no more than in the description of the cardinal case, and the initial steps can be taken without any supposition that they designate numbers or any other objects. Furthermore, the further steps involved in passing to a conception of numbers as objects do not differ in the two cases.

In fact, an informative answer to the question "How many *F*s are there?" places the number of *F*s in the ordering of numbers.[18] A version of this condition has application even in the primitive situation where we do not yet have a conception of numbers as objects, as we have the ability to order the *counters* and learn systems of numerals in an order that is shown by the order in which they are presented or constructed in time. For these reasons, one can say that the cardinal concept of number is inseparable from the ordinal.

§33. Finite sets and sequences

In mathematics, the concepts of finite set and sequence are certainly intimately related to that of natural number. The reader may feel that the exclusion of such objects from our discussion in the last section is artificial or even phenomenologically false. In this section, we will discuss a number of questions concerning these concepts. In particular,

---

[17] Of course it is an important feature of cardinality that it is independent of this order. This is a consequence of the fact that if there is a one-one correspondence of $\{1 \ldots m\}$ and $\{1 \ldots n\}$, then $m = n$. Intuitively, the results of counting are ordinals in the infinite case too, but the fact just mentioned does not generalize to infinite ordinals.

[18] In practice more is demanded than just an answer that uniquely determines the place of the number of *F*s in the order, as is shown by the fact that Arabic numerals are in most cases more informative than 0-*S* numerals would be, even when the numbers are small enough so that giving the latter is feasible, and for very large numbers even Arabic numerals become unsurveyable and tend to be replaced by approximations involving exponentiation. These matters have implications for when someone can be said to know how many *F*s there are. The issues involved were very interestingly explored by Saul Kripke in his Whitehead Lectures at Harvard University in 1992.

in the course of returning to the theme of intuition, we will consider the claim that attributions of cardinal number to sets, or ordinal numbers to sequences, are the most primitive form of attributions of number.

A theory of sets that provides only for finite sets is quite similar to arithmetic. The natural numbers are what is obtained by beginning with 0 and iterating the successor operation an arbitrary finite number of times. Similarly, for a given domain D of individuals, the hereditarily finite sets based on D, abbreviated $HF(D)$, are what is obtained by beginning with the null set $\varnothing$ and iterating the operation of forming from any $x_1 \ldots x_n$, individuals or previously formed sets, the set $\{x_1 \ldots x_n\}$ whose elements are just $x_1 \ldots x_n$. It is well known that for any D, $HF(D)$ is a model of the usual (Zermelo-Fraenkel) axioms of set theory without the axiom of infinity.[19]

The hereditarily finite sets may seem essentially more complex than the numbers, even if D is empty. But the "anadic" operation of forming a set from any finite number of given objects can clearly be replaced by the dyadic one of forming, from a given set x and object y, the set $x \oplus y$ whose elements are just those of x, together with y. $x \oplus y$ is of course just $x \cup \{y\}$, but I am using a different notation because I will present a theory in which this operation is primitive. $HF(D)$ is obtained by beginning with $\varnothing$ and the elements of D and iterating the operation of forming $x \oplus y$ from previously given x and y, that is, of adding an element to a previously given set. This formulation might dispel the impression that the theory of hereditarily finite sets is intrinsically more complicated than arithmetic. In fact, each can be modeled in the other in a straightforward way. Taking the integers as von Neumann ordinals (i.e., $0 = \varnothing$, $Sn = n \oplus n$) yields a model of arithmetic in the theory of HF sets; conversely (for $D = \varnothing$), we can set $\varnothing = 0$ and $x \oplus y = x + 2^y$ (if $y \notin x$), so that if $x_1 \ldots x_n$ are all different and are mapped onto $m_1 \ldots m_n$ respectively, then $\{x_1 \ldots x_n\}$ is mapped onto $2^{m_1} + \cdots + 2^{m_n}$. Using this mapping, Wilhelm Ackermann showed in 1937 that the translations of the axioms of ZF, other than Infinity, are provable in PA.[20]

---

[19] See for example Krivine, *Introduction*, pp. 44–45. I adopt the convention that $HF(D)$ includes D, even though the elements of D are in general not sets. Note that if D is not empty, the axiom of extensionality has to be formulated so as to allow individuals (urelements), as is done in (H1).

[20] "Die Widerspruchsfreiheit der allgemeinen Mengenlehre." It follows that PA and ZF without Infinity are equiconsistent; that is, a finitist proof can be given of the consistency of each relative to the other. A beauty of Ackermann's mapping is that it is onto: every number is the code of a hereditarily finite set; moreover, it does not impose an order on the set.

We cannot expect Ackermann's result to generalize to arbitrary D. It obviously fails if D is uncountable.

§33. Finite sets and sequences

For future reference we will describe a theory of hereditarily finite sets in "arithmetical" style. The theory will have the primitive constant $\varnothing$ and functor $\oplus$, and the membership predicate $\in$. The intention is that the domain should consist of HF sets built up from $\varnothing$ and nonsets. We define $\mathbf{M}x$, 'x is a set', as $x = \varnothing \vee \exists y(y \in x)$. Then the axioms will be

(H1) $\quad [\mathbf{M}x \wedge \mathbf{M}y \wedge (\forall z)(z \in x \leftrightarrow z \in y)] \to x = y$

(H2) $\quad z \notin \varnothing$

(H3) $\quad z \in x \oplus y \leftrightarrow z \in x \vee z = y.$[21]

In addition, there will be a schema of induction:

(H4) $\quad \{A(\varnothing) \wedge (\forall x)(\forall y)[A(x) \to A(x \oplus y)]\} \to (\forall x)[\mathbf{M}x \to A(x)].$

Thus, the induction principle says that whatever is true of the empty set and remains true when an element is added to a set of which it is true is true of all sets.[22]

It is not difficult to see that the usual axioms of set theory (other than Infinity and Foundation) are provable in this theory. (See Appendix 1.) If we take the natural numbers to be von Neumann ordinals, then primitive recursion on natural numbers is a special case of the rule of primitive recursion on sets given in Appendix 1. We obtain another perspective on arithmetic, however, by first introducing a relation, $\sim$, of cardinal equivalence. This can be done in the Cantorian manner, since a one-one correspondence of finite sets is itself a finite set, if one construes a relation as a set of ordered pairs. But it is instructive to observe that the relation $\sim$ can be defined directly by primitive recursion:

$$x \sim \varnothing \leftrightarrow x = \varnothing$$
$$x \sim y \oplus z \leftrightarrow \{(z \in y \wedge x \sim y) \vee [z \notin y \wedge (\exists w \in x)(x - \{w\} \sim y)]\}.$$

We can describe a predicate or functor as "cardinal" if $\sim$ is substitutive in all its argument places. It then seems natural to translate arithmetic into set theory by translating $=$ as $\sim$ and rendering arithmetic functors and predicates by suitable cardinal ones. However, if a term t is translated as t',

---

[21] Note that (H3) has the consequence that if x is not a set, then $x \oplus y = \varnothing \oplus y = \{y\}$. It follows that the antecedent of (H4) is equivalent to $A(\varnothing) \wedge (\forall x)(\forall y)[(\mathbf{M}x \wedge A(x)) \to A(x \oplus y)]$.
[22] Induction principles of this type for the theory of finite sets go back to Zermelo, "Sur les ensembles finis," p. 188. The formulation here derives from Tarski, "Sur les ensembles finis," p. 54. See also Lévy, *Basic Set Theory*, p. 77. Zermelo's concern, however, is to derive it from set-theoretic premises and a definition of finite set.

the idea is that St should be translated as $t' \oplus x$, for any x such that $x \notin t'$. But then the translation contains the extra parameter x, and the idea is that when S is iterated the value of this parameter might change at any stage.

We can, however, introduce *predicates* instead of 0, S, and further arithmetic functors for, say, addition and multiplication. We would have

$$Zx \leftrightarrow x = \varnothing$$
$$Syx \leftrightarrow (\exists z)(z \notin x \wedge y \sim x \oplus z).$$

Addition might then be introduced as follows:

$$Azxy \leftrightarrow (\exists w)(w \sim x \wedge w \cap y = \varnothing \wedge z \sim w \cup y).$$

These relations can easily be seen to be cardinal.[23] It has the difficulty that since the successor of a set, and the sum of two sets, are unique only up to cardinal equivalence, we cannot introduce the usual functors as definite descriptions.

It is natural to implement our idea by thinking of numbers as types, where the tokens are sets and the relation $\sim$ is that of being of the same type. This would be implemented by introducing a functor **C**, understanding **C**x as the cardinal of x, by an abstraction axiom:

$$\mathbf{C}x = \mathbf{C}y \leftrightarrow x \sim y.$$

The extension of the theory by this new functor and axiom is conservative, since we could interpret **C**x as the (unique) von Neumann ordinal y such that $x \sim y$. This axiom is of course just a set-theoretic formulation of Frege's criterion of identity of cardinal numbers.

From an ontological point of view, the theory of HF sets has the complication that it is only by ascent in rank that we obtain larger and larger sets, unless we make some assumption about the domain of individuals. It is instructive to consider whether we can take the same approach to an ordinal conception of number, where instead of an equivalence relation on finite sets we now consider one on finite sequences. But now what is a finite sequence? It is natural to think of a finite sequence as an object s that has a length $lh(s)$ and, for each natural number $i < lh(s)$, an ith term $(s)_i$. Two sequences s and t are the same if $lh(s) = lh(t)$ and for each $i < lh(s)$, $(s)_i = (t)_i$. Such objects might, for example, be functions whose domains are initial segments of the natural numbers. But now if there is even one individual, the finite sequences of individuals are already a model of

---

[23] I am indebted to Warren Goldfarb for proposing this response to the difficulty of the previous paragraph.

§33. Finite sets and sequences

arithmetic, since for any individual x the sequences $\langle\rangle$, $\langle x\rangle\langle x, x\rangle\langle x, x, x\rangle$, ... are all different.[24]

However, this explanation uses the notion of natural number. Is there some way of explaining the notion of sequence so that we could derive the notion of number from it in a noncircular way? Let us call the conception of sequences just mentioned the sequences-as-functions conception. An alternative would be what we might call the sequences-as-tuples conception, where we assume a notion of ordered pair $\langle x, y\rangle$ and an empty sequence $\langle\rangle$. If s is a sequence, the result of adding a term x, which we write s^*x, is $\langle s, x\rangle$. Because of the binary character of pairing, this approach imposes a particular association. A preferable way to think of sequences as tuples is in an inductive-recursive way, as generated from an empty sequence by adding terms successively. For simplicity, let us first consider only sequences of objects from a given nonempty domain D. We use a two-sorted language; I will use the letters p, q, r, s, t as variables for sequences, and u, v, w, x, y, z as variables for individuals.[25] We assume an empty sequence $\langle\rangle$ and an operation of adding a term to a sequence, which we write $*$. We have the elementary axioms:

(S1) $s^*x \neq \langle\rangle$

(S2) $s^*x = t^*y \to (s = t \wedge x = y)$

and an induction principle:

(S3) $\{A(\langle\rangle) \wedge (\forall s)(\forall x)[A(s) \to A(s^*x)]\} \to (\forall s)A(s)$,

where the variable s ranges over sequences. We can again extend the theory by a schema for introducing terms by primitive recursion on $*$. (See Appendix 1.)

Now an equivalence relation $\approx$, which we can read as sameness of length or "ordinal number," can be introduced by inductive conditions,

---

[24] If the domain D of individuals is empty, then there is only one sequence $\langle\rangle$ of individuals, although if one goes on to allow sequences of sequences or sets, one obtains an infinite domain just as in the case of the hereditarily finite sets, in fact already with one ascent, since one can use $\langle\rangle$ instead of the x of the text.

[25] In the two-sorted language, the nonemptiness of D is enforced by using standard quantificational logic.

The same axioms in a one-sorted language yield a theory in which sequences can be terms of sequences. Then we do not need the nonemptiness of D to yield infinitely many sequences; we can take $\langle\rangle$ as the x of the text. This theory is in fact a conservative extension of the theory of sequences of individuals.

for example:

(L1) $\langle\rangle \approx \langle\rangle$

(L2) $s \approx t \to s^*x \approx s^*y$

(L3) $\{A(\langle\rangle, \langle\rangle) \land (\forall s)(\forall t)(\forall x)(\forall y)[A(s, t) \to A(s^*x, t^*y)]\}$
       $\to (\forall s)(\forall t)[s \approx t \to A(s, t)].$

It is proved in Appendix 1 that $\approx$ is an equivalence relation. Now one would like to think of "numbers" as just the sequences of this theory, with the equivalence relation $\approx$ as equality. Addition is just concatenation of sequences, and it and a form of multiplication are introduced by primitive recursions; see Appendix 1.

Nonetheless, there is an important respect in which the theory of sequences remains impoverished so long as we do not use some resources belonging to the concept of number. We would like to be able to talk of the nth term of a sequence, and also of its length. We need to have at our disposal a system of counters, though not yet numbers as full-fledged objects. This fact diminishes the difference between the sequences-as-functions and the sequences-as-tuples conception. Given counters, we can introduce $(s)_n$ and $lh(s)$ by primitive recursions:

$lh(\langle\rangle) = 0; lh(s^*x) = S[lh(s)]$
$(\langle\rangle)_n$ is undefined for all n; $(s^*x)_n = (s)_n$ if $n < lh(s)$, $= x$ if $n = lh(s)$; is undefined otherwise.[26]

Our problem is not that the theory of sequences does not have the resources to provide us with counters or even numbers as objects. If we assume that our language has a name for at least one element of D, call it 1, then there is an obvious translation of first-order arithmetic into the language of sequences, with 0 rendered as $\langle\rangle$ and St rendered as t'^*1, where t' translates t. We can set $(s + t)' = s'^*t'$, and $(s \cdot t)' = s' \times t'$. ($\times$ is the multiplication introduced in Appendix 1.) However, this translation in effect construes numbers as sequences of the form $\langle 1 \ldots 1 \rangle$. It does not capture what I had in mind by understanding numbers as sequences. The problems of translating the language of arithmetic without such a construal are the same as in the cardinal case. The solution I will pursue

[26] We can avoid partial functions by setting the undefined cases equal to a throwaway value (provided we have a name for at least one element of D), say the 1 of the following paragraph. The case where 1 is the genuine nth term of s is distinguished from that in which $(s)_n$ is undefined by the fact that the latter case obtains if and only if $lh(s) \leq n$.

is to think of numbers as types of which sequences are tokens, so that the original idea of thinking of numbers as sequences is compromised. Then the relation $\approx$ is just that of being of the same type. Addition and multiplication of sequences are the operations on tokens that correspond to the usual arithmetic operations understood as applying to types. We can introduce the ordinal number of a sequence by an abstraction axiom:

$$\mathbf{O}s = \mathbf{O}t \leftrightarrow s \approx t.$$

Note that the extension of our theory by $\mathbf{O}$ is again conservative, since we could have defined $\mathbf{O}s$ as the sequence r consisting entirely of 1's such that $r \approx s$. We can now define 0 as $\mathbf{O}\langle\rangle$ and the relation $m = Sn$ as

$$(\exists r)(\exists s)(\exists x)[s = r^*x \wedge \mathbf{O}r = n \wedge \mathbf{O}s = m].$$

§34. Sets and sequences, intuition and number

The purpose of developing these theories is to trace another genetic route to the concept of natural number. Of the two options offered, the ordinal one appears simpler, since one obtains the structure of the natural numbers without ascent in rank. However, perhaps because of a tendency deriving from Frege and Russell to concentrate on the concept of cardinal, writers on the subject have tended to start with sets, and I will concentrate on that case for the moment. Finite sets of individuals, thought of as generated in a step-by-step way such as our formal theory tries to capture, are arguably more concrete than sets in general. One way in which the concept of number might develop is by developing the concept of finite set, in the first instance of individuals, then noting the equivalence relation $\sim$, and introducing cardinals as a kind of types.

One source of the interest of this possible genesis is that some philosophers have claimed that small finite sets of perceptible objects are perceived or intuited. Let us suppose for the moment that that is true; I will examine the question shortly. Then we might also argue that the cardinal numbers of these sets are intuited. What would be involved would be to take in a set, but to take it in as a type, where other sets with the same cardinal are of the same type. Sameness of type would be intuitively verifiable in the step-by-step way by which statements of cardinal number are verified according to §32. Although it would involve another step of abstraction, it seems we might speak of intuition of the correspondence that witnesses the sameness of number, since it is itself a finite set of pairs.

Thus we would have a conception of number according to which some numbers at least are objects of intuition.[27]

What would intuition of a finite set be? Following the Cantorian tradition, we should say that it is a representation of its elements *as a unity*. It goes with the conception of sets as collections discussed in §20. Intuition of a set would be a sort of capstone of intuitions of its elements. The elements would have to be "held together" in a simple case of Kantian synthesis. Thus intuition of a *set* would require a definite conceptualization, for example to distinguish it from intuition of a mereological sum, because it would supervene on intuitions of just the elements of the set, and not, for example, some other objects that occupy the same region of space and time and are perceived along with the elements of the set. But as here conceived, intuition of a set is founded on intuition of its elements in the Husserlian sense already deployed in §28.

It should not be required that the elements of the set should be perceived or intuited all at once; one might intuit a set in a succession of stages corresponding to its generation by adding new elements. Then, of course, one aspect of the holding together is remembering what one has perceived before.[28] But memory is equally essential to perceiving an event that unfolds in time; this is not a disanalogy with perception that should disqualify us from speaking of intuition.[29] It should be pointed out that concepts are involved to the extent that they are needed for

---

[27] In earlier writings I suggested a way in which numbers might be objects of intuition by taking them as "generalized types": numeral-tokens even in *different* numeral systems are of the same type if they "represent the same number" in the sense explained in §32. (See *Mathematics in Philosophy*, pp. 43–47, where, however, the conception is given a modal nominalist interpretation, and "Mathematical Intuition," pp. 163–164.) The present suggestion would imply that such generalized types are objects of intuition. However, the conception seems to me to lose much of its motivation once one sees the relation of alternative numeral systems to our own as a matter of translation. The idea of starting from finite sets or sequences now seems to me of more interest as a possible genesis of the natural numbers. As a final account of the natural numbers, the earlier generalized types conception is open to the structuralist objection that it is just one more *construal* of the natural numbers. What is there about the concept of natural number that makes it true that numbers are these objects rather than some other progression?

[28] Cf. Kant's description of the synthesis of reproduction in the *Critique of Pure Reason*, A100–102.

[29] Still, there are questions about the boundaries. Suppose I am at a family gathering and perceive all at once all the living descendants of my paternal grandparents. Can I bring to bear memories of perceiving the ones who have died and thus come to intuit the set of *all* their descendants? This is strained if, as is true in actual fact, some of them died many years ago. One would not talk of seeing or witnessing an event that was spread out over a number of years, with long discontinuities.

§34. Sets and sequences, intuition and number

identifying the elements. But one need not pick out just these elements by deploying a concept, although once one has done so a concept is thereby "constituted." It would be, roughly, the concept of being this$_1$ or this$_2$ or . . . , where the indices distinguish the uses of the demonstrative 'this' to pick out the elements successively perceived.

One should not expect of such a conception that *any* finite set of in some way intuitable objects can be intuited. The conception already rules out intuition of sets that contain elements that are not intuitable. But clearly, as a practical matter, we are not able to intuit a set of a very large number of elements, even if they can be perceived individually. The problem is the same as that concerning intuiting a string of 10^{100} strokes. But other obstacles can arise that do not come from the scale of our cognitive faculties or the shortness of life. For it to be possible to intuit $\{x, y\}$ it must not just be possible to intuit x and to intuit y; the intuitions must be jointly possible. This may run afoul of the general fact that in order to perceive spatiotemporal objects and events one must reach a certain proximity to them. Suppose, for example, that e and e' are two events such that neither is close to being in the past light cone of the other, so that if an observer is close enough in space-time to one of them to observe it, he cannot possibly observe the other. It follows that he cannot intuit the set $\{e, e'\}$, even though it may be that at some past time he could have moved in such a way as to observe e or e'.[30]

It seems that we might reasonably speak of intuition of finite sets under somewhat restricted circumstances, and this would allow that if numbers are thought of as "types" of which such sets are tokens, then it seems we can also speak of intuition of numbers. This conception still has its difficulties, as we shall see, and it is doubtful that it carries us as far with regard to arithmetic as intuition of strings. Before detailing this, I want to discuss briefly the views of another writer who has approached the concept of number from a standpoint of this general kind.

One thing that is attractive about the idea of intuition of sets is that it makes sense of the obvious fact that sometimes facts about number can be verified by perception, that for certain F and n one can *see that* there are n Fs. Penelope Maddy uses this consideration to argue that suitable finite sets are not only intuited but *perceived*. Maddy's main argument for this claim is based on just the premise I have stated: In her example,

---

[30] More humdrum considerations of a similar kind show that tensed discourse about sets runs into conflict with the principle that if a set exists, then all its elements exist; see chapter 1, note 57.

Steve can, on opening a carton of eggs, see that there are three eggs in it.[31] In a simple case like that, one does not need to count. By contrast, the number of *F*s would not have to get very large before, given a step-by-step verification of the kind described in §32, we would hesitate to say that one sees that there are *n F*s.

Maddy then raises the question what is the bearer of number and considers various candidates (sets, properties, aggregates,[32] Fregean concepts). The ground on which she chooses sets is not phenomenological, as her thesis might lead one to expect. Rather, "we need to look, not to our perceptual experiences, but to our overall theory of the world, and we must ask which of these is best suited to playing the role of the most fundamental mathematical entity."[33] The decision in favor of sets rests on the utility of the concept in setting up a general framework for mathematics, and then on the indispensability of mathematics for science. The conclusion that the perceptual belief that there are three eggs in the carton is a belief about a set depends "on the idea that mathematical entities are indispensable to physical science; if they weren't, there would be no reason to include sets in our overall theory."[34]

One can see that the point of view of Maddy's argument is quite different from that of §32. Roughly, she wants to describe in scientific terms what is involved in seeing that there are three eggs in the carton. The sense in which Steve sees a set is close to that in which he might be said to see a collection of atoms. It is not required that he should identify it *as* a set; in fact, Maddy finds that implausible, on the ground that perceptual experience does not distinguish the different candidates for the bearer of number. This is one reason why she speaks of perception rather than intuition; she relies on the fact that one can see (or hear or smell) an *F* without identifying it as an *F*.

Using neuropsychological theories about the development of the object concept, Maddy engages in a speculation, according to which in

---

[31] *Realism in Mathematics*, p. 58. The ideas go back to her "Perception and Mathematical Intuition."
[32] What we have called mereological sums of objects.
[33] *Realism in Mathematics*, p. 61.
[34] Ibid., p. 62. In subsequent writings Maddy has criticized the "indispensability argument" for mathematical objects; see "Indispensability and Practice" and *Naturalism in Mathematics*. She would not now endorse the argument discussed in the text for the conclusion that there is perception of sets; see *Naturalism in Mathematics*, p. 152 n. 30. An indication that she no longer holds the conclusion, is the remark, in the latter place, that although she still holds that "we have perceptual access to simple numerical facts," she now rejects the conclusion that these facts are facts about sets. But it is not clear to me exactly what she means to retract.

§34. Sets and sequences, intuition and number 209

the course of human development one acquires a neural structure, which might be called a "set-detector," which enables one to acquire perceptual beliefs about sets.[35] It is the activation of this on his encounter with a carton of eggs that enables Steve to see the set of eggs in the carton and to see that there are three eggs there. This is the more fundamental reason for speaking of perception: It is supposed to be a natural faculty that does not depend on any particular conceptualization.

An underlying premise of Maddy's account of these matters is that a statement such as 'there are three eggs in the carton' is a statement about something to which a number is being attributed. It is that which leads to the question whether that something is a set or something else. In our analysis in §32, we did not need any such supposition: One can see that there are three eggs there without seeing any objects except the eggs, the carton, and perhaps other things in the immediate environment. Why should one think otherwise, unless one thinks (contrary to the view of §5) that predicates as such designate, and, moreover, that some cognitive relation to what a predicate designates is necessary for knowing a truth involving the predicate? I do not have a psychological story to rival Maddy's, but it seems to me that the primary elements of such a story would be the capacity to classify what one sees, as expressed in the use of predicates, and to recognize identities and differences.

On this view, seeing that there are three eggs in the carton would be a consequence of seeing that

$$E(\text{this}_1) \wedge E(\text{this}_2) \wedge E(\text{this}_3) \wedge (\forall z)(Ez \to z = \text{this}_1 \vee z = \text{this}_2 \vee z = \text{this}_3),$$

where of course 'Ex' represents 'x is an egg in the carton'. Maddy might object that this would make one's knowledge that there are three eggs in the carton inferential, contrary to psychological evidence.[36] But in avoiding this, how does perception of the set help? On Maddy's view, it seems one has to see *of* the set that it has three elements, and that has the same complexity as the fact that there are three eggs in the carton; indeed, one must see, in addition, that the set one sees is the set of eggs in the carton. Either view can maintain that "inferences" such as arise in a linguistic representation of what is seen take place automatically, at the subpersonal level.

Maddy's view has at least a superficial resemblance to that of Husserl in *Philosophie der Arithmetik*, although Husserl's conception of consciousness of sets is closer to my own description of a possible intuition of them.

---

[35] Ibid., p. 65.
[36] Ibid., p. 60; she cites Kaufman et al.

It is instructive at this point to consider one criticism that Frege makes in his review of Husserl. Husserl suggests that the kind of synthesis involved in intuition of a set is expressed by 'and', so that if one asks what Steve perceives, the reply could be (calling the eggs a, b, and c) 'a and b and c'. Frege objects that we do not ask 'How many are a and b and c?' but rather 'How many eggs are there in the carton?' and the answer is also in that form: 'There are three eggs in the carton.' (Frege seems to prefer, 'The number of eggs in the carton is three.')

> Thus we see that both in the question and in the answer a concept word or a composite concept expression occurs, instead of the 'and' required by the author.[37]

Frege is certainly quite correct in claiming that the typical form of a question about cardinal number, and of a statement of number, involves a predicate. Still, I think that Husserl could reply that, in the case where the answer is verified by perception, just the sort of synthesis is involved as enters into the description of intuition of the set. That is, one identifies *these* objects as the eggs in the carton and observes that there are three of them. But this formulation, which uses the plural, illustrates a problem Maddy herself calls attention to, namely, that one by considerations of this kind still does not get to the concept of *set* rather than another (in this case plurality). Husserl does not distinguish pluralities from what we have called collections, and the observation that the sort of synthesis that the idea of intuition of sets requires is to be found in perceptual verification of numerical claims still does not imply that a set or other such object has to be constituted as an object.[38]

Maddy's answer to this problem, as we have seen, is that it is mathematics and its role in science that leads to the conclusion that it is a set that one perceives in the kind of situation described. Even if one does not reject the indispensability argument in general, one can question this particular application of it. The reason is that a conclusion is being drawn about the organism's basic relation to the environment from the claim that set theory offers the best general framework for axiomatizing classical mathematics. But the mathematics actually applied in the description of the physical environment in which such simple verifications of numerical facts take place, and even the mathematics

---

[37] Review of Husserl, p. 321, my translation. Obviously, I have expressed Frege's point in terms of Maddy's example.

[38] A related point is made against Maddy by Carson, "On Realism in Set Theory," pp. 8–9. Carson's paper discusses Maddy's views about perception of sets in the more general setting of the set-theoretic realism defended in *Realism in Mathematics*.

needed for psychological and neurological theories of how this might be accomplished, can be formulated with other primitives. Therefore, even if classical arithmetic and analysis and the abstract analysis used in quantum mechanics are indispensable for the scientific account of the situation, it does not follow that an axiomatization in terms of sets is indispensable in the same way. It is just not plausible that the formulation in terms of set theory reflects the nature of things to the degree that Maddy's view presupposes. Thus one can ask, concerning Maddy's idea of a neural set-detector, what reason she has for claiming that it detects precisely a *set* rather than detecting one of the other possible bearers of number that she and other writers consider, or simply numerical facts.

The upshot of this discussion is that although it is reasonable to hold that some sort of collective consciousness of the objects falling under a predicate '*F*' is involved in determining by perception what the number of *F*s is, it is not necessary to attribute to the agent perception or intuition of a set as a single object.[39] Maddy, possibly along with the psychologists whom she cites, may be concerned with the question how to describe the awareness of numerical facts by children who have not yet learned language. Here, it seems to me that there is even less reason to suppose that one particular logical regimentation of the situation reflects the nature of the subject's cognitive apparatus in preference to another that expresses the same facts. To the extent that one makes distinctions, surely in this case one should, other things being equal, prefer the least ontologically committal formulation.

§35. Difficulties concerning intuition of finite sets

The objections made in the last section to Maddy's argument for perception of sets also affect the claim that intuition of sets as we described it is necessarily involved in verification of elementary numerical facts.

---

[39] I think the same considerations would apply to the proposal that intuition of sequences underlies the most elementary verifications of numerical facts, as W. W. Tait intimates ("Finitism," §2). Tait is, however, on somewhat stronger ground than Maddy, because he is basing his claim on a view about the nature of the concept of number. If one believes that the concept of number as we know it in mathematics is innate (a view that is not at all absurd), one will naturally interpret in terms of it the most elementary knowledge of facts involving number.

 Tait's view of Number as the form of finite sequences might be interpreted in terms of the generalized-types conception of ordinal number sketched earlier, with arbitrary sequences as tokens. I do not know how well this captures his intention. The conception does not appear in Tait's later papers.

However, one may still be interested in such intuition as a way of giving an intuitive foundation to theories of finite sets and sequences such as those described in §33. The relevance of intuition of sets would surely be increased if it were shown that there is significant intuitive knowledge of them, and, furthermore, if intuition extended to finite sets and sequences of higher rank.

Let us ask first whether it is intuitively evident that $x \oplus y$ is well defined, that is, that a new element can always be added to a given set. In §28, considering the case of strings of strokes, we relied on the fact that for objects of this kind imagination can be a foundation for intuition. To convince ourselves in the same way that $x \oplus y$ is defined, it seems we have to imagine an arbitrary element of HF(D). That clearly constrains D; it must be the sort of domain of which we can imagine an arbitrary element. Then it seems that one can imagine an arbitrary element of HF(D) by imagining a succession of elements of D bracketed in an arbitrary way so that they form ground-level sets, sets of sets, and so on. Then it seems evident that any new such object can be added to a set already obtained.

In the more restricted setting where we consider only sets of individuals, it is sufficient to imagine an arbitrary such set and an arbitrary element of D. In this case, that $x \oplus y$ is always defined does not insure an infinite domain of sets, because $x \oplus y$ differs from x only where $y \notin x$. In particular, if x is already all of D, then $x \oplus y = x$ no matter what y we choose. However, the same picture seems applicable to the case of sequences, and there of course s^*x is always new, even if x already occurs as a term of s.

Nonetheless, our picture gives rise to doubts. First, even if it is accepted, what of the null set? The idea behind our original description of intuition of a set was that intuition of the elements would found intuition of the set. But that leaves intuition of the null set with *no* foundation, but it cannot be ordinary, direct perception. It is tempting to say that the null set is a linguistic fiction, introduced to round out the system. So long as we are concerned only with sets of individuals, this view is viable. But it will turn out that the problem of the null set propagates up through the HF sets.

A second problem concerns the claim that it is possible to imagine an arbitrary set or sequence of individuals, even assuming that one can imagine an arbitrary element of D. This is something more complex than imagining an arbitrary string in the setting of §§27–28. In particular, to see that any string can be extended, we did not need to use the fact that an arbitrary string arises by iterated concatenation of an additional stroke. But it is hard to see what an arbitrary finite set or sequence of individuals would be without some such conception. It follows that what is being

§35. Difficulties concerning intuition of finite sets

imagined is subject to a restriction as to how it has been generated. We need not see sets and sequences in that way; other ways of understanding the basics of set theory are possible. But it is doubtful that any alternative is more favorable to the idea of imagining an arbitrary set.

Another doubt, related to that concerning the null set, arises when one brings in sets of sets. Suppose that I perceive the following array:

Do I intuit (1) six squares, (2) a single set of six squares, (3) three sets of two squares each, or (4) a set of three elements, each of which is a set of two squares? If I simply perceive the array as a Gestalt, that is already taking it in as a unity, but it must fall short of intuition of a definite set. The latter would require something that chooses between the above four possibilities (and other, less perceptually natural ones).

What the different intuitions would have to do, in effect, is to put mental brackets on the array. The absence of any is case (1); in the others we would have:

case (2)   {□□□□□□}

case (3)   {□□} {□□} {□□}

case (4)   {{□□} {□□} {□□}}

The structure we represent by bracketing could be represented equivalently by a tree. We give the tree in case (4):

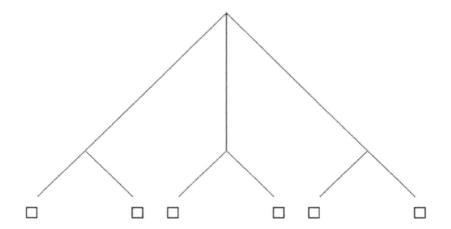

Now in order to "intuit" a set on the basis of perception of its ultimate elements,[40] the subject must impose a bracketing on these elements, which determines the tree of membership relations leading up to the set itself. The bracketing is optional: The subject could have chosen another bracketing leading to an intuition of a different set with the same ultimate elements. But, it will be objected, this makes too much depend on the concept brought to the situation by the subject.

It seems to me that this does show that the analogy with perception is strained in this case in a way that it is not in the case of expressions as types. According to the account of §28, to intuit a type the subject has to perceive or imagine something *as* of a certain type, where the concept of the type is also optional in that the physical tokens can be grouped into types in different ways. The concept of the type serves, however, only as a rule of reidentification, and such a rule is presupposed by any perceptual judgment of the presence of an object. In the case of a set, the "mental bracketing" of course contributes to reidentification, since one would have the same set again only if one had the same ultimate elements bracketed in the same way. But there is a subtle difference. The difference between a token and a type lies basically in the criterion of identity. If a region of space contains an object that is a token of more than one type, then this fact comes to light by comparison with other places and times. But consider the location of the six squares that are the ultimate elements of the sets (2), (3), and (4). Then, in the only sense in which a location could be said to "contain" a set, any location that contains the six squares would contain all three sets (and more). For a set to occur at a location can only be for its ultimate elements to occur there.

The difficulty is reinforced by considering the null set again. It would consist of a bracketing that encloses *nothing* or of a degenerate tree consisting only of the top vertex, with no branching or labeling. Now we have to ask whether, in the intuition of a set, the bracketing itself is intuited. If it is not, then how can there be intuition of the null set, or, for that matter, of pure sets generally? But if it is, then it seems the sensory foundation of this intuition will have to be perception or imagination of some form of representation such as actual brackets or trees. But then, so long as other systems of representation are possible, we cannot say that the configuration of objects and brackets *is* the set, since a configuration with

---

[40] That is, the individuals in its transitive closure. Note that a pure set has no ultimate elements.

§35. Difficulties concerning intuition of finite sets 215

a different form of representation of the bracketing would have an equal claim.

The way in which we have described a putative intuition of sets gives the bracketing a conceptual role, and against the idea that intuition of it is part of intuition of the set is the general consideration that in a perceptual situation involving the application of certain concepts, we do not expect that a linguistic or other embodiment of the concepts should be perceptually present in that very situation. We can reinforce this way of looking at the matter by observing that the theory of hereditarily finite sets over D can be given a straightforward relative substitutional semantics, relative of course to the domain D of individuals.[41] The idea is that given an interpreted first-order language $\mathcal{L}_0$ for which D is the domain, we add terms "for hereditarily finite sets" and variables and quantifiers for them, which, however, we interpret substitutionally relative to D. The terms might be called bracket terms. $\{\}$ is a bracket term, and if $T_1 \ldots T_m$ are terms of $\mathcal{L}_0$ or bracket terms, so is $\{T_1 \ldots T_m\}$. If $X, Y, \ldots$ are the substitutional variables for bracket terms, then the clause for universal quantification in the recursion for satisfaction is:

(5) s satisfies $(\forall X)A$ iff for every bracket term T, and for every s' agreeing with s on variables free in A and assigning a value to every variable free in $A\, X/T$, s' satisfies $A\, X/T$.[42]

This semantics is equivalent to a special case of a natural generalization of the relative substitutional semantics for predicative classes[43] to higher ranks.

It might be argued that this semantics dispenses with the idea of intuition of sets because it offers an ontological reduction, in which sets are eliminated. The discussion of substitutional quantification in general encourages caution about such claims. In a setting like the one we are considering, a pure (not relative) substitutional semantics would be equivalent to an objectual one where the objects are expressions. Such a pure substitutional semantics is clearly possible for the pure hereditarily finite sets, that is, where D is empty. In that case, sequences and

---

[41] In the sense of *Mathematics in Philosophy*, p. 214; cf. p. 67. For details of the present semantics, see Appendix 2.
[42] Bracket terms in this definition are allowed to contain individual variables, but of course not "set" variables. But this clause enables us to interpret formulae containing bound "set" variables, in particular occurring within bracket terms.
[43] *Mathematics in Philosophy*, ibid.

satisfaction drop out and one can simply define truth. We could just as well construe the sets as the bracket terms that designate them.[44]

In the relative substitutional case, clearly the elements of D are left unreduced, whatever one thinks about the controversial issues about the ontology of substitutional theories already alluded to in §32 (p. 195). Thus, in general, one cannot, as in the pure substitutional case, give an equivalent objectual semantics where the objects are the expressions of the substitution class. Rather, the objects must also code assignments to the free individual variables. Thus one can give such a semantics where the objects are ordered pairs of expressions and sequences of elements of D.

But the role of sequences in (5) differs in a relevant way from that in the usual objectual clause for the universal quantifier, which in the underlying language $\mathcal{L}_0$ we might suppose to be:

(6)   s satisfies $(\forall x)A$ iff for every u in D, s_u^x satisfies A,

where $(s_u^x)_i = u$ if x is the ith variable, $(s_u^x)_j = (s)_j$ otherwise. Here sequences play the role of parameters, and the only operations on them that we require are explicitly defined. If we confine ourselves to formulae with a fixed finite set of free variables and bounded quantifier depth, we can eliminate the sequences. This technicality expresses the fact that we interpret a formula with n free variables as an n-place predicate, and a single formula of quantifier depth k involves at most $(n+k)$-place predicates. For this sort of reason, we do not take the usual Tarskian truth definition for a first-order language as imputing to the *object language* an ontology of finite sequences. Where they are required is to obtain satisfaction and truth predicates that cover formulae of unbounded complexity.

We can see the relevance of the difference in the following way. A truth theory in the style of Davidson does not attribute to the object language an ontology of finite sequences of individuals, unless the domain itself is closed under formation of sequences. An individual sentence will be translated into the metalanguage by a sentence containing only quantifiers over whatever the theory interprets the quantifiers of the language to range over. But this could not be done for a truth theory with a relative substitutional quantifier clause such as (5). In that case, the reason is that the substitution class of bracket terms contains terms with any number of free

---

[44] These are of course not unique. But if we fix an ordering $<$ of all bracket terms, we can call a bracket term normal if it is $\{\}$ or $\{T_1 \cdots T_m\}$ where $T_1 \cdots T_m$ are normal and $T_1 < \cdots < T_m$, clearly any pure HF set is designated by a unique normal bracket term. (The ordering need not be of type ω or even a well-ordering.)

§35. Difficulties concerning intuition of finite sets 217

individual variables, and so there is no limit to the number of objects that have to be assigned to interpret an instance of the quantification on the right side of (5), while in the case of (6) at most one more object is involved.

The upshot of this discussion is that if we take the relative substitutional semantics as capturing a speaker's understanding of the language of hereditarily finite sets, or of finite sequences, then we largely remove the motives for characterizing awareness of such sets and sequences as intuition, except in the ground-level case, where it is sequences that are left unreduced. In the semantics, it is clearly sequences as functions that are appealed to, but the work can be done by sequences as tuples in the context of a theory like that sketched in §33 and Appendix 1. This would be a satisfying result if we could put aside our questions about the notion of imagining an arbitrary set or sequence of individuals. If this could be done, then for sequences we would have a situation for intuitive knowledge very similar to what will emerge for strings in Chapter 7.

The manner in which ordinal numbers were introduced in §32 gives them the character of generalized types. So long as their role is just that of numbering sequences of individuals, they can be considered as objects of intuition in the weak sense that individual numbers can be intuited. It follows that there is one route into the concept of number that gives numbers this character. In our discussion of cardinality, however, we saw that regarding number as attaching to a set was not essential to its elementary applications. The same holds for sequences and ordinals. It is not evident that we should see the genesis of number as introducing *bearers* of numbers as objects in such a way that such objects are the foundation of *numbers* as objects. But it does tell us something about the notion of number that a route into it is possible in which this is the case, although its optional character makes it not serve as the basis for an argument like Maddy's for *perception* of sets. And in a fully developed theory of number in theoretical mathematics of course there will be bearers of number.

We have up to now not taken account of a basic feature of the concepts of set and sequence: We allow sets and sequences of objects generally, independently of whether they are in any way intuitively given. In talking of intuitive knowledge about sets and sequences, we have not considered how great a restriction it is that it should be possible to imagine an arbitrary element of the underlying domain D. But it is clearly a restriction.

Note that for the relative substitutional semantics, no use is made of any intuitive character the domain of individuals presupposed for the interpretation of $\mathcal{L}_0$ might have. In order to reinforce the question it raises about intuition of sets, we pose a simple question concerning the axiom of pairing. Given x and y, there is a set whose elements are just x and y. What now is gained if we suppose that x and y are intuitable or that they belong to a domain of which we can imagine an arbitrary element? The substitutional semantics suggests that what is gained concerns not the step from x and y to $\{x, y\}$ but x and y themselves, whose existence is a *presupposition* of that of $\{x, y\}$. The treatment of the concept of object in Chapter 1 indicates how, on a purely formal level, we might have a conception of object not constrained by considerations of intuitability. The relative substitutional semantics sketched here indicates how this would extend to hereditarily finite sets. This might lead to the conclusion that intuition of the elements of a set is quite irrelevant. However, we confront again the fact that sequences of individuals are left unreduced by our relative substitutional semantics.

We need to reflect once again on what we have called the synthesis involved in representing several objects together. Our discussion has brought out that this should be distinguished from taking the objects to constitute a single object, a set as collection. But perhaps we should not see either as particularly bound to intuition of the objects involved. The cases we have considered are all special cases of the principle that any objects can constitute a set. We might see the synthesis involved as a purely intellectual affair that has nothing to do with intuition.[45] On this view, although it could not yield intuition of the set if the elements were not intuited, there is no difference with respect to the transition between the givenness of the elements and that of the set.

§36. Well, then, what are the numbers? Structuralism put in its place

We have considered in §§32–33 two kinds of possible geneses of discourse about natural numbers. On the one, such discourse arises in a series of steps beginning with a situation in which numerals do not function as singular terms, by way of one in which quantification of numeral places is substitutional, and then this quantification becomes objectual and is unified with other quantification. Cardinality is provided for in the language, without the introduction of any objects as bearers of number. Such objects would arise only at a stage at which the schematic character of generalizations about predicates needs to be replaced by regular

---

[45] Compare Kant's remark that "the science of number ⋯ is a pure intellectual synthesis which we represent to ourselves in our thoughts." (Letter to Johann Schultz, 25 November 1788, *Ak*. X 555.)

§36. Well, then, what are the numbers?

quantification over a domain. This development is on this account independent of the development of reference to numbers, although it is no doubt essential to theorizing about cardinality.

On the other type of genesis, sets or sequences are the primary "arithmetical" objects, and numbers arise by abstraction. This account makes the notion of finite ordinal or cardinal the genetically first concept of number. Although this was also true on the first account, that feature of it was independent of the fact that on it quantification "over numbers" is at first substitutional. Although on the second type of account one could at first treat the quantification as substitutional relative to the quantification over sets and sequences, there seems to be little gain in this.[46]

One could easily imagine the first account as part of a model for mathematicians' talk about numbers before the nineteenth century, although a realistic account would have to take in a lot that we have not considered, such as how such mathematicians talked of what we call real numbers and how they related the concept of number to geometry. The account has the feature that numbers first arise as objects of a distinctive type. That means that without being structuralist, the account is congenial to structuralism because it introduces numbers in a way that at least leaves entirely open the question what identities might hold between numbers and other objects.

The question arises how this story fares with respect to a problem that arises in my own treatment of structuralism, as well as in that of other writers, namely, the need to give an instance of the structure, possibly independently of the concept of number or at least cardinality. In §18, it was argued that this is an unavoidable impurity in structuralism, since one must rely at this point on some other domain of objects, in my preferred treatment a domain of quasi-concrete objects and in any event not pure mathematical objects.

If one begins as in §32 with numerical quantifiers, this problem arises as the problem how we insure that it is never the case that $(\exists_n x)Fx$ is equivalent to $(\exists_{Sn} x)Fx$. Given (2a) and (2b), it follows that this cannot happen unless both are false. Whatever we mean by $m = n$, the substitutivity of identity will imply that $n = Sn$ can hold only if $(\exists_n x)Fx$ and $(\exists_{Sn} x)Fx$ are false for all 'F', that is, if the domain of individuals contains fewer than n elements.[47] If the domain is finite, this will be true for sufficiently large n.

---

[46] The abstraction at issue is first-order abstraction, as in Frege's example of the direction of a line. Second-order abstraction would raise additional problems.

[47] Obviously, I have ruled out the interpretation of n as infinite.

Thus, so long as only the elements of an antecedently given domain are numbered, we do not avoid this sort of collapse unless the domain is infinite. Our reflection on the verification in everyday life of simple instances of statements of number does not yield that.

It is hard to see how it could unless the infinity of the domain comes from some other source. Otherwise, on this genetic account, the infinity of the natural numbers does not seem forced on us until we apply numerical quantification to predicates of numbers, that is, after quantification over numbers is unified with other quantification. That appearance is misleading, however, because it is assumed that the character of numbers as cardinals is maintained throughout. But there is no bar to introducing pure arithmetic at some stage, and, for example, assuming $Sn \neq 0$ and the inference from $Sm \equiv Sn$ to $m \equiv n$. (I use '$\equiv$' rather than '$=$' here because I do not want to assume that the relation being introduced is identity.) But once we go objectual, the substitutivity of identity will force different numerals to designate different objects, since if **m** and **n** are distinct numerals, then the class of numerals **k** for which **m** $\equiv$ **k** holds will differ from the class for which **n** $\equiv$ **k** holds.[48]

Does this give an easy route to the infinity of the natural numbers? No, because it clearly presupposes the infinity of the numerals. It depends on intuition just as the proposal of §18 did, but not quite in the same way, since the numerals are not offered as an instance of the structure of numbers. The justification of the assumptions is presumably that we can treat numerals as equivalent in the sense of $\equiv$ only if they are the same, but that can be carried through only if the sequence of numerals is infinite.

In the route by sets or sequences, as it has been set forth in §§33–34, that one can always add additional elements to a set or terms to a sequence is an assumption of the theory of these objects, so that it can be prior to the abstraction involved in the introduction of numbers. We might, however, consider the case where we form only sets of individuals and yet do not assume an infinity of individuals. In that case, of course, there can be collapse of cardinals at a finite point, but it occurs in a slightly different way. If the cardinals are themselves individuals, then the domain must be infinite, just as in Frege's own setting. Let us rather assume that they are a separate sort of objects. If we define 'm is the (cardinal) successor of n' essentially in Frege's way, then, if the domain is finite, there will be a set V of all individuals, which will have a certain cardinality n, but n will

---

[48] It then follows that '$\equiv$' is coextensive with '$=$' on the natural numbers.

§36. Well, then, what are the numbers?

have no successor. There will be no failure of the abstraction axiom, and also no x for which **C**x does not exist.

One might then ask whether treating cardinals as individuals itself offers the "easy route" to the infinity of the natural numbers that we have so far not found. Let us remind ourselves of the framework in which we are working. We have a two-sorted theory, with individuals and sets, and the axioms (H1)–(H4) and the abstraction axiom of §33, where the variables are of the set sort but the terms Cx and Cy are of the individual sort. It is in fact a weaker version of what has come to be called Frege Arithmetic, a theory based on second-order logic in which we add to the language a cardinal number operator $NxFx$, read 'the number of Fs', together with an axiom embodying Frege's criterion of identity for cardinals, called above, following the literature, Hume's Principle. The theory now under discussion could be formulated as a subtheory of Frege Arithmetic by suitably restricting the comprehension axiom and then adding a schema of induction. But it embodies the same fundamental move as Frege Arithmetic, of passing from an equivalence of second-level entities to an identity of ground-level objects. Because in both cases this constrains the domain of individuals to be infinite, this has to be regarded as a substantial assumption, although the resulting theory is still essentially weaker than Frege Arithmetic; it is equiconsistent with first-order arithmetic.[49] Moreover, as we have already treated sets as a separate sort, regarding cardinals as individuals seems inadequately motivated. Once again, we do not have an easy route to the infinity of the natural numbers.

Apart from that, none of the stories we have considered offers an answer to the question what objects the natural numbers are or which of the different construals of the natural numbers with the help of higher-order logic or set theory is correct. That is entirely in accord with the structuralist view defended in §18. However, common to all the versions of both kinds of account is the following: We have a choice as to whether to proceed so that terms and variables for numbers are of the same type as others, and, assuming that mature mathematics does involve regarding numbers as

---

[49] Suppose first that the theory assumes nothing about the number of individuals. Then, assuming consistency, it will clearly have a model with no individuals except the finite cardinals, that is in which the individuals are the natural numbers, and in which all sets are finite. The sets can then be coded as numbers using Ackermann's method.

The result will still hold if we assume countably many individuals that are not cardinals. For the theory will have a model with countably many noncardinals and countably many cardinals. Code the cardinals as 0, 2, 4, 6, $\cdots$ and the noncardinals as 1, 3, 5, 7, . . . Again code the sets by Ackermann's method.

objects among others, at what point to unify the type of numbers with others. This implies that the questions giving rise to the structuralist view of mathematical objects can be postponed until far along in the genesis of discourse about numbers. This holds in particular for the question whether the identity of numbers is bound up with their cardinal or ordinal role: One might imagine their first arising as one or the other but then as cutting loose from this and being understood as a structure that can be *applied* in either way, or in others. The place of a structuralist view of numbers is as an account of numbers in mature mathematics, and we need not assume that such an understanding of the numbers is what the child learns or what our distant ancestors acquired over a long period of time.

§37. Intuition of numbers denied

There is a quite obvious conflict between the structuralist view of numbers and the claim that numbers are objects of intuition in the Hilbertian sense developed in this work from §28 on. The conflict exists for structuralist views in general as canvassed in Chapters 2 and 3, not just for the specific one defended in §18. Being intuited by a certain agent S at a specific time t would be a property additional to structural properties of numbers, even if the numbers are embedded in a larger mathematical structure. Thus, if it is a property of some number n, on the structuralist view it would have to be an external relation such as is involved in applying numbers as cardinals. Such relations, however, have a certain kind of isomorphism-invariance. If the number of Fs is n, then there is a one-one map of the Fs onto the elements $a, fa, ffa, \ldots f^{(n)}a$ of any progression; if there is such a map for one progression, there will be one for any other. But no comparable condition can obtain for being intuited by S at t, for the simple reason that there are progressions whose terms are not objects of the kind that intuition as we have explained it can capture. For example, any structure that is instantiated in such a way that the domain is a set can be instantiated so that the domain consists entirely of sets of very high rank, but, since such sets have a structure too rich to be instantiated in space and time, they cannot be objects of Hilbertian intuition. The same is true of other conceptions of intuition modeled on Kant's. Other examples of nonintuitable objects suggest themselves: other kinds of pure abstract objects, transcendent objects such as God and other objects considered in theology, and theoretical objects in science. Although they are suggestive, I don't want to appeal to any of these: the pure abstract objects

§37. Intuition of numbers denied

because their not being objects of intuition should be the conclusion of an argument, perhaps close to that concerning the natural numbers; transcendent objects because it will not be agreed that they exist; and theoretical objects because, given that they are spatiotemporal, it is at least not obvious that they should not be treated as objects of intuition although it is not practically possible to perceive them.

It is true, as I argued in Chapters 2 and 3, that it is a presupposition of talk about natural numbers that what we are talking about is in some way a possible structure. Moreover, I argued that this possibility is made out by an intuitive model. But the intuition in question need not be intuition of *numbers*, and in the version of §18 it is not. What is asserted to be possible is a structure that is physical, or mental, or intuitive-geometrical, in a way in which, on this structuralist view, numbers are not.

It might be objected that this rejection of intuition of numbers depends on very specific features of the conception of intuition involved, more specific than its generally Kantian or spatiotemporal character. I agree that it depends on the latter, but it does not depend on more specific features of the conception of §28. Elsewhere I have discussed generalized-types conceptions of numbers and indicated that at least a case could be made that there is intuition of numbers as so conceived (see note 27). But it would be incompatible with the structuralist view to accept such a conception as the final conception of what the numbers are. And I think the point can be generalized: So long as the intuition is spatio-temporal, something of the generality of the concept of number will be lost on any conception that admits intuition of numbers. We can, however, leave open whether some essentially different conception of intuition, modeled perhaps on Husserl's or Gödel's, admits intuition of numbers. This question will arise again in Chapter 9.

We should give separate consideration to the question whether, on a structuralist view, any propositions properly about numbers are intuitively evident. This is different from the question to be addressed in Chapter 7, what the limits of intuitive evidence are when one refers to the intuitive model of arithmetic by means of strings. Obviously, however, anything that is not intuitively evident when interpreted with reference to that model would also not be intuitively evident as a proposition about numbers. In particular, that is the case for induction as a general principle. But if numbers are not objects of intuition, it is difficult to see how any propositions about numbers could be intuitively evident in the sense that concerns us. But we can best approach the question by way of a different question, whether the applicability of the concepts of cardinal and ordinal number to objects in general implies that statements about numbers such as '7 + 5 = 12', which are intuitively evident when interpreted with reference to an intuitive model, can still be taken as intuitively evident when this generality is taken into account.

'7 + 5 = 12' implies, by first-order logic with identity, any instance of the schema

(7) $[(\exists_7 x) Fx \wedge (\exists_5 x) Gx \wedge (\forall x) \neg (Fx \wedge Gx)] \rightarrow (\exists_{12} x)(Fx \vee Gx)$.

It should be clear that there will be instances of this schema that are not intuitively evident, once the domain over which the quantifiers range no longer consists of objects of intuition. Does this imply that, in drawing such a consequence, we have come to understand '7 + 5 = 12' so that it is no longer intuitively evident? We have an argument for this conclusion only if we assume that even in this setting *logic* preserves intuitive evidence. But precisely because it is being applied where the domain of quantification has a nonintuitive character, we can question this assumption. Thus the conclusion that the lack of intuitive evidence of such an instance of (7) reflects back, so that '7 + 5 = 12' is itself not intuitively evident, is not forced on us.

The ideas about the genesis of reference to numbers presented in this chapter differ from the versions of structuralism considered in this work, in that according to the genetic accounts some form of reference to numbers is present before the questions giving rise to the abstraction or generalization can even arise. The eventual integration of arithmetic with set theory (and possible highly abstract empirical scientific theories) does bring the numbers into a sort of holistic connection with highly nonintuitive modes of reference and knowledge. It is not clear, however, that statements that were intuitively evident before this expansion of our conceptual resources need to lose that evidence. What is clear from examples like that of '7 + 5 = 12' given above is that their *applications* outside the intuitive sphere will in general not have the same intuitive character. If we model the often-claimed "transparency" of the numbers by our concepts of intuition and intuitive evidence, then it follows that the generality of their application, thus one aspect of what is involved in their status as pure abstract objects, is something added to what is transparent about them. I don't find this an unwelcome result; in fact, some further questions about the transparency of the natural numbers, to be sure not concerning such elementary statements as '7 + 5 = 12', will be considered in Chapter 9.

§38. Appendix 1: Theories of sets and sequences

The first-order theory PS⁻ of hereditarily finite sets sketched in §33 is a theory formalized in first-order logic with identity, with a nonlogical predicate $\in$ of membership and a constant $\varnothing$ for the null set, and a binary functor $\oplus$. The axioms are (H1)–(H3) and the instances of the schema (H4) of induction. We note that the formulation of extensionality (H1) allows for individuals. As it stands the theory says nothing about the individuals, even whether or not any exist. They could be given some structure by adding predicates applying to individuals and, possibly, axioms involving them.

A more restricted theory, of sets only of individuals, is obtained by using a two-sorted language, with variables for individuals and sets. $\varnothing$ will be of the set sort, and $\oplus$ will take a set and an individual as arguments. If we use $a, b, c, \ldots$ for sets, then the axioms take the form:

(H1′) $(\forall z)(z \in a \leftrightarrow z \in b) \rightarrow a = b$

(H2) $z \notin \varnothing$

(H3′) $z \in a \oplus y \leftrightarrow (z \in a \vee z = y)$

(H4′) $\{A(\varnothing) \wedge (\forall b)(\forall y)[A(b) \rightarrow A(b \oplus y)]\} \rightarrow (\forall b)A(b)$.

(These formulations are meaningful for the one-sorted theory if we take $a, b, c, \ldots$ as variables restricted to **M**. Then they offer an axiomatization essentially equivalent to (H1)–(H4).[50]) To allow the case of no individuals, the logic for the individual sort should allow the empty domain. Clearly, without assumptions as to how many individuals there are, the theory will, for each n, have a model with n individuals and 2^n sets.

Let us now consider the one-sorted theory PS⁻.[51] One can straightforwardly see that the axioms of ZF, other than Infinity and Foundation, are derivable. Extensionality and the null set are axioms; the pair set $\{x, y\}$ is $(\varnothing \oplus x) \oplus y$. To prove the sum set axiom, one first shows by induction that any pair a, b of sets has a unique union: If $b = \varnothing$ it is a. If

$$(\forall x)[x \in z \leftrightarrow (x \in a \vee x \in b)],$$

---

[50] (H3′) thus interpreted leaves undetermined what $x \oplus y$ is when x is not a set, so that a strictly equivalent axiomatization would consist of (H1′), (H2′), (H3), and (H4′). On induction, see note 21 above.

[51] For information and references to literature on theories of finite sets, I am greatly indebted to Akihiro Kanamori and Laurence Kirby.

then evidently by (H3)

$$(\forall x)[x \in z \oplus u \leftrightarrow (x \in a \vee x \in b \oplus u)].$$

$\varnothing$ is evidently its own sum set. If $(\forall x)[x \in z \leftrightarrow (\exists y)(x \in y \wedge y \in a)]$, then for any x,

$$\begin{aligned}x \in z \cup u &\leftrightarrow (\exists y)(x \in y \wedge y \in a) \vee x \in u \\ &\leftrightarrow (\exists y)(x \in y \wedge y \in a) \vee (\exists y)(x \in y \wedge y = u) \\ &\leftrightarrow (\exists y)(x \in y \wedge y \in a \oplus u),\end{aligned}$$

so that if z satisfies the condition to be the sum set of a, $z \cup u$ satisfies the condition to be the sum set of $a \oplus u$. Again, the condition implies uniqueness.

Zermelo's separation schema

(AS) $(\exists b)(\forall x)[x \in b \leftrightarrow (x \in a \wedge A(x))]$

is also derivable by induction. For if $a = \varnothing$, one can clearly take $b = \varnothing$.[52] And if (AS) holds for a given a, then if $\neg A(z)$ the same b does for $a \oplus z$; if $A(z)$, then (AS) implies

$$(\forall x)[x \in b \oplus z \leftrightarrow (x \in a \oplus z \wedge A(x))],$$

so that in either case (AS) holds for $a \oplus z$. A similar inductive argument shows that Replacement holds as well.[53] To prove the power set axiom, one first shows by induction that for any x and y, $\{w \oplus y: w \in x\}$ exists. $\{\varnothing\}$ satisfies the condition to be the power set of $\varnothing$. If z satisfies it for a, then $z \cup \{w \oplus y: w \in z\}$ satisfies it for $a \oplus y$. Again by induction, every set has a power set.

Thus, every theorem of ZF (with urelements) without Infinity or Foundation is provable in PS⁻.[54]

These arguments suggest extending PS⁻ by a schema of primitive recursion, for example:

(PR) $\varphi(x_1 \ldots x_n, \varnothing) = \psi(x_1 \ldots x_n)$
$\varphi(x_1 \ldots x_n, a \oplus z) = \chi[x_1 \ldots x_n, a, z, \varphi(x_1 \ldots x_n, a)]$ if $z \notin a$,
$= \varphi(x_1 .. x_n, a)$ if $z \in a$.

---

[52] The restriction of a to **M** is again nonessential; if a is not a set, one can again take $b = \varnothing$.
[53] Cf. Lévy, *Basic Set Theory*, p. 77.
[54] This is stated by Givant and Tarski, "Peano Arithmetic," for the theories without urelements. Some details can be found in Tarski and Givant, pp. 223–226.

§38. Appendix 1: Theories of sets and sequences

However, such a schema is subject to an additional condition to insure that the value of $\varphi(\ldots, a)$ does not depend on the order in which the elements of a have been adjoined.[55] It seems clear that if the functors in the recursion are cardinal in the sense of §33, the condition will be satisfied, but that is a rather special case.

The axiom of Infinity is clearly false in the intended model of hereditarily finite sets based on a domain of individuals (possibly empty, but also possibly infinite). In fact, it fails in a stronger way: It is possible to prove in PS^- that every set is finite, for any one of several definitions of finiteness.[56] This is easy to see by induction for the most familiar one: a is finite if a is equinumerous with a finite von Neumann ordinal, where an ordinal is finite if it and each of its elements is either 0 or a successor.

Foundation clearly holds in the intended model of the theory, not only for the pure hereditarily finite sets but also for the hereditarily finite sets built up from some domain of nonsets. But it is not derivable from (H1)–(H4). Suppose we have a "set" w such that $w = \{w\}$. Treating it as if it were an individual, build up the hereditarily finite sets on it. That yields a model of the theory in which Foundation fails.[57] For future reference, call this model $\mathbf{M}_1$. Let $\mathbf{M}_2$ be the model obtained in the same way beginning with a sequence $a_0, a_1, a_2, \ldots$ such that $a_i = \{a_{i+1}\}$. In $\mathbf{M}_2$, for each i, a_i lacks a transitive closure, so that the sentence TC asserting that every set has a transitive closure fails in $\mathbf{M}_2$. However, $\mathbf{M}_2$ satisfies Foundation. We can assign finite ranks to the sets in the model: $\mathrm{rk}(z) = 0$ if z is $\varnothing$ or a_i for some i, $= \max(\mathrm{rk}(w_i) + 1)$ if $z = \{w_1 \ldots w_m\}$, unless $m = 1$ and $w_m = \{a_n\}$ for $n \neq 0$, in which case $z = a_{n-1}$ and $\mathrm{rk}(z) = 0$. For consider a set $z \neq \varnothing$ in the model. If the minimal rank of its elements is not 0, or if $\varnothing \in z$, then either a minimal rank element or $\varnothing$ has an empty intersection with z. Otherwise, since z is finite, there must be a greatest n such that $a_n \in z$. Since a_{n+1} is the sole element of a_n, $a_n \cap z = \varnothing$. Let $\mathbf{M}_3$ be the model

---

[55] For a discussion of this problem and a necessary and sufficient condition for a more powerful version of primitive recursion, see §4 of Kirby, "Finitary Set Theory."

[56] For several such definitions that arose in early twentieth century work, see my "Developing Arithmetic in Set Theory," §3.

[57] I am indebted to Vann McGee for questioning me about the status of Foundation and suggesting that (H1)–(H4) do not rule out a set that is its own unit set. (Some earlier circulated versions of this work claimed falsely that Foundation is derivable from (H1)–(H4).) Paul Bernays constructs essentially the same models of his own set theory without Infinity, in which Foundation and transitive closure thus fail in just the same way. However, his construction works by modifying Ackermann's arithmetical model. See "A System, Part VII," p. 104 of reprint. I am also indebted to McGee for pointing out that $\mathbf{M}_2$ satisfies Foundation; the proof in the text is based on his sketch. Bernays seems to state the contrary, but the appearance is due to the fact that he states the axiom in terms of classes, so that it is equivalent to $\in$-induction ("A System, part II," p. 19 of reprint).

obtained in the same way again, starting with both w and the sequence of a_i. In $\mathbf{M}_3$, both TC and Foundation fail.[58]

However, this weakness of (H1)–(H4) is overcome if we replace the Zermelian induction (H4) by the stronger

(H4*) $[(\forall x)(\neg \mathbf{M}x \to A(x)) \wedge A(\varnothing) \wedge (\forall x)(\forall y)\{[A(x) \wedge A(y)]$
$\to A(x \oplus y)\}] \to (\forall x)A(x)$.[59]

PS⁻ with (H4) replaced by (H4*) is essentially the theory PS discussed by Flavio Previale.[60] Conversely, (H4*) is derivable if $\in$-induction, or equivalently Foundation and TC, are added to PS⁻.[61]

---

[58] The standard proof of TC requires the axiom of Infinity; see for example Drake, *Set Theory*, pp. 28–29. But it is provable in set theory without Infinity if we have $\in$-induction; see Barwise, *Admissible Sets*, p. 24. Likewise, if $\in$-induction is added to (H1)–(H4), both Foundation and TC are provable.

[59] (H4*) is an adaptation to the situation with urelements of a schema proposed by Givant and Tarski, op. cit. I myself arrived at it by reformulating an equivalent schema proposed in correspondence by McGee. Previale, "Induction and Foundation," and Kirby, "Finitary Set Theory," call (H4) weak induction and (H4*) strong induction.

[60] Op. cit., pp. 216–217. Previale's PS is a conservative extension of what is formulated above, but he also treats $\in$ as defined and derives (H1) from other axioms. (H4*) suggests stronger principles of primitive recursion on hereditarily finite sets, which have indeed been discussed in Rödding, "Primitiv-rekursive Funktionen," and Kirby, "Finitary Set Theory."

To prove TC, apply (H4*) to '$\neg \mathbf{M}x \vee x$ has a transitive closure'. It is trivial if $\neg \mathbf{M}x$, and evidently $TC(\varnothing) = \varnothing$. If x has a transitive closure $TC(x)$, and y has one $TC(y)$, then $TC(x) \cup TC(y)$ is the transitive closure of $x \oplus y$; if y is not a set, $TC(x \oplus y) = TC(x) \cup \{y\}$. By (H4*), every set has a transitive closure.

Now we can derive $\in$-induction. Suppose (1): $(\forall x)[(\forall y \in x)A(y) \to A(x)]$. Following Previale, we write $x \leq y$ for $x \in TC(y) \vee x = y$. (1) clearly implies $A(x)$ if $\neg \mathbf{M}x$ or $x = \varnothing$, so that for these x, $(\forall z)(z \leq x \to A(z))$ holds. Writing this as $A^*(x)$, suppose now $A^*(x)$ and $A^*(y)$. If $z \leq x \oplus y$, then $z \in TC(x)$; or $z \in TC(y)$; or $\neg \mathbf{M}z$ and $z = x$ or $z = y$; or $z = x \oplus y$. In the first three cases, $A(z)$ follows by $A^*(x)$ and $A^*(y)$. In the fourth, if $z' \in x \oplus y$, $z' \in x$ or $z' = y$ by (H3), and in either case $A(z')$ by $A^*(x)$ and $A^*(y)$. But then we have $(\forall z \in x \oplus y)A(z)$, and so $A(x \oplus y)$ by (1). Combining the cases, we have $A^*(x \oplus y)$. By (H4*), $(\forall x)A^*(x)$, from which $(\forall x)A(x)$ follows.

[61] This could be concluded from what is stated in Givant and Tarski. To prove (H4*), we rely on the fact that PS⁻ proves the axioms of ZF other than Infinity and Foundation. But with $\in$-induction and the statement that every set is finite, we can prove that every set has a finite rank. Assign rank 0 to nonsets. Now assume the antecedent of (H4*); call its three conjuncts (i), (ii), and (iii). We can then prove by ordinary induction on n: (iv) $(\forall x)[rk(x) \leq n \to A(x)]$. (i) and (ii) imply that this holds for $n = 0$. Assuming it holds for n, let z be a set of rank $n + 1$. Since z is finite, there is a function f mapping some natural number m one-one onto z, and then there is a function g such that $g(0) = \varnothing$ and $g(k + 1) = g(k) \oplus f(k)$ if $k < m$. $A(g(0))$ holds by (ii). Assume $A(g(k))$ and $k < m$. Since $f(k) \in z$, $rk(f(k)) \leq n$, and so by induction hypothesis $A(f(k))$ holds. Then by (iii) $A(g(k) \oplus f(k))$, i.e. $A(g(k + 1))$, follows. Thus for all $k \leq m$, $A(g(k))$ holds, and since $g(m) = z$, we can conclude $A(z)$, i.e. (iv) holds for $n + 1$.

§38. Appendix 1: Theories of sets and sequences

It might therefore be desirable to replace (H4) by (H4*). However, the system PS⁻ based on (H1)–(H4) has a certain historical pedigree and expresses most directly the intuitions about finite sets developed in §33, as is indicated by the fact that (H4′) is adequate in the case where we consider only sets of individuals.[62]

The theory of sequences of individuals is formulated in §33. Again, we obtain a theory allowing sequences of sequences by using a one-sorted language and a predicate *Seq*. The sequence variables $p \cdots t$ are now understood as restricted to *Seq*.[63] The schema of primitive recursion takes the form:

(PR′) $\quad \varphi(x_1 \ldots x_n, \langle\rangle) = \psi(x_1 \ldots x_n)$
$\quad \quad \varphi(x_1 \ldots x_n, s^*z) = \chi[x_1 \ldots x_n, s, z, \varphi(x_1 \ldots x_n, s)]$.

In this case, the extension is conservative already in the one-sorted theory.[64] Because a sequence is built up in a unique order, the problem for primitive recursion mentioned above does not arise.

The predicate $\approx$ of §33 can be defined by primitive recursion with substitution in parameters. We set $pd(\langle\rangle) = \langle\rangle$ and $pd(s^*x) = s$; then:

(9) $\quad s \approx \langle\rangle \leftrightarrow s = \langle\rangle$

(10) $\quad s \approx t^*x \leftrightarrow s \neq \langle\rangle \land pd(s) \approx t$.

The models **M₁**–**M₃** establish Givant and Tarski's assertion that no two of the theories PS⁻, PS⁻ + Foundation, PS⁻ + TC, and PS are equivalent.

[62] Independently of this issue, it should be pointed out that theories of this kind without induction have been studied and shown to be mutually interpretable with Robinson arithmetic Q. That Q is interpretable in the theory with (H1)–(H3) is announced in Szmielew and Tarski, "Mutual Interpretability"; see also Tarski, Mostowski, and Robinson, *Undecidable Theories*, p. 34. A proof of the interpretability of Q in the set theory is given by Collins and Halpern, "Interpretability of Arithmetic." Another proof of the interpretability of Q in the set theory is given in Montagna and Mancini, "Minimal Predicative Set Theory"; in fact they do without the axiom of extensionality.

Neither Szmielew and Tarski nor Tarski, Mostowski and Robinson assert that the set theory is interpretable in Q (contrary to the statement of Montagna and Mancini, p. 187). A proof can be extracted from Nelson, *Predicative Arithmetic*, Chapters 10–12. It relies on the local interpretability of arithmetic with bounded induction (IΔ_0) in Q (cf. §50).

I am indebted to Allen Hazen for this observation and for most of the references in this note. Whether the last mentioned interpretability result was known earlier I do not know.

[63] The axioms stated do not settle what x^*y is when x is not a sequence. A natural convention would be to have $x^*y = \langle\rangle^*y$ in that case. Then (S1) holds without restriction, but (S2) requires the restriction of s and t to sequences. An alternative would be $x^*y = x$.

[64] In the one-sorted theory, the restriction to sequences is necessary for these equations to work as intended. However, we can accommodate nonsequence arguments by adding an additional clause, say $\varphi(x_1 \ldots x_n, z) = \zeta(x_1 \ldots x_n)$ if $\neg Seq(z)$.

(L1) and (L2) are then obvious. We can then derive (L3). Note first that by sequence induction we have

(11) $\quad s \neq \langle\rangle \to (\exists t)(\exists y)(s = t^*y)$.

In fact, t and y are unique. Suppose now $A(\langle\rangle, \langle\rangle)$ and

(12) $\quad (\forall s)(\forall t)(\forall x)(\forall y)[A(s, t) \to A(s^*x, t^*y)]$.

Then we clearly have $(\forall s)[s \approx \langle\rangle \to A(s, \langle\rangle)]$. Suppose now $(\forall s)[s \approx t \to A(s, t)]$. If $s \approx t^*y$, then by (10), $s \neq \langle\rangle$ and $pd(s) \approx t$. Let s' and x be the t and y given by (11). Then by (S2), $s' = pd(s)$, and hence $s = [pd(s)]^*x$. By (12), $A(s, t^*y)$. Hence $(\forall s)[s \approx t \to A(s, t^*y)]$, and by (S3), $(\forall s)(\forall t)[s \approx t \to A(s, t)]$ follows.

It is then immediate, by taking A in (L3) as $t \approx s$, that $\approx$ is symmetric, and it is also immediate that it is reflexive. We can also derive by $\approx$-induction (L3):

(L4) $\quad s \approx t \to [s = t = \langle\rangle \lor (\exists p)(\exists x)(\exists q)(\exists y)(s = p^*x \land t = q^*y \land p \approx q)]$

and then prove by sequence induction (S3) that $\approx$ is transitive.[65] Clearly we also have from (9), (10), and (11) (or from (L4)):

(L5) $\quad \neg(s^*x \approx \langle\rangle)$

(L6) $\quad s^*x \approx s^*y \to s \approx t$.

Concatenation of sequences is introduced by the following recursion. Abusing notation, I write it again as *.

$$s^* \langle\rangle = \langle\rangle$$
$$s^*(t^*x) = (s^*t)^*x.$$

The standard inductive proof that addition is associative can be used in this setting to show that concatenation is associative.

---

[65] We also have an induction principle for the cardinal equivalence relation $\sim$ similar to (L3):

$\{A(\varnothing, \varnothing) \land (\forall a)(\forall b)(\forall x)(\forall y)[(A(a, b) \land x \notin a \land y \notin b) \to A(a \oplus x, b \oplus y)]\}$
$\to (\forall a)(\forall b)[a \sim b \to A(a, b)]$,

which is similarly derived from the recursion for $\sim$ in §32. It can then be used to prove that $\sim$ is an equivalence relation, as in the text for $\approx$.

§39. Appendix 2: The language of hereditarily finite sets

We also have by sequence induction on t

$$r \approx s \to r*t \approx s*t$$

and by $\approx$-induction

$$s \approx t \to r*s \approx r*t$$

We cannot prove that $s*t = t*s$; that is in general false if the domain D has more than one element. We can, however, prove $s*t \approx t*s$, again by much the same proof as proves the commutativity of addition, using as a lemma

$$s*(t*x) \approx (s*y)*t.$$

Multiplication is defined by the obvious primitive recursion:

$$s \times \langle\rangle = \langle\rangle$$
$$s \times (t*x) = (s \times t)*s.$$

Note that $s \times t$ depends only on the length of t, as it is s concatenated with itself $lh(t)$ times. Multiplication can be proved distributive over addition (concatenation), associative, and commutative in the same manner as this is done in the usual development of arithmetic. Distributivity and associativity are strict identities; commutativity holds only in the sense $s \times t \approx t \times s$.

§39. Appendix 2: Relative substitutional semantics for the language of hereditarily finite sets

Let $\mathcal{L}_0$ be a first-order language, which is interpreted by means of a certain structure with domain of individuals D. Let $x, y, z, \ldots$ be the individual variables, and let $s, t, \ldots$ be individual terms (including variables). I assume a fixed enumeration of the individual variables.

We will add to $\mathcal{L}_0$ some apparatus for talking about hereditarily finite sets and give it a substitutional interpretation relative to D. A language $\mathcal{L}_1$ is described as follows: First we define *bracket terms* inductively as follows:

(a) {} is a bracket term of rank 0.

(b) If t is an individual term, then $\{t\}$ is a bracket term of rank 1.

(c) If T is a bracket term of rank n, then $\{T\}$ is a bracket term of rank n + 1.

(d)   If $\{Z\}$ is a bracket term of rank $n+1$,[66] and T is an individual term or a bracket term of rank $\leq n$, then $\{Z, T\}$ is a bracket term of rank $n+1$. Likewise if $\{Z\}$ is a bracket term of rank $\leq n+1$ and T a bracket term of rank n (or an individual term if $n = 0$).

(b)-(d) are equivalent to

(d')   if $T_1 \ldots T_m$ are individual terms or bracket terms, and the bracket terms are of maximum rank n, then $\{T_1 \ldots T_m\}$ is a bracket term of rank $n+1$ (1 if the sequence contains only individual terms).

Note that bracket terms may contain free individual variables, but no provision has been made for them to contain "set" variables.

We introduce substitutional variables X, Y, Z, whose intended substitution class is the bracket terms. An *open bracket term* is defined by (a)–(d), ignoring rank, with the additional clause:

(e) $X, Y, Z, \ldots$ are open bracket terms.

Atomic formulae of $\mathcal{L}_1$ are atomic formulae of $\mathcal{L}_0$, $t \in T$ where t is an individual term and T an open bracket term, and $S \in T$ and $S = T$ where S and T are open bracket terms. Formulae are composed in the obvious way by sentential connectives and quantifiers of both types.

Now we define satisfaction of a formula of $\mathcal{L}_1$ by a sequence of elements of D, where we consider only formulae of $\mathcal{L}_1$ without free substitutional variables. We will say that a sequence *covers* a formula A iff it assigns a value to every individual variable occurring free in A. We assume that for a formula of $\mathcal{L}_0$, s satisfies A only if s covers A.

(f)   If A is an atomic formula of $\mathcal{L}_0$, then s satisfies A iff s satisfies A in the sense of $\mathcal{L}_0$.

(g)   If A is $t \in T$, then s satisfies A iff s covers A, T is $\{t_1 \ldots t_m\}$ for $t_1 \ldots t_m$ individual terms, and s satisfies $t = t_i$ for some $i \leq m$.

(h)   If A is $S \in T$, where S and T are bracket terms of rank n and m respectively, then there are three cases:

(h_1) $n > m$. Then s does not satisfy A.

(h_2) T is $\{T'\}$, where T' is of rank n. Then s satisfies A iff s satisfies $S = T'$.

(h_3) T is $\{Z, T'\}$, where $\{Z\}$ is of rank $\leq n+1$, and T' is of rank $\leq n$. Then s satisfies A iff either s satisfies $S \in \{Z\}$ or T' is a bracket

---

[66] Clearly that will hold only if Z is a nonempty sequence.

§39. Appendix 2: The language of hereditarily finite sets

term and s satisfies $S = T'$. (The definition is the same for the two subcases of (d) above.)

(i) If A is $S = T$, where S and T are bracket terms of rank n and m respectively, then there are again three cases:

(i_1) $n \neq m$. Then s does not satisfy A.

(i_2) $n = m = 0$. s satisfies A iff S and T are both $\{\}$.

(i_3) $n = m \neq 0$. Then S is $\{U_1 \ldots U_k\}$ and T is $\{V_1 \ldots V_l\}$. Then s satisfies $S = T$ iff for every $i \leq k$ there is a $j \leq l$ such that s satisfies $U_i = V_j$, and for every $j \leq l$ there is an $i \leq k$ such that s satisfies $U_i = V_j$.[67]

(j) If A is $\neg B$, s satisfies A iff s covers B and s does not satisfy B

(k) If A is $B \wedge C$, then s satisfies A iff s satisfies B and s satisfies C.

(l) If A is $(\forall x)B$, then s satisfies A iff s covers A and for every $u \in D$, s_u^x satisfies B. (If x is the jth variable and $j > lh\, s$, we can assume that $(s_u^x)_i = u$ for all i, $lh\, s \leq i \leq j$.)

(m) If A is $(\forall X)B$, then s satisfies A iff for all bracket terms T, and all s' that cover $A\ X/T$ and agree with s on all variables free in A, s' satisfies $B\ X/T$.

Suppose now that we are adding only the language of pure hereditarily finite sets. Then individual terms play no role in the set-theoretic clauses of the definition. In the definition of bracket term, clause (b) is dropped, and clause (d) is replaced by

(d″) If $\{Z\}$ is a bracket term of rank $\leq n+1$, and T is a bracket term of rank $\leq n$ (and one of these inequalities is an equality), then $\{Z, T\}$ is a bracket term of rank $n+1$.

In the satisfaction definition, since bracket terms no longer contain individual variables or other individual terms, clause (g) is dropped and clause (m) is replaced by

(m′) If A is $(\forall X)B$, then s satisfies A iff for all bracket terms T, s satisfies $A\ X/T$.

---

[67] These identities may be identities of individual terms. Note that identities of the form $t = T$, where t is an individual term and T is a bracket term, are not well formed.

In §35 we noted that in a relative substitutional theory of predicative classes, or in the relative substitutional theory of hereditarily finite sets, the truth theory could legitimately be regarded as attributing an ontology of sequences to the object language, as is not the case for a truth theory of the usual sort for a first-order language. In the above definition only clause (m) differs from the usual first-order case in such a way as to encourage that interpretation. But in the simplified situation just considered, where only pure sets are admitted, the replacement of (m) by (m′) means that the definition gives us no reason to suppose that an ontology of sequences is being attributed to the object language.

It should be noted that if the language of hereditarily finite pure sets is interpreted on its own, not as added to a first-order language, then satisfaction plays no role, and we have a substitutional theory of truth of the usual sort.

7 Intuitive Arithmetic and Its Limits

§40. Arithmetic as about strings: Finitism

Although in the last chapter we concluded that natural numbers are not objects of intuition in the sense that was developed in Chapter 5, the system of strings discussed in §28 is still, in some sense, an intuitive *model* of arithmetic. First, it is a structure of the same similarity type as the natural numbers or the positive integers, consisting of an initial element and a unary operation. Second, it consists of objects of intuition in the sense that there is actual intuition of strings sufficiently early in the sequence and it is possible to draw some conclusions about an arbitrary string intuitively. Third, we can easily satisfy ourselves that it satisfies the Dedekind-Peano axioms. These observations raise the second of the two questions posed at the beginning of Chapter 6: How far does intuitive knowledge in arithmetic extend, when arithmetic is understood with reference to this model of strings? This question was effectively already discussed in the Hilbert school.

In Hilbert's papers on foundations in the 1920s and more fully in Hilbert and Bernays's *Grundlagen der Mathematik I* of 1934[1] the method appropriate for metamathematics, what Hilbert called the finitary method, is developed in terms of an interpretation of arithmetic where the objects are strings of just the sort we have been discussing. It is clear from HB I, p. 21, that the authors understand strings (*Ziffern*) to be the *objects* of the theory. But it is not essential that the interpretation be exactly this one;

---

[1] The reader is reminded that *Grundlagen* is cited as HB. Page references are to the second edition, although the passages cited occur in the first edition.

indeed, Hilbert and Bernays write

> In number theory we have an initial object and a process of continuation. Both must be fixed intuitively in a definite way. The particular manner of fixing is nonessential, but the choice once made must be held to for the whole theory. We choose the numeral 1 as initial thing, and appending a 1 as process of continuation. (ibid., pp. 20–21)[2]

We will assume this fixing is done in the manner described in §27. We will not address the question how much generality beyond that our discussion of intuitive knowledge will have. It seems clear that it will have some; on the other hand, a general notion of intuitive model, which would specify that generality, is not likely to be intuitive in the sense that concerns us.[3] The explanation of the finitary method in §2 of HB I contains arguments toward the conclusion that primitive recursive arithmetic, interpreted with respect to the model of strings, is intuitively known.

Of the three claims made at the beginning of this chapter, the first is obvious and the second was argued for in §28 and §29. It seems almost as obvious that the strings satisfy the Dedekind-Peano axioms. But the nature of this conviction is not very easy to analyze. If we accept the conclusion of §29 that the successor operation can be seen intuitively to be well-defined, then it is plausible that the elementary axioms (i.e., the (R1)–(R4) of §31) are intuitively known; this will be argued in the next section. If we understand the strings as what is obtained from | by iterated application of the operation of adjoining one more, then it should be as evident that induction holds for them as that it holds for any structure characterized in this particular way. That is all I wish to claim at this point; that this evidence is intuitive is not part of the claim, and, indeed, the second-order character of the principle of induction should lead us to think it is *not* intuitive. It is most relevant to our concerns to

---

[2] Translations from this work are my own. One might read this passage as saying that the sequence of strings of 1's is an intuitive *model* of arithmetic, clearly one among many possible such. The question would then arise what meaning the notion of model has in the conceptual framework of their metamathematics. It is an indication that in the course of explaining finitary mathematics Hilbert and Bernays engage in reflections that can be formulated mathematically but that are not finitary; cf. note 3.

[3] Concerning the related notion of finitism, about which more later, recent analysts have stressed that they were analyzing finitism from a nonfinitist point of view. It is very doubtful that one could give a real characterization from a finitist point of view, as opposed simply to pointing out that certain modes of reasoning are finitist and others are not. Similarly, one would expect that a characterization of the limits of intuitive knowledge would use mathematical knowledge that is not intuitive.

§40. Arithmetic as about strings: Finitism 237

consider a weaker arithmetic than the usual first-order arithmetic: primitive recursive arithmetic (PRA). This theory is quantifier-free, although its formulae have free variables with a generality interpretation. The formulae are equations of terms and combinations by the usual operators of propositional logic. Terms are composed from variables, 0, S, and functors introduced by what we might call the rule of primitive recursion. This allows us, given functors ψ, χ of n and $n+2$ arguments respectively ($n \geq 0$),[4] to introduce a new functor φ of $n+1$ arguments and to assume the equations

(PR) $\varphi(0, a_1 \ldots a_n) = \psi(a_1 \ldots a_n)$,

$\varphi(Sb, a_1 \ldots a_n) = \chi[b, \varphi(b, a_1 \ldots a_n), a_1 \ldots a_n]$.[5]

Logic consists of the obvious rules concerning identity, classical propositional logic (but see §42), and a rule for substituting terms for variables.

The aim of this chapter is to determine how far intuitive knowledge in arithmetic extends, when it is understood with reference to an intuitive model like that of strings. However, it is helpful for this inquiry, and also of interest in itself, to keep in mind issues about Hilbert's conception of the finitary method. Gödel remarked that "finitary mathematics is defined as that of intuitive evidence,"[6] and some such conception of the epistemological character of the finitary method did belong to the outlook of the

---

[4] This formulation presupposes that functors can be introduced by explicit definition, which can be reduced to particular schemata as in Kleene, *Introduction to Metamathematics*, pp. 219–221. Otherwise arbitrary terms composed from already introduced functors need to be allowed in (PR). I use the term 'functor' simply for singular terms with empty argument places; I do not wish to assume that they designate *functions*; compare the use of the word 'predicate', in particular in chapter 1, where it is not assumed that predicates designate properties or relations.

[5] The term "rule of primitive recursion" has its point, because in a formalism with function variables, we could formulate a single pair of axioms

$R(0, a, f) = a$
$R(Sb, a, f) = f[b, R(b, a, f)]$,

which will do the work of the rule and might be called the axiom of primitive recursion. Proof-theoretic work shows that the relation of the rule and axiom of primitive recursion is closely connected with that of the rule and axiom of induction.

[6] "Über eine bisher noch nicht benützte Erweiterung des finiten Standpunktes," p. 240, my translation. (I use 'evidence' to translate the German *Evidenz*, although a neologism like 'evidentness' might better avoid the misleading connotations of the English word.) In Gödel's 1972 English version of the paper, "On an Extension of Finitary Mathematics which has not yet been Used," what corresponds to the quotation in the text is the remark that "finitary mathematics is defined as the mathematics of *concrete intuition*" (p. 272, emphasis Gödel's).

Hilbert school, although it is not stated quite so directly in their writings. A certain kind of avoidance of reference to the infinite is also part of the method, and this might be motivated and considered independently of conceptions of intuition, as it is in some more recent writings on finitism (see later).

Gödel's remark can be unpacked as two theses:

(F1) If a proposition has been proved by the finitary method, then it is intuitively evident.

(F2) If a proposition is intuitively evident, it can be given a finitary proof.

It is useful to consider these theses in connection with two theses that together constitute a mathematical characterization of finitism:

(F3) Proofs in PRA are finitist; hence any theorem of PRA is finitistically provable.

(F4) If a proposition in the language of primitive recursive arithmetic is finitistically provable, then it is a theorem of PRA.

(F3) is clearly expressed in writings of Hilbert and Bernays; an analysis of finitism that did not yield it would be hard put to it to show that it was faithful to Hilbert's intentions. In discussions of the subject it has been noncontroversial. This is, however, partly a terminological matter, since arguments have been given against accepting primitive recursion in general, in some cases on grounds akin to those underlying the finitary method. Views that set the limits of mathematics (or of mathematics with a specific kind of methodological superiority) more narrowly than finitism are usually called "strict finitist," and I will follow that usage.

(F4), by contrast, is controversial as a claim about the finitary method as intended by Hilbert. It is favored by many of Hilbert's general remarks about the method. However, Richard Zach has pointed out that Ackermann's first consistency proof of 1924 uses a function enumerating the primitive recursive functions, which cannot be available in PRA.[7] Under the impact of Gödel's incompleteness theorem, Bernays fairly quickly became convinced that an extension of the finitary method was needed to prove the consistency of classical first-order arithmetic (PA).[8] It

---

[7] "The Practice of Finitism." Such a function can be proved total in PA, but the proof requires induction on a predicate with two nested quantifiers.

[8] I use this now usual designation for classical first-order arithmetic although it is not entirely happy.

§40. Arithmetic as about strings: Finitism

was not claimed that the proofs obtained in the 1930s by Gentzen and then Ackermann were finitist, and there has been general agreement that they are not. That still leaves a substantial intermediate area, where analyses of finitism have differed. Kreisel in 1958 proposed an analysis according to which finitism goes substantially beyond PRA and claimed that it is possible to codify the totality of finitist methods in a system such that one can give (in PRA) a proof of the consistency of PA relative to it.[9] By contrast, two different analyses offered by W. W. Tait in 1967 and 1981 yielded defenses of (F4).[10] Because Tait has offered the most developed defense of it and it has come to be identified with him, I will refer to (F4) as *Tait's Thesis*.

A mark of finitist argument is that it should not involve what Gödel calls "abstract concepts." It may not be entirely clear what makes a concept abstract. But higher order and intensional notions are evidently excluded. Thus Tait does not allow the finitist to use the general notion of function,[11] and Gödel mentions among abstract objects those of meaningful

---

I have not been able to determine exactly when Bernays came to this view. It is not expressed in HB, although that could well have been in deference to Hilbert, who says in his brief preface that Gödel's theorem only shows that for further consistency proofs "one must exploit the finitary method in a sharper way than is necessary in considering the more elementary formalisms" (p. vii).

The view that finitary methods are insufficient to prove the consistency of PA is expressed, with a very slight hesitation, in Bernays's well-known "Sur le platonisme dans les mathématiques," p. 68 (trans. pp. 270–271). (This paper was presented as a lecture in 1934.)

Bernays's slight hesitation in 1934 arises from the fact that he claims only empirical certainty for the claim that "elementary combinatorial" proofs can be formalized in a theory like PA. But he is somewhat more emphatic on this point in the immediately following lecture "Quelques points essentiels de la métamathématique," p. 88. Bernays, however goes on to discuss the characterization of the "finitary point of view." He mentions the fact that some had not distinguished this point of view from intuitionism, and evidently does not think the explanations given by Hilbert precise enough to exclude this. However,

> Nonetheless, in metamathematical proofs one has always confined oneself to a narrower framework, in view of the tendency natural to elementary evidence. One remained in the domain of reasonings that can be formalized without using bound variables. It is by this limitation that we fell into the difficulties mentioned. In fact, our thesis that one can formalize in $\mathfrak{N}$ [i.e., PA] every proof of an arithmetical theorem that conforms to the finitary point of view is only valid if this point of view is interpreted in the restricted sense. (p. 90, my translation)

Nothing he says is clearly incompatible with the view that the "restricted sense" still allows some recursions going beyond PRA, such as that whose use by Ackermann was noted by Zach.

[9] "Ordinal Logics and the Characterization of Informal Concepts of Proof."
[10] "Constructive Reasoning" and "Finitism."
[11] "Finitism," pp. 23–24.

statement and proof, where he has particularly in mind proofs as understood in the foundations of intuitionistic logic.[12] I will take for granted that some such idea is needed for a constructive understanding of the language of first-order logic. Hilbert's description of finitism allows only for a very limited use of quantification, so that quantified statements effectively cannot enter into logical combinations.[13] Because the analysis of finitism is not my concern, I will take for granted that the notions Gödel and Tait exclude from finitary mathematics are rightly excluded from it.

Gödel proceeds to argue that abstract notions are necessary in order to prove the consistency of PA and that propositions involving them cannot be made intuitively evident in the required sense. Then from (F1) it follows that this consistency cannot be proved finitistically. Gödel does not mention Tait's thesis, although he seemed to adhere to it in two lectures of the 1930s.[14] I will not comment on Gödel's argument in a more systematic way, although it is worth noting that, in its surface structure, it avoids appeal to a thesis like (F4) at the cost of using the presumably controversial epistemic notion of intuitive knowledge. Tait's analysis of finitism leads to the same conclusion without using this notion.

Focusing, as is my intent, on intuitive knowledge, the relation of (F1) and (F3) is an issue independent of those concerning Tait's thesis. Together they imply that proofs in PRA convey intuitive knowledge. This is rather explicitly argued in the writings of Hilbert and Bernays. One can combine them to obtain the following thesis:

> *Hilbert's Thesis.* A proof of a proposition according to the finitary method yields intuitive knowledge of that proposition. In particular, this is true of proofs in primitive recursive arithmetic.

---

[12] "Über eine bisher noch nicht benützte Erweiterung," pp. 240, 244.
[13] "Über das Unendliche," p. 173, trans. p. 378.
[14] "The Present Situation in the Foundations of Mathematics," pp. 51–52; "Vortrag bei Zilsel," p. 92. However, that this is so for the first of these passages has been challenged, with strong arguments, in Wilfried Sieg's introductory note to the correspondence between Gödel and Jacques Herbrand in the former's *Collected Works*, vol. V. In footnote 4 of "Über eine bisher noch nicht benützte Erweiterung" commenting on Kreisel's analysis, Gödel treats it as capturing an extended sense of finitism. That might indicate that he still held Tait's thesis for the proper sense of finitism. The footnote is modified substantially in "On an Extension of Finitary Mathematics" (see p. 274). Gödel there seems to prefer a later discussion of the matter by Kreisel in "Mathematical Logic," pp. 168–173, 177–178. To determine Gödel's views about finitism in the post-war period would require closer analysis of these writings as well as correspondence with Bernays in the period from 1965 to 1972 than we can undertake here. See Solomon Feferman's introductory note to that correspondence, Gödel, *Collected Works*, volume IV.

§40. Arithmetic as about strings: Finitism

Our discussion of intuitive knowledge in arithmetic will simultaneously be a consideration of arguments for Hilbert's Thesis. Some of the arguments will come directly from the writings of Hilbert and Bernays.

First, however, I wish to document the attribution to Hilbert of this thesis as well as (F1) and (F2), attributed to the Hilbert school by Gödel. The latter was, after all, not a member of the Hilbert school, and Tait and Kreisel, to whom I have also referred, are of a later generation than any of Hilbert's collaborators. But although Hilbert was no systematic philosopher, I don't think there can be much doubt that finitary mathematics, as he understood it, was to be based on intuition in such a way that it would be reasonable to characterize a finitist proof as yielding intuitive knowledge in some sense. Hilbert, however, stressed the role of intuition as providing the *objects* of finitary mathematics. Thus consider the following famous passage:

> As a condition for the use of logical inferences and the performance of logical operations, something must be already given to our faculty of representation, certain extralogical concrete objects that are intuitively present as immediate experience prior to all thought. If logical inference is to be reliable, it must be possible to survey these objects completely in all their parts, and the fact that they occur, that they differ from one another, and that they follow one another, or are concatenated, is immediately given intuitively, together with the objects, as something that neither can be reduced to anything else nor requires reduction.[15]

Hilbert's term 'concrete' (*konkret*) gives rise to confusion. Because he makes clear that he is primarily thinking of symbols of a formalized language and goes on to talk of them as types, his objects are abstract rather than concrete in the sense of §1, and as this distinction is made in Anglo-American discussions of nominalism, unless perhaps he intends the talk of types as a *façon de parler* to be reduced to talk of tokens.[16] Otherwise,

---

[15] "Über das Unendliche," p. 171, trans. van Heijenoort, p. 376. These remarks are repeated almost verbatim in several papers of Hilbert, going back to "Neubegründung der Mathematik (Erste Mitteilung)," p. 163. (This observation is made in Majer, "Geometrie und Erfahrung.")

[16] This is suggested in one place, "Neubegründung der Mathematik," p. 163 n. 1, where Hilbert says that he calls "signs of the same shape (*Gestalt*) 'the same sign' for short," thus intimating that being the same sign is thought of as an equivalence relation of tokens rather than strict identity. There is a question whether Hilbert could still maintain this in "Über das Unendliche," where he insists that nothing infinite is to be found in the physical world, in particular not a potentially infinite sequence of tokens. He could, however, hold some form of modal nominalism. Neither he nor the often more explicit Bernays addresses this issue.

the objects he calls concrete are quasi-concrete in the sense of §7, that is, it is intrinsic to them to have instantiations in the concrete in the narrower sense.[17] They might similarly be called quasi-spatiotemporal, but it would not be misleading to call them simply spatiotemporal, as one understands geometric figures to be spatial. In any case, Hilbert's term evidently has an additional meaning of something like "perceptual," as is shown by its association with intuitive givenness, for example, in the following brief characterization of the finitary method:

> What is characteristic of this methodological standpoint is that considerations are put forth in the form of *thought experiments* on objects that are assumed to be *concretely present* (*konkret vorliegend*).[18]

In finitary arithmetic, Hilbert understands the objects to be numerals, strings of occurrences of '1'. Thus, according to him, the objects of finitary arithmetic are objects of intuition. Exactly how Hilbert understood the term *Anschauung* is a large question, since he often uses it with a quite broad meaning, in keeping with the unphilosophical use of mathematicians.[19] Nonetheless, the concept of intuition underlying his account of finitism can be described more precisely, in particular so as to give a reasonably clear sense to the claim that finitary proofs yield intuitive knowledge. I will assume in the subsequent discussion that the latter notion is based on the conception of intuition laid out in §28. That seems to me reasonably faithful to Hilbert. The conception certainly has certain essentials in common with his: Its objects are quasi-concrete and spatiotemporal; its scope is a minimal generalization of that of ordinary perception so as to take in abstract objects such as sign configurations and geometric figures.

---

[17] In the corresponding place in "Neubegründung der Mathematik," Hilbert uses *diskret* instead of *konkret*. Clearly the objects he has in mind are discrete. But nothing obviously follows about what he means by *konkret*. Hilbert certainly had a conception of geometric intuition, of which the objects would presumably be continuous. Cf. Majer, "Geometrie und Erfahrung" and its revised English version "Geometry, Intuition, and Experience: From Kant to Husserl."

[18] HB I, p. 20, emphases in the text. Bernays, in a paper roughly contemporaneous with "Neubegründung," speaks of the domain of the "concrete-intuitive," within which foundational considerations are to be developed. See "Über Hilberts Gedanken zur Grundlegung der Arithmetik," p. 18. It may be significant that in the 1972 English version of Gödel's "Über eine bisher noch nicht benützte Erweiterung," *Anschauung* is rendered as "concrete intuition"; see, for example, the passage quoted in note 6.

[19] Consider, for example, Hilbert and S. Cohn-Vossen, *Anschauliche Geometrie*. Here intuition, being geometrical, is still not too far removed from Kant's conception, but the authors are not attempting to base an epistemological account of geometry on it.

§40. Arithmetic as about strings: Finitism

Likewise, we of course understand intuitive knowledge as in §29. In the case of singular propositional knowledge, it will typically be based on intuition of the objects the proposition is about. This is clearly in accord with the usage of the Hilbert school. Although I have not found the terms *anschauliche Erkenntnis* and *anschauliche Evidenz* in Hilbert's papers, they do occur in Bernays's writings of the time of his close collaboration with Hilbert.[20] The appeal to thought-experiments in the account of the finitary method in Hilbert and Bernays can be seen as an attempt to convey intuitive knowledge. In particular, in actual application it serves to convey intuitive knowledge that is general with respect to strings. The intuitive knowledge or evidence referred to by Bernays (see note 20) was clearly intended to comprehend general propositions about such objects as strings.

Our description of PRA implies that Hilbert's Thesis follows from the following subtheses (interpreted with respect to the intuitive model of strings):

(I1)   Successor can be seen intuitively to be well defined.

(I2)   The elementary successor axioms can be known intuitively.

(I3)   In each case of introduction of a function symbol by primitive recursion, if the assumed functions have been intuitively seen to be well defined, then this is so of the new function introduced, in such a way that the recursion equations are known intuitively.

(I4)   Logical inference preserves intuitive evidence.

(I5)   Inference by induction preserves intuitive evidence.

(I1) has effectively been argued for in §29. (I2) will be taken up in §41. (I4) will be the subject of §42. (I5) and (I3) will be addressed in §43 and §44 respectively. §45 will consider what conclusion we can draw about Hilbert's Thesis and the limits of intuitive knowledge in arithmetic, discussing along the way both the truth and the relevance of the claim that every string can be intuited.

---

[20] In "Über Hilberts Gedanken," p. 15, Bernays describes his inquiry as whether it is possible to give a foundation for the "transcendent assumptions" of mathematics by means of "primitive anschauliche Erkenntnisse." Bernays' most important philosophical statement before the impact of the incompleteness theorem is "Die Philosophie der Mathematik und die Hilbertsche Beweistheorie." Certainly an idea of intuitive knowledge underlies the discussion there of the finitary standpoint, which at one point is described as the "Standpunkt der anschaulichen Evidenz" (p. 40).

§41. The elementary axioms

In the usual formalizations of arithmetic, PRA included, it is not explicitly stated that 0 designates something or that Sn is defined for each value of n. Moreover, the variables are understood to range over numbers, so that the predicate 'x is a number' is not expressed in the formalism; if need be, it could be rendered as '$x = x$'. For our purposes, however, we do need an understanding of these terms that is not incompatible with the intuitive character of the simple propositions concerning them that the language of PRA does express.

Given that the initial string is an object of intuition there should be no difficulty in using a term designating it in a theory purporting to be intuitive knowledge. On the present interpretation '0' designates the initial string. If, as in some formalisms for talking about partial functions, we have a "predicate" that would be read "is defined," '0 is defined' surely expresses intuitive knowledge. Following a standard usage, we will write 't is defined' as $t\downarrow$.[21] However, this consideration applies not to PRA but to a possible object language that allows nondesignating singular terms.

In §29 I argued that we can see intuitively that any string can be extended by an additional stroke. If we accept that conclusion, it follows that given our intended interpretation of S, the terms that can be used in a theory consisting of intuitively known propositions are closed under S. We could express this by saying that if t is defined, then so is St. One might discern an existential quantifier in "can be extended" (as well as in the definedness predicate). I do not need to quarrel with this. My use of it is compatible with Hilbert's understanding of it: It can be cashed in by an explicit expression such as $s|$ for the immediate extension of a string s, and this in turn can be cashed in by intuition, and moreover I have not used it in logical combinations. In formalisms with a definedness predicate, it is envisaged that it will be so used, for example

---

[21] See for example Beeson, *Foundations of Constructive Mathematics*, pp. 97 ff. Such a predicate is analogous to the existence predicate in systems of free logic. (Cf. the discussion of E-logic and E^{+-}-logic in Troelstra and van Dalen, I, 50–52, and Feferman, "Definedness.") It is not our intention, however, to interpret statements involving a term t where $t\downarrow$ is false with reference to an outer domain of nonexistent objects. In fact, Beeson's logic of partial terms and developments of it assume that even equations of undefined terms are false. Such legislation is quite right for the intended application, but my own preference in free logic more generally, and its application to modal logic, is not to legislate in this way but rather to leave options open: Whether a predicate can be true of a "nonexistent object" is not something that *logic* should decide.

§41. The elementary axioms

in the statement that if $t\downarrow$ then $St\downarrow$; questions about this will arise later, in §44.

Our entitlement to use the expressions '0' and 'St' for given t presupposes not just existence but uniqueness. That this is intuitively known is part of what is being claimed when it is said that the intuition in question is intuition of *types*. '0' can designate a string on our intended interpretation only if it is of the same type as |. What we require is that this type be intuitively recognizable. As we saw in §28, this does not mean that there cannot be obstacles to determining of a given token that it is of that type. However, if we have before us a type of the same formal language, it should be recognizable whether or not it is | or some other type, and it seems clear from the explanation given there that it is; for example, in the latter case it will contain an additional occurrence of |.

Concerning the functionality of successor, there is the complication that it does depend on the fact that our language contains only a single symbol. If we consider a symbolism with just two symbols, say '—' in addition to '|', and again the successor of a string is the result of adding one more symbol on the right, then successor is no longer unique; for example the string '| | — |' will have the two successors '| | — | |' and '| | — | —'. But evidently in the one-symbol case there is just one possibility of adding another symbol (type) to a string.

Peano's first two axioms, and the introduction rules (R1) and (R2) of §31, have the additional content that 0 is a *number* and that the successor of a number is a number. In the present context, these simply express the way we are interpreting the language of arithmetic. As items of intuitive knowledge they do have a degenerate character: The role of intuition is simply to supply a denotation to the singular terms involved.[22] We shall encounter similar degeneracy in other cases. This is hardly a problem for Hilbert's Thesis.

We have at least this minimal role of intuition in the case of the other two elementary axioms codified in (R3) and (R4) in §31, that the successor of a number is never 0 and that successor is one-to-one. The first one says that the result of adding | to a string is never simply |. This is evident if one represents to oneself the process of adjoining another |. It could not be of the *type* of | because it would contain more than one occurrence of |.

---

[22] This seems to be one of the things that troubles James Page in "Parsons on Mathematical Intuition." See "On Some Difficulties," Section II.

In the present context the third axiom can be formulated as:

(A3) If strings x and y have the same successor, then $x = y$.

This is slightly more complex than it seems because it is a statement about types: It implies about tokens that if tokens a and b have successors of the same type, then a and b are themselves of the same type. Now, what is the "successor" of a string x but something obtained by attaching one more stroke to x? This "attaching" takes place in the space outside that occupied by x, so that if x differs from y, then there is no way in which one could, by adding a stroke to each of x and y, obtain identical results. If we think of sameness of type in terms of one-one correspondence, then clearly a correspondence between s and t can be extended if a | is added to each, by making these |s correspond to each other.[23] Unlike the uniqueness of successors, the uniqueness of predecessors is a general feature of strings and does not depend on the fact that our alphabet consists of a single symbol.

In all these cases, it might be questioned that the statements involved are intuitively known on the ground that they are really analytic and obtain by virtue of the concepts involved or by virtue of the meaning of such terms as 'stroke', 'string', and 'attach'. I have no investment in conceptions of analyticity, and I do not insist that this cannot be so in any of these cases. The most that would follow, however, is that the intuitive knowledge involved is degenerate in the sense mentioned earlier. Any sense in which it is analytic that | exists and that the operation of attaching an additional | is defined would be dependent either on intuitive givenness of the sort I have attempted to describe, if one understands them along the lines proposed in §28.

It is still clear that the successor operation is exhibited clearly enough in intuition so that we cannot deny that these elementary axioms are intuitive on the sort of grounds on which we would deny it for the general principle of induction and presumably for those applications of induction or recursion that involve abstract concepts, however we draw the limits of the intuitive.

The reader may legitimately feel that more should be said about equality of strings. The idea seems natural that in our simple case equality is

---

[23] One might see a difficulty in a case where with one of the strings, the additional | is already there but has not been counted as part of the string, and it happens to belong to the other string. But note that for a one-one mapping of A onto B to be extendible to one of $A \cup \{a\}$ onto $B \cup \{b\}$ by mapping a onto b, it is sufficient that $a \notin A$ and $b \notin B$. This clearly obtains in the present case; the worry arose because $a \in B$.

"generated" by the two principles that two strings each consisting of a single symbol are equal, and equal strings have equal successors. Applied directly to types, however, these principles are trivial and by themselves give rise only to self-identities. This ceases to be so, however, when they are combined with recursion equations for additional operations.

In spite of the general inadequacy of the nominalist view, two corresponding principles about the notion of being of the same type are informative and describe a canonical procedure for determining sameness of type, modulo the problem of vagueness discussed in §28. They can be stated as follows:

(1) If a and b are each single stroke-tokens, then a and b are of the same type.

(2) If a and b are of the same type, and c results from a by attaching a stroke token on the right, and d results from b by attaching a stroke token on the right, then c and d are of the same type.

§42. Logic and intuition

We now turn to subthesis (I4), that logical inference preserves intuitive knowledge. But note that the logic allowed in PRA is limited to elementary reasoning about identity and classical propositional logic.

As we indicated in §24 and §29, one could stipulate that intuitive evidence should not depend on reasoning or inference at all. In fact, we do not use the word 'intuition' in a way that would violate this stipulation, but in talking of intuitive knowledge or intuitive evidence we do. Clearly, that was the intent of Hilbert and other theorists of mathematics, and such a stipulation would make the notion of intuitive evidence inapplicable to any whole mathematical theory that includes the results of proof. But then some basic forms of reasoning are going to have to be allowed as preserving intuitive evidence, and some will count as logical. The general principle should be that what is allowed is to be absolutely basic to reasoning in general, or to reasoning about objects or about the particular kind of objects with which we are concerned.

A principle that should be considered is that if an inference from, say, premisses A and B to a conclusion C is to preserve intuitive evidence, then the inference should itself be a matter of intuitive insight or be built up from a sequence of inferences each of which is. We might call this principle "Descartes' principle," as in the *Rules for the Direction of the Mind* he demands of deduction that the elementary inferences should be intuitions. One might put this by saying that 'if A and B, then C' must be intuitively known. However, that is not right, because if one has that knowledge, and the knowledge that A and that B, the step to C is still an inference, which one would naturally describe as a conjunction-introduction to attain 'A and B' and then modus ponens to attain C. So, in order to avoid the famous Lewis Carroll regress, one cannot escape the problem of the status of inferences, distinguished from that of propositions, with respect to intuitive knowledge. Descartes' principle is probably counterintuitive in general, but it is particularly so when interpreted in terms of the present conception of intuition and applied to logical inference. One reason is surely the generality of logic, which extends to propositions that involve reference to objects that are not objects of intuition and concepts that are not in any reasonable sense intuitive.[24]

Logical inference has the character that, generally, no new intuition of objects is needed to attain the conclusion that was not already needed to attain the premises. There may be exceptions, such as disjunction-introductions where the new disjunct contains reference to objects not referred to in the premiss. But in these cases, the intuitions would be at most presuppositions of the use of certain parts of the conclusion. It would be reasonable to say that intuitive knowledge is *not* preserved if a new disjunct is introduced that is in some way not intuitive in its content.

No formulation of PRA can do without simple rules concerning identity: instances of $a = a$ and the substitutivity of identity. These would be reasonably understood as minimal inferences for reasoning about objects, at least mathematical objects. In particular, the most elementary computations proceed by these rules. Given that the intuitive knowledge in question rests on intuition of objects, if we allow any extension at all of intuitive knowledge by reasoning, such basic reasoning concerning identity would have to be allowed. In fact, however, their consequences are quite trivial until they are combined with induction and recursion.

Already this admission has the consequence that there will be degenerate cases of intuitive knowledge, for example, statements of the form $t = t$, where t designates an intuited object, and the general statement $a = a$.

---

[24] We have not tried to introduce or explain a notion of intuitive concept, but if one is clear about what is an abstract concept in the context of Gödel's discussion of finitism, then an intuitive concept could simply be one that is not abstract. It is still not clear what would be the place of logical notions in such a distinction. We would not want to say that identity, the conditional, conjunction, and disjunction are abstract, so long as they are applied to atomic sentences that do not contain abstract concepts. By contrast, quantification does appear to take us outside the realm of the intuitive.

§42. Logic and intuition

Logic does not rest on intuition in the way that perceptual judgments rest on perception. In both these cases, intuition plays a certain role in seeing them to be true: in the former because it is intuition that gives the object that is said to be self-identical,[25] in the latter because intuition is involved in understanding the range of the variable and thus in its coming to have a definite sense. (As a pure law of logic, it is schematic.) Still, it would not be right to say that it is by intuition that we know these statements, since they are applications of more general rules that are not restricted in their application to objects of intuition.

Why should the use of propositional logic be held to preserve intuitive knowledge? The question now arises how we understand the connectives. Hilbert and Bernays thought of them as truth-functional. The facts they relied on are the following: Any atomic formula of PRA has the property that for any assignment of numerals to its free variables, it can be decided by computation. (This clearly depends on the computability of primitive recursive terms.) But then for a compound formula, a truth-value can, again, be computed for any assignment of numerals to the variables. Hence they described true generalizations in this language as "verifiable."[26]

This computability of the logical connectives when used with the language of PRA is certainly hepful in convincing oneself that PRA accords with general demands of constructivism. In fact, there is a stronger result than this brief exposition suggests: PRA can be formulated as an equation calculus, in which the only logic is reasoning about identity of the sort discussed earlier.[27] (Such formalisms often are called "logic-free," but for our present purposes identity counts as a logical notion.). Then propositional connectives obeying the classical rules can be introduced by contextual definition. The use of primitive recursion to this end is minimal: The only recursions that are not essentially definitions by cases are those for addition and subtraction. In §44, I argue that one can see intuitively that these are well defined. Thus the reduction of classical propositional logic to the

---

[25] I leave aside the question whether $t = t$ is to be allowed as true in cases where t does not designate anything. No such case arises in the language of PRA, and in the logic of partial terms with the definedness predicate, only $t{\downarrow} \to t = t$ is assumed.

[26] The concept is introduced in HB I 237. Note that in the case of formulae with free variables but without quantifiers, which is the case that concerns us, a formula is said to be verifiable if "it can be shown" that every result of replacing the variables with numerals is a true formula. It seems likely that the authors had in mind that this could be shown by a finitistic proof.

[27] See, for example, Goodstein, *Recursive Number Theory*.

equation calculus is not affected by the doubts expressed there about the intuitive character of primitive recursion in general.

Nonetheless, one has a more natural formulation of PRA if one takes the connectives, or at least some of them, as primitive. Then the reliance of Hilbert and Bernays on their computability introduces the mathematical "can" at a point where we wish to avoid it. It may seem that it is unavoidable, for how can we justify the use of classical logic other than by the observation that the truth-functions are computable? We can, however, formulate PRA with positive propositional logic (i.e., the negation-free fragment of intuitionistic propositional logic), derive intuitionistic logic using the definition of $\neg A$ as $A \to 0 = 1$, and then prove any instance of the law of the excluded middle. (See Appendix.) The use of primitive recursion in this derivation is again minimal, essentially the same as is used in reducing propositional logic to the equation calculus. One might still not be entirely satisfied with it because of the use of the definition of $\neg A$; for example, it leaves the statement that 0 is not equal to 1 trivial. We can, however, think of $\neg A \leftrightarrow (A \to 0 = 1)$ as an axiom schema.[28]

There would still be questions about the understanding of the connectives in this logic; the usual intuitionistic explanation fits neither with finitism nor with our own concern with intuitive knowledge.[29] But it should be clear that what is assumed, the basic logic of identity and minimal propositional logic, is minimal for reasoning about objects. It doesn't have the implications that quantificational logic, particularly classical, was found to have in application to an infinite domain of objects. The reduction of this application of logic to an equation calculus will show that it does not go beyond the intuitive if we accept the conclusion that this is true of the use of identity, 0, S, and the minimal amount of recursion involved. Possibly some other understanding of the connectives, perhaps starting from the intuitionistic explanation and reducing the notion of proof involved to consider only "intuitive" proofs, would also be satisfactory. At all events, the case that this weak logic takes us beyond the intuitive has not, so far as I know, been made.

---

[28] Note, however, that if instead $\neg(0 = 1)$ is an axiom, the left-right direction of this biconditional is an instance of the characteristically intuitionistic schema $\neg A \to (A \to B)$, although the right-left direction is a simple *modus tollens*.

[29] It is worth noting, as Grigori Mints pointed out to me, that several writers of finitist inclinations preferred the formulation of PRA as an equation calculus. See Goodstein, *Recursive Number Theory*, Skolem, "Begründung der elementaren Arithmetik," and Lambek and Scott.

§42. Logic and intuition

One might then ask about bounded quantification. Bounded quantification can be defined in PRA; for example, as for any formula A there is a term t_A such that $A \leftrightarrow t_A = 0$ is provable, evidently $\forall x < a\, A(x)$ is equivalent to $\Sigma_{x<a}\, t_{A(x)} = 0$. Given the bounded least number operator, also definable in PRA, one could also define $\forall x < a\, A(x)$ as $\mu x < a\, \neg A(x) = a$.[30] Whether either iterated summation or the bounded least number operator is intuitive is a somewhat more delicate question, which I will defer to §44.

By contrast, Hilbert's remarks about quantification in finitist mathematics would suggest that already reasoning in which unbounded quantifications enter into logical combinations, even within the limits of intuitionistic logic, does not preserve intuitive knowledge. Classically, in using quantifiers over a domain, we assume we have a definite enough conception of the domain so that quantification applied to predicates that have a definite truth-value for each element of the domain will yield predicates or sentences that themselves have a definite truth value (possibly for given values of free variables). In the case of the natural numbers, Hilbert and most of his contemporaries understood that to amount to presupposing the natural numbers as an infinite totality, something that would be incompatible with finitism. Apart from that, there is no reason to think that any insight we might have to that effect is intuitive in the sense that concerns us now. The usual constructive understanding assumes for the truth of $\forall x \in N A(x)$ a construction or proof that, for each particular number n, yields a proof of $A(n)$, but as the constructivist tradition has made clear, there is no reason to assume that either there is such a construction or there is a counterexample.

This point affects only classical logic, and Hilbert's claim has been understood to imply that even the intuitionistic use of unbounded quantifiers would take us beyond the intuitive.[31] We have no reason to believe that proofs in the sense of the standard interpretation of intuitionistic logic (what is called the BHK interpretation[32]) are objects of intuition.

---

[30] I assume, with Kleene, *Introduction to Metamathematics*, p. 225, that $\mu x < a\, A(x) = a$ if there is no $x < a$ such that $A(x)$.

[31] In point of fact it was so understood with a certain wisdom of hindsight, since, as is well known, the distinction between intuitionistic and finitist methods was not clear to researchers in foundations before the early 1930s, in particular before Gödel's incompleteness theorems and his relative consistency proof of classical to intuitionistic first-order arithmetic in "Zur intuitionistischen Arithmetik und Zahlentheorie." Gödel was perfectly clear about the distinction by 1933 ("The present situation"), Bernays by 1934 ("Sur le platonisme").

[32] After Brouwer, Heyting, and Kolmogorov. See Troelstra and van Dalen, I, 9–10.

Some will have to contain higher-type operations. Consider, for example, a statement of the form

$$\forall x \exists y Axy \to \forall u \exists v Buv.$$

A proof of the antecedent will yield a function giving y in terms of x; a proof of the statement will therefore have to yield, *given* such a function, a function giving v in terms of u. Nothing that has been said in our earlier discussions, or in writings in the tradition of the Hilbert school, gives any reason to think that a proposition of that form is intuitive enough in its content so that reasoning concerning it can yield intuitive knowledge.

It does not follow, however, that no alternative understanding of the language of first-order arithmetic will yield such a result. Indeed, there are mathematical theories with full classical logic that have been shown by proof-theoretic arguments to be conservative extensions of PRA; it was in effect already shown by the work of the Hilbert school that this is true for the result of adding classical logic to PRA, preserving the restriction that induction is to be applied only to quantifier-free formulae.[33] More recently, this has been shown for a second-order theory WKL_0 that is capable of proving a significant part of the elements of analysis.[34] If Hilbert's Thesis could be shown to be true, such proofs could be used to obtain interpretations of theorems of these systems that would show them to be intuitively known if so interpreted. Although there is no reason to believe that such interpretations cannot in principle be obtained, their concrete shape remains to be delineated. They would not, however, be the straightforward naive interpretation of the statements of the theory.

§43. Induction

Induction and recursion will naturally be seen as the crux of the issue surrounding Hilbert's thesis, since it is they that cash in our conception of generalization about the natural numbers, and we do not have nontrivial

---

[33] One can allow it for formulae containing bounded quantifiers or even Σ_1^0 formulae, i.e., those obtained by beginning with quantifier-free formulae and applying bounded quantifiers and possibly unbounded *existential* quantifiers.

[34] See Simpson, "Reverse Mathematics" and *Subsystems*, ch. IV. Ch. VIII §3 of the latter gives a model-theoretic proof of the conservativeness result, originally proved by Harvey Friedman. A proof-theoretic proof is given in Sieg, "Fragments of Arithmetic."

§43. Induction

arithmetic without that. Earlier,[35] I made the suggestion that induction should be regarded as conservative of intuitive evidence because of its analogy with a logical principle. One can see this analogy from the way formal arithmetic was characterized in §31, with (R1) and (R2) as introduction rules for the number predicate and the rule of induction (R5) as elimination rule. The formal analogy of induction to a logical principle is very strong, and it is exploited in proof-theoretic studies of systems containing arithmetic. Is it clear, however, that it is more than a formal analogy?

We cannot make for induction the claim we made for logical reasoning in the previous section, that it is basic for reasoning in general, or for reasoning about objects. It has rather the character of being basic to reasoning about objects in a particular domain, the objects obtained from an initial element by arbitrary finite iteration of a given operation. In the case whose intuitive character is at issue, Hilbert's model of strings, the initial element is an object of intuition, and it is intuitively known that the operation (adding a stroke to a string) is well defined. If we admit some conception of the domain of strings as intuitive, it seems we ought to admit induction as preserving intuitive knowledge.

I say "some conception," because we can certainly characterize the domain in a way that would have no claim to be intuitive, for example by a Fregean definition using full second-order logic. Even the idea that induction should hold for *any* predicate that we can precisely understand introduces what the participants in the discussion of finitism and intuitive evidence would have regarded as abstract concepts. What is claimed by the finitist is that instances of induction, provided that their content does not involve any nonfinitist notions, become evident one at a time. This is a natural way to reconstruct Hilbert's position, and it is at least implicit in the discussion of Tait.[36] This is in keeping with (I5): applications of induction *preserve* intuitive evidence; there is no claim to the effect that a principle of induction is intuitively evident.

It follows that it would not be right to say that we understand clearly what it is to iterate a successor operation a finite number of times and *then* see that induction is true of the objects resulting from such iteration beginning with 0.

---

[35] "Intuition in Constructive Mathematics."
[36] "Finitism," esp. section IV. The reader should not deduce from this convergence that Tait would approve of my own use of the notion of intuitive knowledge.

§44. Primitive recursion

The discussion so far has made the case for Hilbert's thesis rest to a large extent on recursion, that is, on subthesis (I3). For even if we grant that inference by induction preserves intuitive evidence, this will not lead us very far in arithmetic unless we have some of the function symbols that in PRA are introduced by primitive recursion. What we have called the rule of primitive recursion allows us, given functors ψ, χ of n and $n+2$ arguments respectively ($n \geq 0$), to introduce a new functor φ of $n+1$ arguments and to assume the equations

$$\varphi(0, a_1 \ldots a_n) = \psi(a_1 \ldots a_n),$$
$$\varphi(Sb, a_1 \ldots a_n) = \chi[b, \varphi(b, a_1 \ldots a_n), a_1 \ldots a_n].^{37}$$

Let us consider first a degenerate case, where χ does not depend on its second argument, and the second equation thus takes the form

$$\varphi(Sb, a_1 \ldots a_n) = \chi[b, a_1 \ldots a_n].^{38}$$

To convince ourselves that φ is well defined, we only need to convince ourselves that every number is either 0 or of the form Sb. In the case of the intuitive model that concerns us, this is immediate from the way in which it has been explained. Every "number" is either '1' or has been obtained from it by a succession of adding a '1' on the right, and that is the successor operation. We don't have to understand "a succession of" such operations to see that every number is either the initial one or a successor.

This trivial case, however, leaves out precisely the feature of primitive recursion that gives it its power, that it enables us to iterate a given function and cashes in the idea that the numbers are obtained by arbitrary iteration of the successor operation. The first nontrivial case is that of addition, which is equivalent to iteration of the successor operation beginning with an arbitrary a.

At this point, let us return to Hilbert and Bernays. They do give what would be intuitive arguments for the well-defined character of primitive recursive functions, both in general and in particular cases such as addition. Let us first consider what they say about addition and multiplication.

---

[37] ψ and χ may of course be explicitly defined from functors introduced earlier; cf. note 4. On the term 'rule of primitive recursion', see note 5.

[38] This "degenerate" recursion schema is reducible to a single case: If we set $\xi(0, a, b) = a$ and $\xi(Sc, a, b) = b$, then we can define the functor φ of the text explicitly as $\xi[b, \psi(a_1 \ldots a_n), \chi(b, a_1 \ldots a_n)]$.

§44. Primitive recursion

I will shift ground in one respect, and, following them, treat the sequence of positive integers beginning with 1 rather than the non-negative integers beginning with 0; that is more natural for an interpretation in their intuitive model, since then '1' designates itself.

Then addition reduces simply to concatenation of strings, and one can hardly doubt that it is intuitive in the required sense. The basis recursion equation is replaced by the identity $a + 1 = a + 1$, so that the only recursion axiom that is needed is

$$a + (b + 1) = (a + b) + 1,$$

and that is certainly intuitively obvious if we think of addition as concatenation of strings considered as spatial configurations.

Concatenation does not take place by adding on to the string a one term of b at a time; rather, one can see that $a + b$ is defined without relying on the buildup of b by a sequence of steps. I think it is for this reason that Hilbert and Bernays claim that the general associative law is "immediate from the definition of addition" (p. 23); in particular, we do not need induction to see its truth.[39] That reinforces the interpretation of them as thinking of addition in the way I am proposing, although their own explanation (p. 22) is a little obscure. It will be less clear that addition is intuitive if we explain $a + b$ as what is got by iterating the successor operation beginning with a instead of 1. That we understand an operation that effectively does that is a conclusion.

On this view, addition is a special case, in that iteration is not directly involved in understanding it. Therefore, its intuitive character does not show that the same will obtain for primitive recursion in general. But let us go one step further and consider the case of multiplication. Hilbert and Bernays again give a special explanation, which appeals to a kind of replacement:

> Multiplication can be defined in the following way: $a \cdot b$ designates the numeral that one obtains from the numeral b by, in its construction, always replacing 1 by the numeral a. (p. 24)

This is a persuasive picture, but it amounts to saying that the same procedure of generating strings can be carried out with the role of '1' replaced by that of a. It may be most persuasive, however, if we suppose that every string can be intuited, an assumption that I have wanted to avoid.

---

[39] In this respect, they contrast the associative with the commutative law.

This assumption does not seem to me essential. What is really involved is that we imagine an arbitrary string replacing '1' in the construction of b; because it is a bounded spatial configuration, the case is just a more complex version of what is involved in seeing the truth of (I1).

Even if you are not persuaded by this, I ask you to grant for the sake of argument that multiplication is intuitively well defined. Unlike the explanation we gave of addition, this is an entering wedge for the idea that iterating an intuitively well-defined operation will yield an intuitively well-defined one. But the situation is rather special, because what happens at each stage is the same, namely, the insertion (in effect construction) of a. In particular, there is a bound, given in advance, on the complexity of what is added to the construction at each stage.

These considerations don't easily generalize, and it is particularly with respect to exponentiation that questions have been raised. Already in 1934 Bernays questioned whether the evidence that $67^{257^{729}}$ exists is intuitive.[40] Subsequent writers have expressed doubts about exponentiation on strict finitist or other grounds.[41] These doubts are reinforced by the acceptance of polynomial-time computability as a criterion for feasible computability, so that, in general, computation of the exponential function is not feasible.

Although I will return to the case of exponentiation, it is not clear that we can deal with it by considerations specific to the case at hand, as we did with addition and multiplication. Let us return to the general case. To make Hilbert's thesis plausible, we need to convince ourselves that primitive recursion introducing a new functor yields intuitively evident equations, provided that the defining equations associated with the functors on the right-hand side of the equations are intuitively evident. That is to say that the thesis of the conservativeness of intuitive evidence extends from induction to recursion. But this is a larger step than one might at first think, just because primitive recursion introduces a new function symbol, one might say: a new concept. The analogy between induction and a logical principle does not seem to be of any help at this point.

---

[40] "Sur le platonisme," p. 61, trans. p. 265.

[41] Bernays was doubting not exponentiation but the intuitive character of the evidence that it is well-defined. But cf. van Dantzig, "Is $10^{10^{10}}$ a Finite Number?" For more recent instances, see Nelson, *Predicative Arithmetic*, and Isles, "What Evidence is there that 2^{65536} is a Natural Number?"

§44. Primitive recursion

Hilbert and Bernays do claim that primitive recursion is finitistically acceptable and offer an argument to this effect (pp. 25–27). The core of it goes as follows: Suppose (taking the case without parameters) a functor φ has been introduced by the equations

$$\varphi(1) = a$$
$$\varphi(n+1) = \psi(\varphi(n), n).$$

Then, for a given string m, one can by a sequence of m steps obtain an outcome which is either a or a term of the form

$$\psi[\psi(\ldots\psi(a, 1), \ldots, m-2), m-1]$$

which no longer contains φ.

> With this we have reached a computable expression, for ψ is to be an already known function. One has to carry out this computation from inside out, and the numeral obtained as its output is to be assigned to the numeral m. (p. 26)

But this seems to be an argument by Σ_1-induction: A computing procedure is described, and it is shown that it terminates. The key step is to show that in computing $\varphi(m)$ for a given numeral m, one can by a succession of steps replace it by a term that does not contain φ. As a proof, it can be structured in different ways; the one that seems closest to the surface of the text would argue that if $m = 1$, this replacement is trivial, and if it can be carried out for m, then it can be carried out for $m + 1$. Once one has the φ-free term, moreover, there is still the task of computing it, and the depth of nesting of ψ in it is precisely m: Another induction seems involved in seeing that its computation terminates.

One also might see the matter the following way: The recursion equations describe a procedure for computing φ. For $m = 1$, the procedure simply outputs a. For $m + 1$, the procedure is to compute $\varphi(m)$, with result say k, and then compute $\psi(k)$. We can easily prove that the computation terminates, by induction using the predicate 'the computation of $\varphi(m)$ terminates', which is essentially '$\varphi(m)$ is defined', ('$\varphi(m){\downarrow}$') or '$\exists k[\varphi(m) = k]$'. To avoid making these statements trivially true, the language has to allow nondesignating singular terms.[42] One can see the considerations as a description of how, parallel to the construction of a string or other numeral m, one constructs the value of $\varphi(m)$. But we haven't been given a noncircular way of showing that the construction must work, unless we give

---

[42] Cf. §41, in particular note 21.

ourselves something equivalent to Σ_1-induction. Only the argument itself shows that the use of definedness or the existential quantifier satisfies Hilbert's condition that existential quantification be cashable by explicitly giving the instance making the statement true. So our problem has not been advanced.[43]

It is worth pointing out, however, that in "Die Philosophie der Mathematik und die Hilbertsche Beweistheorie" (see note 13), Bernays gives the same "replacement" account of multiplication and attempts to generalize it to exponentiation, in a context in which he points out that exponentiation quickly gives rise to numbers whose representation by numerals of the Hilbertian canonical form is not practically possible:

> Here also there are bounds for the executability of repetitions both in the sense of actual representability and in the sense of physical realization. Consider the example of the number $10^{10^{10}}$. We can reach this number in a finitary way as follows: We start from the number 10, which, in accordance with one of the normalizations given earlier we represent by the expression
>
> 1111111111.
>
> Now let z be any number which is represented by a corresponding expression. If we replace in the above expression each 1 by the expression z, then again a number expression arises, as we can make clear intuitively, which for communication we designate with "$10 \times z$". Thus we obtain the process of multiplying a number by 10. From this we obtain the process of passing from a number a to the number 10^a, in that we let the number 10 correspond to the first 1 in a and, to each attached 1, *the process of multiplying by 10*, and continue until we are at the end of the expression a. The number obtained by the final process of multiplying by 10 is designated by 10^a.
>
> This procedure offers no difficulty from the intuitive point of view.[44]

In the case of exponentiation, each 1 in the construction of a is replaced by a "process." It is hard to see how this description can do what Bernays apparently wants, since processes are not clearly objects of intuition. Possibly, he means that what the 1 is to be replaced by is *the result of multiplying by 10* what one obtained at the previous stage. But then the argument is again either just a description of a computing procedure or an

---

[43] These formulations were suggested to me by Tait, "Finitism," §4. It's not clear to me whether Tait would disagree with my comments on it, apart from his skepticism about the notion of intuitive evidence.

[44] "Die Philosophie der Mathematik und die Hilbertsche Beweistheorie," pp. 38–39, my translation, emphasis added.

§44. Primitive recursion 259

argument involving Σ_1-induction, since without an inductive hypothesis one does not have such a "result."

There is really nothing specific to exponentiation in Bernays's argument; one could replace multiplication by any function that has been seen to be well defined and argue that this will obtain for its iteration. It seems likely that Bernays came to think his 1930 argument unclear; possibly, he saw the general argument for primitive recursion in Hilbert and Bernays as replacing it.

I conclude that neither Hilbert nor Bernays offers a noncircular argument for (I3) and therefore that they do not offer a convincing argument for Hilbert's thesis. It seems to me very doubtful that this obstacle can be overcome. If it could, by an argument of the kind given by Hilbert and Bernays, one would expect the argument to be convertible into a proof that, say, exponentiation is well defined, using hypotheses essentially weaker: unproblematic primitive recursive functions and induction. Unless one allows unbounded Σ_1-induction, one will not get the result: If the functions that one allows are all polynomially bounded, and only Δ_0 induction is allowed, then one will not be able to prove that exponentiation is everywhere defined.

Before we assess where we stand, we must consider a loose end from §42. There we pointed out that one might argue for the claim that, in the language of PRA, classical propositional logic preserves intuitive knowledge by formulating PRA as an equation calculus and introducing the connectives by contextual definition. However, this definition uses addition and subtraction to reduce equations $s = t$ to the form $\phi(s, t) = 0$. The same device is needed if we begin with minimal logic and undertake to prove instances of the law of excluded middle; see the Appendix. That addition can be seen intuitively to be well defined was argued earlier. The case of subtraction is more like that of multiplication, in that, in order to obtain $a - b$, it seems we cannot avoid proceeding step by step, for example, dropping strokes from the end of a as we add them in the course of generating b. However, the two procedures are so closely parallel that one can hardly deny that if a is given and b has been generated then so has $a - b$.[45] Certainly, if we accept multiplication as intuitively well defined, then we must accept subtraction. But the case of subtraction is easier, involving

---

[45] If all the strokes of a have been removed before the generation of b is complete (i.e., if $a < b$), then we set $a \dotminus b = $ the initial element. So we obtain what is sometimes called "truncated subtraction" $a \dotminus b$. If we set $pr(0) = 0, pr(Sn) = n$, then $x \dotminus 0 = x$ and $x \dotminus Sn = pr(x - n)$. The notation is from Kleene, *Introduction to Metamathematics*, p. 223.

as it does the dismantling of a string in parallel to the construction of another.[46]

§45. The limits of intuitive knowledge

The discussion of primitive recursion produces a sort of stand-off, familiar to those of us who try to understand the justification of mathematical induction. If one accepts the rule of primitive recursion as preserving intuitive knowledge, it is not clear that this commits us to accepting as intuitively known anything that can be seen not to be such on other grounds, unless one demands that the objects involved be practically intuitable. The iterability of procedures that we can see to lead to a result is, plausibly, part of our notion of numbers as the result of iterating the successor operation beginning with 0, or of Hilbert's strings as the result of iterating the putting on of an additional '1' beginning with '1'.

By contrast, the conclusion that even exponentiation must be regarded as everywhere defined in a genuinely intuitive arithmetic does not seem to be forced on us. It seems open to us to hold that what is intuitively known in arithmetic is limited to a fragment of primitive recursive arithmetic that does not include recursions for any function that grows as fast as exponentiation. The arguments to the contrary canvassed in the previous section turned out to be circular.

On balance, I am inclined to the negative conclusion, that the arithmetic that is intuitively known does not include exponentiation. This conclusion is encouraged by the fact that exponentiation is not feasibly computable, as that notion is understood by most writers on the subject; in particular, it is not polynomial-time computable. Exponentiation is also not predicative according to the criterion of Nelson's *Predicative Arithmetic*, which will be discussed in Chapter 8. Another more plausible characterization of what is there called "strict predicativity" does allow exponentiation as strictly predicative, but then it is a serious question whether iterated exponentiation is such and whether what is strictly predicative allows more than Kalmár elementary functions. Thus it seems that in any case acceptance of Hilbert's Thesis would imply that intuitive knowledge in arithmetic goes beyond the strictly predicative.

Rejection of Hilbert's Thesis would still leave open the question whether the functions that can be seen intuitively to be well defined have

---

[46] Cf. the remark about comparison of length (and thus order) in HB I, p. 22. $a \leq b$ can be defined as $a - b = 0$, if subtraction is defined as in the previous note.

§45. The limits of intuitive knowledge 261

weaker closure properties, such as being closed under *bounded* primitive recursion. The latter would imply that bounded quantification can be introduced into intuitive arithmetic, a plausible result. I am far from sure about questions of this kind. It is possible that the notion of intuitive knowledge is not precise enough to decide them. They also could depend on the answers to difficult mathematical questions. There are quite straightforward questions about restricted recursions whose answers are not known; for example, it is not known whether the polynomial-time computable functions are closed under bounded recursion.

So far, we have discussed these issues without attending to questions about the *possibility* of intuition. In particular, we have not made any assumption to the effect that every string *can* be intuited. If the possibility in question is practical, in particular, if what is at issue is actual human ability, then this assumption is false, unless we revise our mathematics to conform to strict finitism, so that the answer to the objection that a string of length 2^{65536} cannot be intuited would be that we have no reason to believe that there is any such string. It is also not likely that we can construct a real, physical machine that would have analogous capacities, such as computing exponentiation by 2 for the argument 65536.

Now let us suppose that we do allow ourselves to understand the 'can' in terms of abstract mathematical possibility.[47] The considerations that show successor to be intuitively well-defined show that if a string can be intuited, so can its successor. (This is plausibly true also for practical possibility.) One can then argue by induction that every string can be intuited. This might be questioned on the ground that similar reasoning leads to sorites paradoxes.[48] Given that the context is supposed to be mathematical possibility, I don't think that is a convincing objection. But it does suggest that there is something trivial about the conclusion. It is the concept of number, by way of an application of induction, that tells us that every number can be intuited, rather than some independent insight into what is intuitable yielding a domain of intuitable objects. But just for

---

[47] I don't use here the term "in principle" because it is ambiguous. Possibility in principle may be abstract mathematical possibility in the sense in which I am understanding it, or it may be constrained by other considerations, for example basic physical considerations. It is with the latter understanding that R. O. Gandy argues that there are finite bounds to what inscriptions can in principle exist and limits to what can in principle be known. If his concern had been with intuition, he would very probably have said the same about what can in principle be intuited. See "Limitations to Mathematical Knowledge."

[48] There are, of course, different views about what part of the reasoning leading to sorites paradoxes should be rejected. I am claiming that in the present situation the appropriate response is to embrace the conclusion that every string can be intuited.

this reason, I don't think it likely that a convincing argument for the falsity of the thesis will be forthcoming. Objections in the literature to the claim that every string can be intuited seem either to understand the modality in terms of human ability or to amount to a rejection of the application of mathematical modality in combination with epistemic notions. The former objection is based on a misunderstanding.

The second objection, that combination of mathematical modality with epistemic notions (or perhaps other non-mathematical notions) is questionable, is one that I think should be regarded seriously. I have given a fuller consideration of some issues surrounding it elsewhere.[49] But if we do allow ourselves to reason mathematically about the possibility of intuition, then it is easy to prove 'A string of length $t + 1$ can be intuited' for every term t of PRA and possibly for more.[50]

Does this give rise to an argument for Hilbert's thesis? Suppose s and t are such that I can intuit a string of length s, and I can intuit a string of length t. Then, surely, if I can calculate out the terms, then by comparing them I can determine whether $s = t$ is true. It would follow that any closed formula of PRA can be decided in an intuitive way. But what follows about intuitive knowledge of generalizations? The most this argument could show is that if a formula of PRA is true for all values of its variables, then, in each particular case, this can be known intuitively. But this does not yield intuitive knowledge of the generalization. Even if this obstacle can be circumvented, I see no way to get around the fact that what these considerations yield is the *possibility* of intuitive knowledge, according to a rather liberal kind of possibility. But Hilbert's thesis concerns actual intuitive knowledge, at least given a proof in PRA. Even if the questions about the ideas involved can be resolved, the case for Hilbert's thesis does not seem to be materially advanced. More generally, the relevance to questions about what we know intuitively of questions about what intuitions are possible "in principle" is not clear.

§46. Appendix

For the conventional formulation of PRA, we can assume that the logic allowed is minimal propositional logic, with $\neg A$ defined as $A \to 0 = 1$. To obtain intuitionistic logic, it then suffices to derive $0 = 1 \to A$ for any A.

---

[49] "What Can We Do 'In Principle'?".
[50] "Finitism and intuitive knowledge," §7. The conclusion is quite trivial unless we recognize the context in which t occurs as intensional; see ibid., pp. 267–268.

§46. Appendix

If A is atomic it is an equation $s = t$. We introduce by primitive recursion a function symbol φ satisfying $\varphi(0) = s$ and $\varphi(Sn) = t$. Clearly $0 = 1$ implies $\varphi(0) = \varphi(1)$, whence $s = t$.[51] The other cases are shown in the usual way by induction on the construction of A. Furthermore, since $0 = 0$, $0 = 0 \vee \neg(0 = 0)$. From the axiom $\neg(Sx = 0)$, we have $Sx = 0 \vee \neg(Sx = 0)$. So by induction,

(13) $\quad x = 0 \vee \neg(x = 0)$.

Now using the lemma

(14) $\quad x = y \leftrightarrow (x-y) + (y-x) = 0$,[52]

we can infer $x = y \vee \neg(x = y)$ by substituting $(x-y) + (y-x)$ for x in (13), so that we have $A \vee \neg A$ for atomic A. By induction on the construction of A, we can derive excluded middle for any A.[53] And because we have intuitionistic logic, we can also derive $\neg\neg A \to A$.

---

[51] I owe this device to A. S. Troelstra. φ has as parameters the free variables of s and t.
[52] Regarding subtraction see the end of §44.
[53] Suppose $A \vee \neg A$ and $B \vee \neg B$. To see $(A \to B) \vee \neg(A \to B)$, assume first A. If we assume B, $A \to B$ and hence $(A \to B) \vee \neg(A \to B)$ follow. If we assume $\neg B$, assume $A \to B$. Then B follows, which is a contradiction, whence $\neg(A \to B)$, whence $(A \to B) \vee \neg(A \to B)$. By $\vee$E we discharge the assumptions of B and $\neg B$. Assume now $\neg A$. By intuitionistic logic $A \to B$ follows, whence again $(A \to B) \vee \neg(A \to B)$. By $\vee$-elimination, we can discharge the assumptions of A and $\neg A$. The case of disjunction is similar, and the others are simpler.

8   Mathematical Induction

§47. Induction and the concept of natural number

Writers on the foundations of arithmetic have found it difficult to state in a convincing way why the principle of mathematical induction is evident. We have presented in earlier chapters different formulations of the principle, as a second-order statement with different interpretations in §§10–11 and as a rule in §31. What we have in view continues to be induction on natural numbers; however, the discussion applies to any structure whose domain consists of the objects obtained from an initial one by iterating a unary operation.[1]

A classic view on the question, first developed by Frege and Dedekind and adopted by Whitehead and Russell in *Principia Mathematica*, is that induction is a consequence of a definition of 'natural number'. The comprehension principle of second-order logic implies that, if we define a simply infinite system or progression as in §11, then induction holds in every simply infinite system. The matter can be presented a little more simply. Suppose we have 0 and S, in the context of classical logic so that the existence of 0 and the defined character of S are presupposed. Then we can define

(1) $Na \leftrightarrow (\forall F)\{[F0 \wedge (\forall x)\ (Fx \rightarrow F(Sx))] \rightarrow Fa\}$,

and induction in the form

(2) $\{A(0) \wedge (\forall x)[Nx \rightarrow (A(x) \rightarrow A(Sx))]\} \rightarrow (\forall x)(Nx \rightarrow A(x))$

follows easily.

---

[1] For much of the discussion it is not essential that the operation be one-one and exclude the initial element from its range; of course, if one of these fails, the structure involved will be isomorphic to a finite initial segment of the natural numbers with the "last" number having as successor either itself or some earlier number.

§47. Induction and the concept of natural number

According to this proposal, induction falls out of an explanation of the meaning of the term 'natural number'. But this way of viewing induction is independent of the particular proposal and even of the use of second-order logic, which one might object to at the beginning of a theory of natural numbers, for example, on the ground of its impredicative character. (Impredicativity will be a theme later in this chapter, in particular in §50.)

An alternative is to undertake to capture by rules the idea that the natural numbers are what is obtained by beginning with 0 and iterating the successor operation, again taking these as given in the sense of presupposing some interpretation of them.[2] Clearly, this yields what were called in §31 the introduction rules for N. In spite of their overfamiliar character, I repeat them here:

(R1) $\quad N0$

(R2) $\quad \dfrac{Nx}{N(Sx)}$

To cash in the idea that the numbers are what is obtained by iterating the successor operation, we understand (R1) and (R2) as the canonical way of arriving at statements to the effect that something is a natural number; it is only by virtue of them that something is a natural number. Their role is like that of an inductive definition. To (R1) and (R2) might be added a gloss such as, "Nothing is a number except by these rules," often called an extremal clause. That implies that N is *minimal* so that the introduction rules hold. Suppose for example that N and N' are both closed under the rules but N' properly includes N. Then it won't be true that N' has only elements that are there by virtue of the rules, since everything the rules require to be there is already in N.[3]

---

[2] According to George and Velleman, "Two Conceptions," this is a "bottom-up" characterization and is to be contrasted with the "pare-down" characterization embodied in the explicit definition in the style of Frege.

[3] I have tried to avoid introducing at this point the idea of *finite* iteration of the successor operation. But suppose that N is the result of iterating S an arbitrary finite number of times, while N' is obtained by some transfinite iteration of S, in both cases beginning with 0. Then my claim is that N' does not satisfy the extremal clause. This is shown by the fact that transfinite iterations require some additional provision for limit stages of the iteration.

It seems to me that, contrary to the suggestion of George and Velleman ("Two Conceptions," p. 313) the question whether transfinite iteration of S is admissible (as a constructivist might deny) is irrelevant. Ordinary induction and primitive recursion articulate what finite iteration *is*; if one is going to consider transfinite iteration, to

But now suppose $A(a)$ is a predicate for which the introduction rules hold:

$A(0)$
$A(x) \to A(Sx)$.

Then $A(a)$ must be true of any natural number. But that is just the induction principle, which, as in §31, takes the form of an elimination rule:

(R5) $$\dfrac{A(0) \quad \begin{array}{c}[A(a)]\\ \cdot\\ \cdot\\ \cdot\\ A(Sa)\end{array} \quad Nt}{A(t)}$$

There is a very natural picture that arises here: If x is a number, then, by the introduction rules, we reach x by beginning with 0 and taking a succession of steps from y to Sy. Then, by a parallel succession of steps, we can show that $A(y)$ holds for each y figuring in the construction, and therefore that $A(x)$ holds. In fact, for each x, we can construct a formal proof of $A(x)$ by beginning with $A(0)$ and building up by modus ponens, using $A(x) \to A(Sx)$. As a *proof* of induction, this is circular: the "construction" of x by a *succession* of steps is itself inductively defined, and it is by a corresponding induction that it is established that A holds at each point in the construction. Nonetheless, it is still useful for metamathematical arguments concerning induction in formalized theories, and it is no worse than arguments for the validity of elementary logical rules.

It might be better to view this idea as relating induction and recursion: A recursion is given for a function $p(n)$ that, for any n, gives a formal proof of $A(n)$ given proofs of $A(0)$ and of $A(m) \to A(Sm)$ for arbitrary m. The same idea is expressed in a more abstract way in the treatment of induction in constructive type theories, where conditions are laid down for an object to be a proof of a proposition of a given form. Consider the case of a formula $A(a, n)$ with a single parameter a, which we assume to

articulate it one needs additional apparatus. This also might be presented by an inductive definition, which will have its own extremal clause. This theme will be pursued in §51.

§47. Induction and the concept of natural number

be of the type N of numbers. Then a proof of $A(a, 0)$, for arbitrary a, will be a function g that, for any $m \ \varepsilon \ N$, yields a proof of $A(m, 0)$; given the treatment of the conditional in such theories, a proof of $A(a, x) \to A(a, Sx)$ for arbitrary a and x will be a function h that, for any $m, x \ \varepsilon \ N$, and proof p of $A(m, x)$, yields a proof of $A(m, Sx)$. Then if we set

$$f(m, 0) = gm$$
$$f(m, Sn) = h[m, n, f(m, n)],$$

then for any m and n in $N f(m, n)$ will be a proof of $A(m, n)$.[4]

Clearly, what we have given here is not a complete explanation of the concept of natural number. One omission is corrected by referring to §31: the remaining two Peano axioms, presented there as the rules (R3)–(R4). Even accepting the open-endedness of the notion of a well-defined predicate, it will be protested that we have said too little about what counts as such. No general rule, such as that a predicate is to be true or false of every number, is likely to be helpful until a body of mathematics is built up. 0, S, identity, and basic propositional logic give us something but too little. What is decisive is admitting some primitive recursions. The procedure of constructive type theories, of treating primitive recursion as primitive, where the values of the function introduced may be of an arbitrary type, is powerful but requires a specific logical framework. Another possible route is that addition and multiplication may be assumed (possibly on the basis of the intuitive ideas of §44) and further recursions obtained by the use of quantificational reasoning, in the way familiar from PA. These two routes do not exhaust the possibilities. But in the end the concept of natural number cannot determine what counts as a well-defined predicate, because in one way or another, either through application or through the further development of mathematics, relations of numbers to objects of other domains will enter in.

There are a number of questions about how to understand these rules and the justification of induction offered here. One that arises both for our procedure and the Fregean one is: What is the range of the *first-order variables*? In typical instances of inductive definitions, all the definition aspires to do is to define a predicate of objects of a *previously* given domain, and so we can take first-order quantifiers to be understood in some independent way. Frege himself seems to have assimilated his own

---

[4] Cf. Tait, "Finitism," p. 32. In the situation of note 3, it seems that if x is reached from 0 by a transfinite succession of steps, then, similarly, a proof of $A(m, n)$ can be obtained by a transfinite recursion. But again, some provision will have to be made for limit stages.

case to this one in assuming that first-order quantification is quantification over *all* objects, in an absolute sense independent of the particular context of inquiry. I don't want to assume that; indeed, I would argue, and have argued elsewhere, that there is no such domain.[5] Should we still interpret our explanation of the notion of natural number as picking out the natural numbers from a previously given domain? One might reply that our conception of the natural numbers is that of a *structure* and therefore does not give individual identities to the objects playing the role of 0, 1, 2, and so on; therefore, there should be no unique answer to the question from what domain the natural numbers are picked out or even whether there is one. The generality with which we have proceeded is in keeping with this structuralist view and, therefore, cannot exclude the case where there is a previously given domain: Some instances of the structure of natural numbers are substructures of other structures, and we might describe such substructures by inductive definitions. In the interesting cases, the other structures will be of other types, such as structures of sets.

However, to assume that this is always the case is to assume that some infinite structure is given to us independently of our knowledge of the kind of structure the natural numbers instantiate. What could that structure be? In §12, it was questioned whether, without already having some mathematics presupposing an infinity of objects, we could regard the physical world or the human mind as offering such an infinity. In some constructions of mathematics, set theory is assumed, and the problem is to single out an instance of the natural number structure in the universe of sets. That fits the objective of developing arithmetic in axiomatic set theory. Whether it gives the most fundamental understanding of mathematical induction and its validity can be and has been questioned, but I shall not pursue the matter here because the procedure of set theory is to use an explicit definition of more or less Fregean type. However, a question of a similar nature will be pursued in §50 in connection with Feferman and Hellman's work.

If there is no such previously given infinite structure, then it is as if we had arrived at the concept of natural number by pulling ourselves up by our conceptual bootstraps, so as to understand the notion of some

---

[5] Essays 8 and 9 of *Mathematics in Philosophy*. The matter is still controversial; see Cartwright, "Speaking of Everything" and more recent literature such as Williamson, "Everything." Different points of view, including my own, are expressed in the essays in Rayo and Uzquiano, *Absolute Generality*.

§47. Induction and the concept of natural number

such structure and convince ourselves of its possibility without having in advance the conception of a domain of objects from which the objects of the structure are picked out. But that this is possible was in effect argued in presenting the Hilbertian intuitive model of arithmetic in §§28–29. So let us assume for a moment that the above explanations refer to the model of strings, so that '0' designates '|' and 'S' expresses the operation of adding a '|' on the right. The intuitive verification that every string can be extended tells us that 'S' is defined for the arguments that concern us. But it does not depend on insight into the specific totality of such strings; this is shown by the fact that in the sense in which a new '|' can be added to any string, it can be added to any bounded geometric configuration. On this reading, then, the variable in (R2) is more inclusive than a variable over strings, but I do not want to say that it ranges over a more inclusive *totality of objects*. There would not be a convincing answer to the question what that totality is. We could specify the range as something like "object given in space or time." But this very general rubric might go with quite different ways of *individuating* such objects. The generality is akin to the kind of generality Husserl called "formal," characteristic of formal logic.[6] In a sense, we used free variables with a generality interpretation without yet knowing what they ranged over. Then the considerations beginning with the idea that the strings are what is obtained by beginning with '|' and iterating the operation of adding another '|' are what explains the domain of our first-order variables, what gives us a definite domain of quantification. This view would be an instance of the idea, found in other approaches to constructive mathematics, that numbers are a type. That should lend some pedigree to the idea that the natural numbers, or some instance of the structure that is basic, do not need to be understood by picking them out from a more comprehensive domain of objects.

A second question concerns the schematic character of the induction rule. Here we have formal generality in almost exactly in Husserl's sense. Induction holds for any well-defined predicate. In thus referring to arbitrary predicates, the statement of the rule makes no assumptions about what counts as a predicate. It has the same purely formal character as the principles of predicate logic itself. As such a generalization about predicates, the rule is not a generalization over a given domain of entities and could not be, because it is not determined what predicates will or can be constructed and understood. Of course, this does not depend just on arithmetic. Predicates will be admitted that essentially involve

---

[6] *Ideen*, §13.

other notions, certainly notions arising in the further development of mathematics but even nonmathematical notions, at least provided that they are precise.[7] The inescapable vagueness of the principle of induction understood in this way was already remarked on in §5. There, we mentioned a way of shunting this vagueness elsewhere: replacing the rule by an axiom, that is, a single statement expressed by quantification over sets, properties, or Fregean concepts. Such a procedure, though it is appropriate for many purposes, is not suitable for the explanation of the number concept by rules that we have been engaged in, precisely because it presupposes a domain of objects from which the natural numbers are picked out.

This understanding of induction implies that the applicability of the rule is not limited to predicates defined in some particular first-order language such as that of first-order arithmetic. But we must not take it as implying the unavoidability or even the legitimacy of full second-order logic, even given the claim of the impredicativity of induction to be discussed in §50. Given a domain D for individual variables, full second-order logic is naturally understood by taking second-order variables to range over all subsets of D. In any case, it requires that the "predicates" of objects in D should be closed under second-order quantification. Nothing we have said implies this; indeed, our picture should suggest rather the opposite, that the "predicates" talked of in the induction rule are a quite open-ended and indefinite totality, depending on linguistic and conceptual resources of whose limits we have no real conception. That would raise a question whether second-order quantifiers, in particular their use to define new predicates as in second-order logic, can have a definite sense. The implication of our earlier discussions of second-order logic is that any such sense that would license impredicative logic must derive from the concept of set.[8] That such second-order logic is not forced on us at this point is shown by the fact that a mathematics that assumes the concept of natural number but from there on is entirely predicative is perfectly coherent.

A third question is whether and in what sense induction is an analytic or conceptual rule or truth. The Fregean treatment makes this so in a definite sense, expressed by Frege's own definition of an analytic truth as

---

[7] I do not take a stand on whether that is required for a predicate to be well-defined for this purpose. It has been argued that induction is not applicable to vague predicates, and indeed such an application is part of the reasoning leading to the sorites paradox.

[8] In addition to §§11–12, see *Mathematics in Philosophy*, pp. 216–217.

§47. Induction and the concept of natural number

one whose proof requires only logic and definitions. But this conclusion requires accepting second-order logic as logic, and one could still have the reservation that the existence of 0 and the defined character of S are presupposed; in Frege's own development the proof of these required his criterion of identity for numbers ("Hume's Principle"), which he in turn derived from the disastrous axiom V. But even if that is not brought in, a more schematic version of induction turns out to be analytic if the Fregean definition is applied to second-order logic without any further principle giving the existence of any numbers.[9]

The explanation of the number concept by rules makes induction follow from an explanation of that concept; it is certainly in some sense "conceptual." It is so in a negative sense, in that it is not intuitive in the sense of §29. But beyond this negative point, what more is said than that if we understand the concept of natural number, instances of induction will be evident to us? (The general rule may not be, without introducing the additional notion of predicate and some semantic reflection.) We don't see what else this evident character could rest on than our understanding of the concept of number. But that does not imply that we have said anything very positive about what it *does* rest on.

In §43, I argued that in the context of finitist arithmetic induction preserves intuitive knowledge, making use of its analogies with logical rules, in particular its serving as an elimination rule for N. The limitations of that argument should be clear from §44, where difficulties arose about the intuitive character of primitive recursions beginning with exponentiation. If we give ourselves the logical apparatus needed to reduce recursion to induction, then induction comes to have the marks of a synthetic principle. Primitive recursion gives us the ability to represent larger and larger numbers, so that it quickly becomes not feasible to represent them as canonical numerals built up from 0 and S or even as Arabic numerals. Any argument to the effect that terms built up from functors introduced by primitive recursion are reducible to canonical numerals will require equivalent recursion or induction.

This point is reflected in the reasons that have been offered for rejecting primitive recursion in its full generality, usually that it admits functions that are not feasibly computable, but also that it admits impredicativity. Is such a rejection compatible with our own claim that induction holds for any well-defined predicate? If one allows, for primitive recursive terms, the predicate '$f(n, a)$ is defined', then one can prove by induction

---

[9] This was effectively proved by Frege himself in *Begriffsschrift*.

that primitive recursive functions are defined for all arguments. But the strict finitist who rejects primitive recursion may well deny that such a predicate is well-defined. Admitting it is admitting at least a restricted form of induction on existentially quantified formulae, and the latter seems to have been excluded from finitary reasoning already by Hilbert. It's also possible that a strict finitist might reject it on grounds of vagueness. There seems to be no contradiction or patent absurdity in the position of someone who refuses to accept enough recursion and induction to obtain even primitive recursive arithmetic. But the most plausible line of objection is not to induction in the schematic form in which we have preferred to state it but rather to the admission of predicates or functors that give greater expressive power, in particular more powerful instances of induction.[10]

§48. The problem of the uniqueness of the number structure: Nonstandard models

Independently of the issues about reference to natural numbers and the issues about structuralism discussed earlier in this work, we have to acknowledge that it is a strongly held intuition that the natural numbers are a unique structure. This is evidenced by the fact that we typically speak of *the* natural numbers. What I have in mind is the view that the natural numbers are at least determinate up to isomorphism: If two structures answer equally well to our conception of the sequence of natural numbers, they are isomorphic. I will call this latter thesis the Uniqueness Thesis. Is this something one can sensibly doubt? In this section I shall consider some grounds that might be offered for such doubts, in particular one that would occur to anyone familiar with basic mathematical logic: the existence of nonstandard models of arithmetic.

An initial, intuitive ground for questioning the Uniqueness Thesis is that the explanations of the notion of natural number contain elements of vagueness. The explanation favored in §47 has this character, because it relied on induction as an inference that could be made with

---

[10] The reliance on induction in explaining the concept of natural number is a characteristically modern stance. One might well ask what explanations of the number concept were available in the many centuries during which arithmetic was part of mathematics but the principle of mathematical induction had not yet been formulated. (Thanks to Hilary Putnam for reminding me of this.) This is obviously a large historical question that I cannot deal with here. One might ask how clearly such explanations would have differentiated the *natural* numbers from other number systems.

§48. The problem of the uniqueness of the number structure

any well-defined predicate, without the prospect of specifying exactly what the range of such predicates is. The same is true of the cruder characterizations that are often given, that the natural numbers are what is obtained by beginning with 0 and iterating the successor operation *forever*, or an *arbitrary (finite) number* of times.

It would be hasty to conclude from this that the notion of natural number is itself vague, in the sense that it is not objectively determined what is a natural number. That would amount to saying that there are "borderline cases" of natural numbers, objects of which it is not definitely true that they are natural numbers but it is not definitely false either. Such an admission would be at first sight fatal to the Uniqueness Thesis, as it allows that one might make the notion of natural number more precise in different ways; for a given borderline case, one including it as a natural number and one excluding it, with the two possible more precise versions having an equal claim to be the natural numbers. It seems unlikely that the different ways of making the notion more precise would all lead to isomorphic structures.

If one has some experience with the foundations of mathematics, one will find something unreal in the idea that 'natural number' might be a vague predicate like 'heap' for which we might have *given* objects such that the meaning of the word does not guide us to a decision as to whether the word applies to the object or not. But we do have to consider whether non-standard models present us with just such a situation. Moreover, the idea that the notion of natural number might be vague in some more subtle way was advanced some years ago by Michael Dummett, in two articles that are perhaps the most instructive in the literature on the questions raised about the Uniqueness Thesis, although they are not the main theme of either.[11] Dummett's views will occupy us later.

It might seem that doubts about the Uniqueness Thesis can be dismissed at the outset on the basis of a theorem of Dedekind already mentioned in §10: Any two simply infinite systems (or progressions in the terminology of §31, going back to Russell) are isomorphic.[12] It will surely be agreed that the natural numbers are a progression. Dedekind's theorem will occupy us in §49, and it does indeed buttress the Uniqueness Thesis. Still, one should be wary of any claim that a mathematical result settles a philosophical question. In particular, we have to remember that no demonstration is better than the assumptions on which it rests. In §10, with reasonable faithfulness to Dedekind, we assumed the set-theoretic

---

[11] "The Philosophical Significance of Gödel's Theorem" and "Platonism."
[12] *Was sind und was sollen die Zahlen*, para. 132.

conception of structure, and that is the setting in which Dedekind's theorem is typically proved. It is a commonplace in the foundations of mathematics that the idea of natural number is more elementary than that of set, at least if the concept of set is to give rise to full-blown set theory, as embodied in ZF and related theories. Although perhaps in the end we should not be bothered by the use of less elementary means to prove the uniqueness of the natural numbers, we cannot just accept them without any examination.[13]

A technical development that might give us pause about the uniqueness of the natural numbers is the already mentioned existence of nonstandard models of arithmetic. (In fact, it reflects the vagueness mentioned above of the notion of predicate in explanations appealing to induction.) For, it might be said, our concept of natural number cannot do more than determine what is true about the natural numbers. But even if we suppose given the set of all these truths, there will still be nonstandard models, in which the truths all hold (and the falsehoods all fail to hold), but which are not isomorphic to the natural numbers. Perhaps it is better to put the matter more neutrally: There will be models, not isomorphic to each other, which are still models of all true sentences of arithmetic.

This observation leads to a reason for questioning whether the uniqueness of the structure of natural numbers is really assured by Dedekind's theorem. For the logical considerations behind the existence of nonstandard models are quite general, and apply also to set theory. Thus, of whatever set theory in which we have proved Dedekind's theorem, there will also be nonisomorphic models. And nonisomorphic models of set theory can give rise to nonisomorphic models of arithmetic. Consider now two models $\mathbf{M}_1$ and $\mathbf{M}_2$ of set theory, and let ω_1 and ω_2 be their sets of natural numbers. Dedekind's theorem is a theorem of set theory; hence it is true in each of $\mathbf{M}_1$ and $\mathbf{M}_2$. But what that tells us is that within $\mathbf{M}_1$ any structure satisfying the conditions for being a progression is isomorphic to ω_1, and within $\mathbf{M}_2$ any structure satisfying the conditions for being a progression is isomorphic to ω_2. But it does not tell us that ω_1 is isomorphic to ω_2; indeed, as non–well-founded models of set theory can be constructed (shortly, we will indicate how), they need not be isomorphic.

I have not said what a "sentence of arithmetic" is, and of course those familiar with the technical background will remind me that the existence

---

[13] The reader may be inclined to observe that Dedekind's theorem can be proved in a very weak theory of sets and classes, such as the EFSC of Feferman and Hellman, "Predicative Foundations." We will take account of this fact in §49 and §50.

§48. The problem of the uniqueness of the number structure 275

of nonstandard models rests on fundamental properties of *first-order* logic. But the difficulty still arises if we take the language of arithmetic to be higher order. For then we can still have nonisomorphic "general models," and the question will arise on what grounds we can exclude those that are nonstandard (perhaps now in the sense usual for higher-order logic).[14] General models arise when the language of higher-order logic is interpreted in the natural way in the language of first-order set theory, and one looks at arbitrary models of the latter; thus, the existence of nonisomorphic general models will follow from the fact that in different models of set theory, the structures of the natural numbers can be different.

A first answer could be that the *distinction* between standard and nonstandard already offers the basis for the "exclusion" of the nonstandard models. One characterization of a standard model of arithmetic would simply be as one that is isomorphic to the natural numbers. That is unhelpful, because if the Uniqueness Thesis should be false, we would be unjustified in speaking of "the natural numbers." A second characterization (closer to the logical one) would be as one in which induction holds for *all* subsets of the domain. But then we have returned to the use of set theory, which was just what made us uncomfortable about simply appealing to Dedekind's categoricity theorem. The observation about models of set theory would suggest a kind of relativism, akin to the relativism about cardinality that Skolem argued for on the basis of similar logical considerations: What we have in mind by "natural number" is relative to the underlying set theory. The language of set theory admits different interpretations, which give rise to different sets of natural numbers that are not even isomorphic structures. A model with more sets may allow the exclusion as nonstandard of certain objects that are integers in a smaller model, just as it may "collapse cardinals" by introducing one-to-one correspondences that were not to be found in the smaller model.

It might be helpful at this point to consider an example. If we made the assumption, much stronger than the one we are considering, that there is a unique universe of pure sets, then that should determine uniquely what sentences of the usual first-order language of set theory are true. Then

---

[14] A model of arithmetic is said to be standard if it is isomorphic to the natural numbers. For any set of sentences in a second-order language, a model is said to be standard if the domain of the second-order variables is the set of *all* subsets of (and relations on) the domain of the individual variables. (The same idea applies to logics of third and higher order, which, however, we do not need to consider.) If the Dedekind-Peano axioms are expressed in second-order form, then if a model of them is standard in this second sense it is standard in the first sense (by Dedekind's theorem). The Submodel form of the Skolem-Löwenheim theorem (see below) implies that the converse does not hold.

these sentences will constitute a set, since there are only denumerably many of them. Now let Σ be this set of sentences. (We do not really need the strong assumption: All we need to know about Σ is that it is a complete theory and that it contains those sentences that we accept as axioms of set theory, say, those of ZFC plus whatever large cardinal axioms we find acceptable.) Now consider how we might construct a model that is nonstandard with respect to arithmetic. Let $N(x)$ be whatever we have chosen as the natural number predicate of the language; of course Σ will contain a sentence saying that $N(x)$ has a set Z as its extension. Let us now add to the language a countable sequence of new constants c_0, c_1, c_2, ..., and add to Σ the sentences $N(c_n)$ and $c_{n+1} < c_n$. (in the order of the natural numbers, easily definable in arithmetic) for each integer n. It is easy to see that the augmented set (call it Σ') is also deductively consistent, since only finitely many of the new sentences could be used in deriving a contradiction. Thus by the completeness theorem, Σ' has a model **M**, which will clearly be nonstandard, since it contains an infinite descending chain of natural numbers, which implies the existence of an infinite descending $\in$-chain.[15] We could have added to Σ' all instances of replacement in the extended language, so that the model **M** will not exhibit failures of any of the usual second-order principles (including induction) in the augmented language.

However, it is still the case that this construction witnesses the fact that the model is nonstandard. Of course we are able to say of it that it contains an infinite descending sequence of natural numbers. But we would like to say that it fails to satisfy one of the basic facts about natural numbers that we cited, that induction fails to hold for any well-defined predicate. Counterexamples can easily be given, though some tempting ones are only apparent: 'There are finitely many $x \in Z$ such that $x < c_0$' is of course true in the model. What we need to generate genuine failures is a name of the *set S* consisting of the "numbers" denoted by all the c_i. S is a non-empty set of numbers; a simple application of induction yields that S has a least element. But that is clearly false in **M**, so that the instance of induction that implies it must be false as well. S is definable if we add to the language the predicate "x is the Gödel number of a sentence belonging to Σ'".[16]

---

[15] If the numbers are represented by finite von Neumann ordinals, so that $<$ is $\in$, evidently this will be an infinite descending $\in$-chain. But however the numbers are represented, it is easy to see by Dedekind's theorem that **M** will contain such a chain and so is not well-founded.

[16] For example, if the nonstandard model is constructed in the Henkin way, the set of sentences true in the model, in a language with a new list of constants d_0, d_1, d_2, ... denoting all the elements of the domain of the model, is defined in the language of set

§48. The problem of the uniqueness of the number structure 277

Thus, for any complete extension of ZF in the usual language of set theory, we have a (non–well-founded) model in which the natural numbers are nonstandard. It is by having the truth predicate for the model with which we start that we recognize the new model **N** as nonstandard; after all, we are able to say of **N** that there is, in it, an infinite descending $\in$-sequence. Because the sentence 'There is no infinite descending $\in$-sequence' is true in **N** (being a theorem of ZF), we are able to recognize a discrepancy between the interpretation of such statements with reference to the model **N** and what they mean as we use them in the *construction* of **N** (which we are here representing as a set-theoretic construction). (We don't need semantic reflection on that use; we simply are in a position to make statements of the form 'p and "p" is not true in **N**'.) There are instances of mathematical induction in the language of set theory augmented by the above predicate that fail in **N**.[17]

In talking about the Löwenheim-Skolem theorem and the philosophical claims that have been made on the basis of it, we have to keep in mind the two versions of the theorem. I will call them the Constructive and the Submodel versions. The Constructive version is the one that is a corollary of the strong completeness of first-order logic. One begins with a set of sentences that is deductively consistent (such as the set Σ' mentioned earlier) and constructs a countable model of it. It is only the Constructive version that directly gives rise to a model that is nonstandard with respect to the integers (in other words, not an ω-model). The Submodel version begins with a given model and concludes the existence of a countable elementary submodel, but if the original model is an ω-model, then the submodel is as well. It also should be remarked that the manner in which the axiom of choice is used in the proof of the Submodel version means that the submodel is not *constructed*; rather, we obtain a proof that a countable elementary submodel *exists*. We are not, however, able to define the domain and the $\in$-relation of the submodel even relative to the domain and $\in$-relation of the given model, only relative to those and a choice function on the nonempty subsets of the domain of the original model. This fact, however, may serve to reinforce rather than to call in

---

theory augmented by the above predicate, say by a predicate $\Delta(x)$ of the augmented language. The domain of the model consists of equivalence classes of constants of the expanded language, where $s \sim t$ holds if and only if $\Delta(s = t)$. Then the set $\{c_0, c_1, c_2, \ldots\}$ is just the set of equivalence classes $[s]$ such that $\exists n \in \omega \, \Delta(s = c_n)$. (I write formulae as arguments of Δ where strictly the arguments should be codes of them such as Gödel numbers.)

[17] Strictly, **N** expanded by a relation corresponding to the added predicate.

question the skepticism about the idea of "absolute" uncountability that, from Skolem onward, has been inferred from the Löwenheim-Skolem theorem. Once one has shown the existence of the countable model, the question will arise: How do you know that it was not that model that you were talking about in the first place? That formulation of the question, however, is too crude: What we prove, after all, is the existence of a *countable* submodel, and from this fact it follows, at least, that its domain is not the universe. Rather the question has to be something like this: How do you know that in your use of the language of set theory, in particular in the proof of the Submodel version of the Löwenheim-Skolem theorem with which we have been concerned, the "universe" of sets is not, from a "higher" perspective, countable? In other words, how do you know that there is not a more generous reading of the language of set theory, in which our present universe is simply a countable Skolem hull?[18]

I will not attempt here to answer that skeptical question directly. What I have said about the difference between the Constructive and Submodel versions of the Löwenheim-Skolem theorem already implies that the Submodel version is not enough to encourage skepticism about the uniqueness of the natural numbers. By contrast, although the Constructive version makes it easy to construct models of set theory in which the integers are nonstandard,[19] the above remarks show that from the construction one can already see that the model is nonstandard, either from a set-theoretic or an arithmetical point of view. Hence it seems to lend no support to the view that all models of arithmetic are on the same footing.

Still, one might be worried because of a certain asymmetry between being standard and being nonstandard. We can identify a model as nonstandard because we find, in a more comprehensive language than that of the theory with which we began, an instance of induction that fails in the "natural numbers" of the model. But to say that a model is standard is to say that *all* instances of induction hold. It may be questioned whether that statement has a definite meaning. One can say instead that induction holds for all *sets*; that is, any set a containing 0 and containing Sn whenever it contains n, contains all natural numbers. But whether such quantification over all sets has a definite meaning is precisely what is being questioned by the relativistic point of view about set theory. It thus

---

[18] These skeptical questions might be a reply to the argument on pp. 107–108 of Hart, "Skolem's Promises and Paradoxes."

[19] Or, of course, to construct non-standard models of arithmetic directly.

seems that to convince ourselves that the notion of "the standard model of arithmetic" is unambiguous, we would have to answer set-theoretic relativism.

It seems, then, that we have a rather unsatisfactory situation: The positive reason we might have to believe that the natural numbers are *not* unique is quite unconvincing, because when we produce a model of number theory not isomorphic to the "natural numbers" with which we begin, it counts as a model only because of a limitation on the means of expression of the language of the theory for which a model has been constructed. By contrast, the Löwenheim-Skolem theorem seems still to cast doubt on whether we have really "captured" the "standard" model of arithmetic.

§49. Uniqueness and communication

Let us now try to make a new start with our problem. A first, and very helpful, suggestion was made some years ago by Michael Dummett. In essence, Dummett's observation is that in order to formulate the idea of models of arithmetic, standard or not, one must *use* the expression "natural number" or some other (such as "finite") interdefinable with it:

> It is, however, circular to think that since what we mean when we speak of the natural numbers cannot be fully explained by reference to the incomplete formal characterisation, it must therefore be explained instead by reference to the conception of the standard model. For this conception must be given to us by means of some description, and this description will itself make use either of the notion of "natural number" or of some closely related notion such as "finite."
>
> Within any framework which makes it possible to speak coherently about models for a system of number theory, it will indeed be correct to say that there is just one standard model, and many nonstandard ones; but since such a framework within which a model for the natural numbers can be described will itself involve either the notion of "natural number" or some equivalent or stronger notion such as "set," the notion of a model, when legitimately used, cannot serve to explain what it is to know the meaning of the expression "natural number."[20]

---

[20] "The Philosophical Significance of Gödel's Theorem," pp. 191–192, 193.
  Although the context is a discussion of Gödel's incompleteness theorem, Dummett's argument is aimed at the idea that it is "by reference to the standard model" that the notion of natural number is understood. Skolem would perhaps have thought it an

The second of these quotations appears to say that the apparently unhelpful answer to our question, which we considered early on, that a nonstandard model can be seen to be such because it is not isomorphic to *the natural numbers* is in fact the best answer that can be given, and therefore, presumably, a sufficient one. We must then, it seems, look to the use of the term 'natural number', as well as those for particular numbers, to distinguish it from other mathematical terms that designate kinds of structures rather than unique structures and thus to convince ourselves of the Uniqueness Thesis. The fact that the numerals are singular terms is of less relevance than appears at first sight. For on any "structuralist" view of arithmetic, whether or not it supposes the numbers to be unique as a *structure*, terms like '0', '1', and other more complex ones will be relative in the sense that they do not have a reference that is fixed by more than the relations of the structure. On the "eliminative structuralist" views considered in chapter 2, such terms will have an implicit parameter for a realization of the structure of the natural numbers, either presupposed or quantified over. It has to be a further question whether all such realizations are isomorphic.

One might take to be embodied in our usage simply the *assumption* that there is one sequence of natural numbers.[21] This seems to me pretty clearly not a satisfactory position. For suppose we describe some other system of objects that answers reasonably well to the idea of natural number; let us call them 'pseudo natural numbers'. Then there should be some reason, connected with the use of the term 'natural number',

---

entertainable view that it was by reference to a "standard model" of axiomatic set theory that the notion of *set* is understood, and have seen his considerations based on the Löwenheim-Skolem theorem as arguments against that view. It would be a question worth discussing to what extent Dummett's observation could be carried over to this case.

One might compare Dummett's observations with the following earlier remark of Bernays:

> That the unreality (*Uneigentlichkeit*) of certain structures of the uncountable is so much more noticeable to us than the unreality that already lies in the conception of the number series as a structure, surely rests on the fact that our concept of a formal theory already intends the same sort of unreality as that of the number series. ("Bemerkungen zum Paradoxon von Thoralf Skolem," p. 118, my translation)

[21] Dummett sometimes gives the impression that that is his intention. I am inclined to think this impression false, as Dummett also engages in reflections on the principles of arithmetic, but it is not clear how he would use them to answer doubts about the Uniqueness Thesis.

§49. Uniqueness and communication

why these objects are not the genuine article, or indeed why they are not what we were talking of all along when we talked of natural numbers.

At this point one would naturally go further into the use of the language of natural number. Given the specific problem, it is the principles of arithmetic, beginning with the elementary Peano axioms and induction, that are immediately relevant. At this point we can consider again Dedekind's categoricity theorem. Two considerations seem to suggest that it will not be helpful: the fact that, in spite of Dedekind's theorem, first-order arithmetic is not categorical (i.e., the existence of nonstandard models already discussed in §48), and the fact that we obtain categoricity by considering the induction principle in the context of *standard* second-order logic. Although Dummett suggests that we have to understand the induction principle in this way in order to appeal to it in communicating our understanding of the notion of natural number,[22] I shall argue the contrary. But I shall follow Dummett in focusing on the problem of communication concerning the natural numbers.

If we look a little more closely at Dedekind's theorem and its proof, we see that it is essentially *first-order*. Although it can be proved in a weak theory of sets and classes,[23] I shall present it as a schematic theorem about a pair of number predicates. Suppose our language contains a singular term '0', a one-place functor 'S' and a predicate 'N', and also another such triplet '$0''$', 'S''', 'N'''. Suppose that the elementary Peano axioms hold for each. In keeping with Skolem's recursive arithmetic, we can introduce by primitive recursion a functor f, with arguments n such that Nn, and values x such that $N'x$, such that $f(0) = 0'$ and $f(Sn) = S'(f(n))$. To see that f defines a one-one mapping of the Ns onto the N's, we need only first-order instances of induction.[24] By (trivial) N-induction we have $Nn \land n \neq 0 \to (\exists p)(n = Sp)$, so that if $n \neq 0$, $fn = f(Sp)$ for some p, and so $fn \neq 0' = f0$. Thus if $fn = f0$, then $n = 0$. Suppose now Nn and (*): $(\forall m)[Nm \to (fm = fn \to m = n)]$. If $fm = f(Sn)$, then $fm = S'(fn)$. Then $fm \neq 0'$ and so by the above $m \neq 0$. Thus, for some

---

[22] Dummett, "Platonism," esp. p. 210; cf. the discussion in my "The Uniqueness of the Natural Numbers," pp. 33–34. The suggestion, however, may arise from the specific context of his paper, which argues against a platonistic theory of mathematical intuition.

[23] Feferman and Hellman, "Predicative Foundations," theorem 5, p. 8. I prefer not to introduce the complication of classes into the present discussion. Feferman and Hellman's formulation will be discussed in §50.

[24] By N-induction we mean the formulation (R5) of the rule of induction given in §31 and §47. N'-induction is the same rule with 0, S, N replaced by $0'$, S', and N'. Note that trivial N-induction proves $Nn \to N'(fn)$.

p, $m = Sp$, and $fm = S'(fp)$. But then $fn = fp$, and by (*) $n = p$, and so $m = Sn$. Thus we have proved

$$Nn \to \{(\forall m)[Nm \to (fm = fn \to m = n)]$$
$$\to (\forall m)[Nm \to (fm = f(Sn) \to m = Sn)]\},$$

and by N-induction, f maps the Ns one-one into the N's.

Note that this first part of the argument uses only N-induction and primitive recursion introducing a functor defined on the Ns. To see that f is onto the N's, however, we use N'-induction on the predicate

(3) $\quad (\exists m)[Nm \wedge fm = n]$.

Clearly, for $n = 0'$ we can take 0 as m; if m witnesses it for a given n, then Sm will witness it for $S'n$.

How is it possible that the proof of Dedekind's theorem can be first-order, although first-order arithmetic is not categorical? What is essential to the proof that the two structures are isomorphic is not any rich system of subsets of the one or the other, but rather that it is possible to apply induction and recursion within one of them in a way that involves reference to the other.[25] The usual language of first-order arithmetic only talks about the natural numbers; the functors and predicates operate within the natural numbers. Thus it does not contain the predicates induction on which would show that some independently described structure is isomorphic to it.

Although this formulation of Dedekind's theorem shows that it does not require a strong second-order or set-theoretic apparatus, it is not clear that it gets us over the difficulty for the Uniqueness Thesis posed by nonstandard models. We have assumed a first-order logical setting. So long as the language remains first-order, even if it is augmented by further predicates and functors, a consistent set of sentences in it will have models

---

[25] Our discussion has been carried out in a classical framework. However, a categorial characterization of the natural numbers due to F. W. Lawvere gives rise, as Tait has observed ("Against Intuitionism," n. 12), to a natural formulation of Dedekind's theorem in an intuitionistic typed theory of constructions. The manner in which the natural numbers are introduced in this logical setting makes particularly clear what is involved: with the type N of natural numbers goes primitive recursion with values of an arbitrary (and therefore possibly other) type A. For a full treatment see Martin-Löf, *Intuitionistic Type Theory*, pp. 71–76.

Either in this setting or in the classical one of the text, however, the role of primitive recursion makes the argument incompatible with strict finitism. And, indeed, it has been observed that from a strict finitist point of view the Uniqueness Thesis should not be expected to hold. See Gandy, "Limitations to Mathematical Knowledge," p. 140.

§49. Uniqueness and communication

in which the two number series will have nonstandard (albeit isomorphic) models. What has been accomplished for our purpose by formulating Dedekind's theorem in this way?

Before I undertake to answer that question, I will first pose a new difficulty for the Uniqueness Thesis. Suppose that we imagine two speakers who both use the first-order language of natural numbers. The first, Michael, uses 0, S, and N, whereas the second, Kurt, uses $0'$, S', and N', both, presumably, building up from these by primitive recursion and other devices. We might ask how one could come to know that his "numbers" are isomorphic to the other's. Let's imagine that Michael attacks this problem by a method approximating to that of Donald Davidson's "Radical Interpretation." He is able to identify $0'$, S', and N' as a singular term, a one-place functor, and a one-place predicate respectively. He will observe that Kurt "holds true" the elementary Peano axioms, and in keeping with the method, he will attribute to Kurt classical first-order logic. Let us assume that there is no reason why Michael should not apply the principle of charity and conclude that the elementary Peano axioms, as affirmed by Kurt, are true. Kurt also accepts N'-induction on any "well-defined predicate." We will extend our supposition and allow Michael the conclusion that any such induction affirmed by Kurt is true.

There might be reasons of a different kind from those mentioned so far, for example, about the reference of individual numerals, why Michael should not interpret 'N''' as used by Kurt as simply meaning 'N'. Suppose he is at least reserved about that question and incorporates Kurt's numerical vocabulary into his own rather than interpreting it by expressions he already has. Will he be able to convince himself that the two number systems are at least isomorphic? He will be able to introduce the functor f by primitive recursion and see that it maps the Ns into the N's; those steps are simply applications of primitive recursion and induction that he already accepts, to be sure open-ended as to vocabulary. For the same reason, he will be able to carry out the first part of the argument for Dedekind's theorem and show that f is a one-one mapping of the Ns into the N's. His application of charity will be enough to yield the conclusion that $0'$, S', N' is a model of arithmetic as he understands it.

Carrying out the second part, however, meets an obstacle. Michael will have observed that Kurt accepts induction on any well-defined predicate; he may even interpret some expression by Kurt of the intention to accept any such induction, even for predicates not yet formulated in his language. But it seems that in order to conclude that N'-induction on the predicate (3) above falls under some such general intention, he would have to regard (3) as a well-defined predicate *according to Kurt*. This would follow if N is such and f a well-defined functor. But what in the procedure of radical interpretation would allow Michael to treat them in that way? Couldn't Michael regard Kurt's vocabulary as too restricted to take in N (and therefore as not giving even the occasion for introducing a functor by primitive recursion on N)? That would allow him to take Kurt's N's as the numbers of a nonstandard model. At all events, it appears that he does not have grounds for ruling this out.

Let us suppose further that Kurt also undertakes to interpret Michael and goes through steps corresponding to those we have just described in Michael's interpretation of Kurt. Thus, Kurt will take 0, S, N into his own language and admit that the elementary Peano axioms and instances of induction affirmed by Michael are true. He will be able to define by primitive recursion a functor $g: N' \to N$ that he will be able to prove maps the N's one-one into the Ns. It might seem that the fact that he will find Michael holding true sentences embodying his interpretation of *himself* will enable him to see that g is onto. That is, Michael has admitted N' into his own vocabulary, so that it seems that Kurt can conclude that it is a well-defined predicate according to him and apply the induction needed to see that g is onto. However, the question arises whether Kurt is entitled to identify the N' that he takes over in his interpretation of Michael with N' as he himself previously used it. If he can't just assume this, he should distinguish them, as, let's say, N' and N'_M. He has, from his interpretation of Michael, an $f: N \to N'_M$, which he can compose with g to see that the N's are included in the N'_Ms. But to see that the composition of g and f is onto, he will have to apply induction on N'_M to a predicate involving his own N', and then the same question will arise: What entitles him to treat that as a well-defined predicate according to Michael?[26]

It seems that so long as we follow the paradigm of radical interpretation, the interpretation of someone's discourse about the natural numbers will not be constrained to make the agent's numbers isomorphic to the interpreter's, because the interpreter has a more expressive language in general, and we haven't found a reason why he must assume that his own number predicate falls in the scope of the agent's

---

[26] Hartry Field considers a somewhat similar situation involving two speakers' languages for set theory. Although the issue he poses explicitly is whether the terms for 'set' are *coextensive*, he states that his considerations "undermine recent attempts to get categoricity . . . results by 'essentially first-order' means"; among these he counts the argument of "The Uniqueness of the Natural Numbers." See "Are our Logical and Mathematical Concepts Highly Indeterminate?" pp. 421–422.

§49. Uniqueness and communication

induction.[27] But the manner in which we have deployed the idea of interpretation will strike the reader as unrealistic as a picture of how Kurt and Michael might understand each other in actual conversation. The procedure of radical interpretation is described as a procedure for constructing a *theory* of what the language used by another, in the typical case a member of another linguistic community, expresses by its sentences. It would be beyond the scope of this work to examine whether a speaker's understanding in normal conversation is best represented as the construction of a theory, particularly a global theory of the sort we have envisaged. At all events, a plausible alternative would be that speakers take each other's words at face value and respond to them without theorizing about what they mean unless difficulties arise that such reflection could help to resolve. This would be in line with the view, which I have defended elsewhere, that language as used is prior to semantic reflection on it.

Does this alternative point of view lead us anywhere on the Uniqueness Thesis? It might seem that it leads either to a begging of the question or to the conclusion that there is no question there to be considered. The attribution to Kurt and Michael of different number predicates was obviously somewhat artificial. Two English speakers would both use the expression 'natural number' or, if no confusion with other number systems threatened, simply 'number'. It's reasonable to assume that it has the form of a one-place predicate. (We assumed that the interpreter in the situation discussed above could conclude that.) Let's assume that about Kurt and Michael. Then they will treat each other's predicate 'number' as "meaning the same thing," at least so long as they do not come to disagree about the principles of arithmetic. Because it has the form of a one-place predicate, without an additional argument place that might possibly be filled by different "number sequences,"[28] it seems that they are just talking of the objects that are natural numbers.

Within this stance of taking each other's words at face value, however, they may still be able to inquire as to the justification of their use of such a one-place predicate, and even minimize the effect of disagreements

---

[27] Field offers a similar argument to much the same conclusion in the Postscript to "Which Undecidable Mathematical Sentences have Determinate Truth values?" Where we seem to disagree is whether that is the end of the matter. The remarks about communication in "The Uniqueness" were meant to respond to objections like Field's; I trust what is offered below will be found more convincing than Field found my earlier treatment.

[28] As would be the case with general types of structure where no one thinks there is (up to isomorphism) a unique instance, such as groups or fields.

of a certain restricted kind. Suppose that each of them teaches the development of arithmetic in set theory, but that Kurt prefers the natural numbers as finite von Neumann ordinals, whereas Michael prefers the Zermelo numbers.[29] Let's suppose they are rather literal-minded about this, so that each one takes the number *n* to be *identical* to the corresponding set. Then they can't be using 'natural number' as predicates with the same extension.[30] In what sense can they continue to take each other's words at face value?

Let's suppose that they have often conversed about number theory independently of its construction in set theory. They agree in accepting the elementary Peano axioms and induction. Let's assume that each can convey to the other his intention to accept induction for any well-defined predicate and to allow the introduction of functors by primitive recursion.[31] If they discover their disagreement about the identity of individual numbers with sets, they will have to distinguish their number predicates, perhaps regarding them as in some way indexical. But nothing else will be disturbed; they can continue their conversation about pure number theory and about higher mathematical constructions built on it. But if they abandon this restriction, they will be in a situation something like that of our original version of the example, in which they have two different number predicates. But if we allow each simply to understand that of the other, then each will be able to take the other's into his own language; for example, Michael might, to avoid confusion, use the expression 'number$_K$', and Kurt might use the expression 'number$_M$'.[32] Each might raise the question whether the numbers are isomorphic to the "subscripted" numbers and use the Dedekindian argument to answer the question in the affirmative. This will put a limit on the disagreement they have got into by too literally identifying numbers with sets, since they will not only agree in what sentences of pure number theory they accept but will see that their number structures are isomorphic.

---

[29] We might imagine them to be the teachers of Paul Benacerraf's Johnny and Ernie; see "What Numbers Could not Be," p. 278.

[30] Presumably, however, they could just observe how little they have to go on to fix the identity of individual numbers and conclude that they should not assume that their number predicates have the same extension.

[31] One might show that one accepts a term as a well-defined predicate by accepting induction on it or on something constructed from it.

[32] Each will, as in the earlier discussion, also have to distinguish his own terms for individual numbers from the other's.

§49. Uniqueness and communication

We now have the situation that was lacking when we viewed Michael's understanding of Kurt as a case of radical interpretation; namely, he will take his own number predicate as a well-defined predicate according to Kurt, and so he will allow himself to use it in induction on Kurt's numbers. That will enable him to complete the proof that his own numbers are isomorphic to Kurt's; clearly, the situation is symmetrical between them. It is important to observe that the conclusion that their number sequences are isomorphic does not depend on any global agreement between them as to what counts as a well-defined predicate. We should not expect such agreement, as it is not necessarily determined what will count for either as a well-defined predicate. Also, the question of a background set theory (or some functional equivalent) within which the argument is conducted does not arise. Their conversation proceeds with given predicates, and although induction and primitive recursion are assumed, nothing from which they might be derived plays any role.

We can imagine the dialogue between Kurt and Michael extended to a larger community. With the means brought to bear so far, any pair will be able to see that their numbers are isomorphic. Some means of generalizing predicate places will of course come into play as soon as the demand is made to see this in general. If it takes the form of an appeal to classes, first-order comprehension is sufficient, as the version of Dedekind's theorem given by Feferman and Hellman illustrates (see note 23).

Where does this leave us with respect to the Uniqueness Thesis? No proof has been given that the "intended model" of either Kurt or Michael is the standard one. Although if they can take each other's words at face value in the way described, they will be able to convince themselves that their number sequences are isomorphic, it does not follow that they are standard. The discussion of the question from the point of view of radical interpretation seems to show that an interpreter coming from outside can interpret them so that the arithmetical truths of each, including those relating the "numbers" of each to those of the other, are all true from the interpreter's point of view, and yet their number series are nonstandard, albeit essentially the same nonstandard model. This interpreter, however, will use his own notion of natural number, or some equivalent, to describe his model. It seems that his interpretation can survive only as long as he does not enter into the sort of dialogue with Kurt or Michael that they have had with each other.

I think the lesson we can draw from these considerations is that the point of Dummett's observation that the notion of natural number must be used in the construction of models of arithmetic is that, in the end, we have to come down to mathematical language as *used*, and this cannot be made to depend on semantic reflection on that same language. We can see that two purported number sequences are isomorphic without strong set-theoretic premisses, but we cannot in the end get away from the fact that the result obtained is one "within mathematics" (in Wittgenstein's phrase). We can avoid the dogmatic view about the uniqueness of the natural numbers by showing that the principles of arithmetic lead to the Uniqueness Thesis, but this does not protect the language of arithmetic from an interpretation completely from outside, that takes quantifiers over numbers as ranging over a non-standard model. One might imagine a God who constructs such an interpretation, and with whom dialogue is impossible.[33] But so far the interpretation is, in the Kantian phrase, "nothing to us." If we came to understand it (which would be an essential extension of our own linguistic resources) we would recognize it as unintended, as we would have formulated a predicate for which, on the interpretation, induction fails.

This way of looking at the situation will remind the reader of Putnam's approach to "Löwenheim-Skolem" predicaments in "Models and reality." He calls his standpoint "non-realist semantics":

> From that standpoint, it is trivial to say that a model in which ... the set of cats and the set of dogs are permuted ... is "unintended." ... Such a model would be unintended *because we don't intend the word 'cat' to refer to dogs*. (p. 24)

Putnam's nonrealist semantics amounts at this point (not necessarily everywhere in his recent writings) to what Quine in his earlier discussion of ontological relativity called "acquiescence in our mother tongue."[34] It is a long way from this modest nonrealism to anything the tradition would recognize as idealism, and even to the thesis of Dummett's that truth-conditions cannot be the basic terms by which the meaning of the expressions of a language is explained.

Apart from the question whether it is persuasive in its own terms, this argument leaves a number of questions, some of which have been addressed in recent literature. Many will find the conclusion weak, indeed,

---

[33] Such a God might formulate a definite idea of our "total science" or "total belief" and therefore an idea of what *could* be a well-defined predicate for us; he could then reason in the manner of Putnam's "Models and Reality." A point at which I may differ with Putnam is that I don't think *we* have such an idea.

[34] "Ontological Relativity," p. 49. Such a stance appears to allow some semantic reflection, such as Putnam allows himself in the quotation in the text. The nature of this reflection, and how far it can go, is a large question which I do not pursue here.

§49. Uniqueness and communication 289

that seems to be the view of those who have discussed the matter most extensively: Field, Vann McGee, and Shaughan Lavine.[35]

Field makes clear that he is not satisfied with the conclusion argued for above, and he offers a proposal for reaching a stronger one.[36] The idea is roughly that the physical universe singles out a standard model for arithmetic, or, in the context of Field's discussion, makes definite what sets are "genuinely" finite. He proposes a "cosmological hypothesis," that time is infinite and Archimedean, and argues that, if it is true, the notion of finiteness can be defined in such a way that in allowable models of the resulting theory, what sets are finite will be invariant. He is at pains to point out that the particular hypothesis he lays out is not essential; a number of other such hypotheses would accomplish the same end.

One's first reaction is to protest that any indeterminacy infecting the mathematical conception of finiteness could be expected to infect the physical conception as well. Field himself supposes his cosmological hypothesis embedded in a certain physical theory S, which will have non-standard models. Field's Archimedean hypothesis, more precisely stated, is that any set of events that has an earliest and a latest member, and such that any two of its members occur at least one second apart, is finite. Now, of course, in a nonstandard model the finiteness of such a set can be witnessed by a mapping onto the integers less than some nonstandard integer, and Field's infinity condition will insure that this happens in some cases. So if in general the existence of nonstandard models threatens the idea that 'finite' is a determinate concept, why does it not do so in this case?

Field takes this possibility to be excluded by an assumption of the determinateness of the *physical* vocabulary:

> More precisely, I will take an interpretation or model of our own language to be "unallowable" or "objectively unintended" unless the extension of 'cow' in the model includes precisely those members of the domain of the model that are cows; and similarly for every other physical predicate in the language.[37]

---

[35] For Field see notes 27 and 28 above. Vann McGee, "How We Learn Mathematical Language," and "Truth by Default"; Shaughan Lavine, *Understanding the Infinite*, ch. VII §4, and *Skolem was Wrong* (unpublished). McGee concentrates on set theory, and that is the principal concern of Lavine as well. I am not certain that McGee's publications are addressed to our question, and I do not discuss them here.

[36] "Are our Logical and Mathematical Conceptions Highly Indeterminate?" pp. 416–418; "Which Undecidable Mathematical Sentences have Determinate Truth-values?" pp. 340–342 of reprint.

[37] "Which Undecidable Mathematical Sentences," p. 340.

Concerning a nonstandard model of S, Field claims that a set of events whose "finiteness" is witnessed by a nonstandard integer will either have to contain objects satisfying 'events' that are not events, or pairs e, e' of events satisfying 'earlier than' or 'at least one second apart' such that e is not earlier than e', or e and e' are not one second apart. Consider an 'event' in the sense of the model that unfolds at a stretch of time in the nonstandard part. Field's assumption is that that is not an event.

I find it hard to see how someone could accept that assumption who does not already accept some hypothesis that rules out nonstandard models as unintended on mathematical grounds. If our powers of mathematical concept formation are not sufficient to do the latter, then why should our powers of physical concept formation do any better? We are not talking about events in a commonsense context, but, rather, in the context of a physical theory developed in tandem with sophisticated mathematics. Field may wish to insist on this interaction and to claim that our mathematical conceptions themselves get their sense and justification in part from their application in the sciences. But the manner in which he places the weight of securing determinacy on the physical vocabulary, so that the mathematical vocabulary gets its determinacy at second hand, does not entirely square with this. It is more likely that he privileges the physical and believes that it can give determinacy to our concepts through causal relations or some other "externalist" means, whereas mathematical structure does not do this. The issues such a view would raise have to be beyond the scope of this work.[38]

An approach toward a stronger conclusion that promises not to lead into issues of this kind is that taken by Shaughan Lavine in his manuscript *Skolem was Wrong*.[39] The idea on which we have proceeded, that the principle of mathematical induction is entirely open-ended, so that induction should hold for any predicate that we might come to understand and is not restricted to a fixed language, has been modeled in mathematical logic by the device of schematic theories. Reviving an older approach to axiom schemata, the schematic letter in the induction schema (and other principles such as separation and replacement in set theory) is taken into the object language along with a substitution rule. In logical work, such as Solomon Feferman's on reflective closure and unfolding, it has been

---

[38] I will only comment that Hilary Putnam has argued in many places that the notion of causality will not do the kind of work Field seems to be requiring of it.

[39] In the context of relativism about set theory, the idea is sketched in *Understanding the Infinite*, pp. 228–240. I thank Lavine for permission to cite his unpublished work.

§49. Uniqueness and communication

investigated what will result when, beginning with a theory formulated in this schematic form, we follow out to some natural closure the process of expanding the formal language, together possibly with such principles as go with the expansion.[40]

Lavine states as a theorem the categoricity of what he calls the "full schematic theory" of first-order arithmetic. One begins with PA, formulated in the above schematic way with the induction schema in the object language. PA(+), full schematic arithmetic, is then PA with instances of induction allowed for any expansion of the language. Lavine states as a theorem the categoricity of PA(+). The first proof that he gives is model-theoretic and, he states very clearly, not satisfactory for the philosophical purpose. It proceeds in effect by supposing that PA(+) has a nonstandard model and showing that, if one introduces a predicate I true of its standard part, then induction will fail in the model for some predicate containing I.

Already at this stage, however, the question arises how we are to understand PA(+). Lavine offers mathematical results, such as the above categoricity theorem, that contain generalization over languages. Such generalization is familiar from mathematical logic and does not differ essentially from other forms of mathematical generalization. But in its familiar form it is accompanied by some mathematical definition of language, so that "language" is a predicate of some mathematical theory, usually straightforwardly formalizable in some manner or other. However, Lavine makes a distinction between talk of *any* expansion and talk of *every* expansion, and says of a statement about the former that it is a full schematic statement that lets us infer, given an expansion, that it has the property stated. The idea apparently is that there is something schematic about such a statement. For the present, I will leave the matter at that, but it seems that PA(+) is not as precisely defined as mathematical theories generally are.

For a more satisfactory approach to our problem concerning the natural numbers and related problems concerning set theory, Lavine uses the theory of inductive definitions. He considers the theory with two natural number predicates. A predicate defining the isomorphism of the Ns and the N's can be introduced by a positive inductive definition. Given the full schematic character, induction involving it, on either the Ns or the N's, is permitted by the theory. Formulae witnessing the isomorphism are

---

[40] See for example Feferman, "Reflecting on Incompleteness." Lavine gives his own explanation of what he calls a full scheme in "Something about Everything."

provable. Except for using a binary predicate rather than a unary functor, this is essentially the same version of Dedekind's theorem that we have deployed.

There is a difference that Lavine apparently thinks makes his result stronger in a way relevant to the philosophical issue. Lavine describes it by saying that my version[41] only applies to two particular arithmetical languages, whereas his are stated for arbitrary pairs of languages. It is not easy to see what this difference amounts to. Considered as a purely syntactic theorem, it is hard to see any difference at all: After all, I have considered a purely hypothetical case. Lavine indeed suggests that it is this purely syntactic understanding that he himself relies on. Then he would be quite right that there is no background set theory presupposed. The theorem could be formalized in a theory of pure syntax that can describe the predicate calculus and theories formalized in it. Such a theory would not be identical to primitive recursive arithmetic but would be reducible to it.

By itself that does not promise to lead to a stronger result than we have claimed, or even as strong a result: After all, our own argument used, besides the mathematical statement, considerations concerning communication and understanding. What prevents the syntactic theorem from being interpreted with reference to a nonstandard model? What seems to do the work is the kind of generality involved in Lavine's understanding of PA(+). Lavine raises precisely the problem that bedevils the application of all such theorems: If we interpret the result with respect to some model, it shows that the structures for the natural numbers in that model are isomorphic, but it does not prevent interpretation with respect to different models that give rise to nonisomorphic structures of natural numbers.

I think the reply he has in mind is something like this: Suppose you present me with an interpretation of my talk of natural numbers (and different domains of "natural numbers"), in which precisely the feared pathology occurs. My idea was, however, that induction holds for any predicate, no matter how the language is expanded. But in order to interpret me as talking of a nonstandard model, you have to produce an expansion of the language that gives your game away, because you will have to describe your interpretation and thus introduce a predicate, to be sure not one I formulated or necessarily could have formulated, for which

---

[41] I.e., that of "The Uniqueness," pp. 34–35, which does not differ in any relevant way from the version of the theorem above.

induction fails for my numbers. But PA(+) allows induction on this predicate, so that what you have produced is not a model of PA(+). I won't expand this reply further because I don't really disagree with it. I do think, however, that it requires that you enter into the sort of dialogue with me that we considered above. If instead of "you" we have the sort of God I invoked before, then my means of defeating the proposed interpretation do not get off the ground.[42]

§50. Induction and impredicativity

The derivation of induction from an explicit definition of the natural number predicate yields induction for all predicates in the language with which one is working only if one has the equivalent of full second-order logic.[43] I shall consider later in this section what one might accomplish with less. But unless one accepts some substantial restrictions, if one proceeds in this way the impredicative character of second-order logic already infects the natural number concept.

In 1963 Michael Dummett made the following remark:

> ... the notion of 'natural number', even as characterised by the formal system, is impredicative. The totality of natural numbers is characterised as one for which induction is valid with respect to any well-defined property, where by a 'well-defined property' is understood one which is well-defined relative to the totality of natural numbers. In the formal system, this characterisation is of course weakened to 'any property definable within the formal language';

---

[42] As my references to other writers indicate, the issues about uniqueness discussed here have also been discussed with reference to set theory. A prerequisite for the kind of approach taken here would be an essentially first-order version of the "quasi-categoricity" theorem of Zermelo, "Über Grenzzahlen und Mengenbereiche," which states that of any two standard models of second-order ZF, either they are isomorphic or one is isomorphic to the sets of rank $< \kappa$ for some strongly inaccessible κ in the other. Such a version exists and is elegantly formulated in Lavine, *Skolem was Wrong*. It would take us too far afield to go into the question whether an argument of the type offered here can be extended to set theory.

[43] A theorem of John Myhill shows that if in a ramified second- or higher-order logic one defines the natural numbers in Frege's way with the second-order quantifier having a definite level or order, then there will be instances of induction of higher level that will not be provable. See Myhill, "The Undefinability of the Set of Natural Numbers," p. 21. Essentially Myhill's result was stated in 1959 by Wang, with only the briefest indication of a proof, in "Ordinal Numbers and Predicative Set Theory," p. 642. Kurt Gödel had observed that Russell's attempt to prove the contrary in Appendix B of the second edition of volume I of *Principia Mathematica* is fallacious. See "Russell's Mathematical Logic," pp. 135.

but the impredicativity remains, since the definitions of the properties may contain quantifiers whose variables range over the totality characterised.[44]

Dummett, discussing Gödel's incompeteness theorem, was thinking of the natural numbers as characterized by a formal system such as PA. But his point is surely that a characterization of the natural numbers that included induction as part of it will be impredicative; this does not hold just of the characterization by an explicit definition in the context of second-order logic or set theory.[45]

What might be meant by a property that is "well-defined relative to the totality of natural numbers"? I'm not sure that Dummett has in mind more than a well-defined property of natural numbers, that is, what is expressed by a predicate that is significant for natural number arguments. But it seems that what creates the worry about impredicativity is those predicates containing quantifiers over numbers. Because the number concept is characterized as one for which induction holds for any well-defined predicate or property, there is impredicativity if those involving quantification over numbers are included, as they evidently are.

It might seem that on this view impredicativity would arise very early in the development of number theory, as soon as induction is applied to predicates containing predicates or functors introduced by recursion,

---

[44] "The Philosophical Significance of Gödel's Theorem," p. 199. In 1983 I noticed this remark when "The Impredicativity of Induction" was largely complete. Edward Nelson's statement of a very similar point in *Predicative Arithmetic*, pp. 1–2, seems entirely independent both of Dummett's remark and my paper, though possibly not of the paper of Myhill cited in note 43.

Commenting on the result of that paper, Myhill remarks that if one's "mathematical philosophy" is classical,

> one would reason that the result shows that impredicativity is present in mathematics from the very beginning, i.e., the natural numbers, and that consequently any philosophy of mathematics which repudiates impredicative definitions *ipso facto* repudiates mathematics itself. (op. cit., p. 27)

Taken strictly, this claim is very questionable, since a classical view of mathematics hardly requires that the notion of natural number be understood by an explicit definition. If one replaces 'impredicative definitions' by the vaguer term 'impredicative notions', then an argument for Myhill's claim would require something like the (earlier) observation of Dummett's elaborated in the text.

By contrast, Myhill seems to think the constructivist could avoid this conclusion. The view defended here is that he could do so only at the price of a dogmatic view of the clarity of the notion of natural number and the evidence of mathematical induction. However, such a dogmatic view could plausibly be attributed to Poincaré and possibly also Brouwer.

[45] So many different characterizations by set theory are possible that we should not take for granted that any such characterization will share the evident impredicativity of the Fregean one. We will consider some examples later.

§50. Induction and impredicativity

because the presupposition that they are defined already involves the general concept of natural number. At all events it would arise as soon as induction on predicates involving quantifiers over numbers is used. If we allow a rule for introducing functors by primitive recursion, that will not be as soon as one might think, given how much arithmetic can be done in PRA. These remarks leave somewhat unclear when arithmetic becomes impredicative according to this view, but we can leave this question to return to it later.

A more urgent question is whether the explanation of the number concept by rules gives rise to the same impredicativity. To address this question, let us first go back to something very basic. Why is it that impredicative specifications of sets are said to be so? Consider, for example, the set-theoretic version of the Fregean definition of 'natural number'. According to it x is a natural number if it belongs to every *set* that contains 0 and is closed under successor. Consider now the set N_0 of x such that x is a natural number, defined as we have just done. Clearly N_0 falls in the range of the quantifier; it even contains 0 and is closed under successor. Our specification of the set N_0 defines it in relation to a totality to which it belongs. Now clearly impredicativity is a property of the specification or definition of the set N_0. In particular, we don't need to suppose that the definition of 'natural number' gives the meaning of that phrase in any strong sense; 'natural number' doesn't figure essentially in the matter at all. We also don't have to suppose that the specification gives the meaning of the designation 'N_0' of a certain set, since the impredicativity is a property of the specification.

Above, we have effectively extended the designation 'impredicative' from the specification of the set N_0 to the definition of the *predicate* 'natural number'. It's hard to see how anyone could quarrel with this, as in order to apply the definition one will need to pass from predicates to sets that are their extensions. However, impredicativity is still a property of the definition and not, for example, of the set.

In Dummett's remark quoted earlier, it is said that the *notion* of natural number is impredicative. He may be attributing impredicativity to the concept of natural number as such, or to the meaning of the predicate 'natural number'. I am more interested in a weaker thesis that follows from this, that certain explanations of the predicate 'natural number', in general not explicit definitions, are impredicative. That moves less far from the original character of impredicativity as a property of definitions. It might be claimed in addition that the explanations in question are the philosophically most attractive ones, or even that no explanation is in sight that is not impredicative. That was my own view in earlier writing.[46] The standard for an informal explanation to be impredicative or not are not so clear as they are for an explicit definition, so that we can't expect the issue to be quite so clear-cut as it is for explicit definitions or specification of sets by abstracts.

Let us first consider Dummett's case. He assumes that induction – either in the general form we have considered in §47 as holding for any well-defined predicate, or in the more restricted form referring to any predicate definable in the language of a formalism for arithmetic such as PA – is directly part of the explanation of 'natural number'; that is, the predicate is explained as one for which every relevant instance of induction holds. Then, because induction is to be applied to predicates containing quantifiers over natural numbers, the sense in which this explanation is impredicative is pretty clear. In §47, induction, with its generalization over "predicates," was used in an informal explanation of the predicate 'natural number'; this is the first of the two cases of Dummett's remark just identified.

Dummett's stronger thesis that the "notion" or "concept" of natural number is impredicative poses to someone who doubts it the challenge to give an explanation of the concept that passes muster as predicative. One might reject the demand for an explanation at all. I don't see how one can be satisfied with that; in particular, it seems to involve rejecting the demand for an account of the evident character of induction. A more interesting program would be to try to give such an explanation that would be predicative. Some recent work in logic can be viewed in that light.

Before going into that, we should dispel a potential confusion about predicativity. Both in the early discussions of impredicative definitions and in much later logical work on predicativity, a predicative conception has been allowed to assume the natural numbers. Poincaré, who was the first to use the notion of impredicativity for critical purposes, certainly allowed himself that; indeed, what he was most concerned to criticize was reductive accounts of natural number by means of logic. Hermann Weyl, who in *Das Kontinuum* argued that some standard reasoning in analysis should be rejected on grounds of impredicativity, distinguished what he considered the "extensionally definite" character of the concept of natural number from the lack of such definiteness of the concept "property of natural numbers," from which he inferred the lack of definiteness of the

---

[46] In both the 1983 and 1992 versions of "The Impredicativity of Induction."

§50. Induction and impredicativity

notion "set of natural numbers."[47] The work on the analysis of predicativity in the late 1950s and early 1960s assumed the natural numbers; for example, formalisms characterized as predicative were allowed to have induction for any predicate expressible in the formalism. At the beginning of his classic paper on the analysis of predicativity, Solomon Feferman characterizes "the predicative conception" as holding that "only the natural numbers can be regarded as 'given' to us. . . . In contrast, sets are created by man to act as convenient abstractions (*façons de parler*) from particular constructions or definitions."[48] If the predicative conception is understood in that way, of course there is no room for the idea that the concept of natural number, or some particular characterization of the natural numbers, might be impredicative.

This suggests that we should make a distinction in the concept of predicativity between predicativity given the natural numbers, and predicativity understood more narrowly, without allowing at the outset full induction or instances of it that might be claimed to be impredicative. It is the former notion that was at issue in the work on the analysis of predicativity of the 1950s and 1960s. The earlier history may have left an ambiguity on just this point, as Poincaré insisted on the irreducible character of the concept of natural number, whereas Russell followed Frege and Dedekind in offering a reduction of induction to a definition. Weyl effectively followed Poincaré on this point. It is obviously in the context of the narrower notion that the thesis of the impredicativity of ordinary induction, or of the notion of natural number, arises. I will call predicativity in this narrower sense strict predicativity. That there is such a distinction should be uncontroversial. It should be understood as not taking as given not just the natural numbers but also any other notion for which a similar question of predicativity arises, unless, of course, it can be dispelled. But we

---

[47] On this view of Weyl, see my "Realism and the Debate on Impredicativity, 1917–1944." For a fuller account of the approach of *Das Kontinuum* and its significance, see Feferman, "Weyl Vindicated."

[48] "Systems of Predicative Analysis," pp. 1–2. Feferman appears to echo the famous aphorism of Kronecker, that God created the natural numbers and everything else in mathematics is the work of man. Bernays remarks that the point of view of *Das Kontinuum* fits better with this aphorism than did Kronecker's own view, which is closer to finitism ("Die Philosophie der Mathematik und die Hilbertsche Beweistheorie," p. 41 n.).

It should be remarked that in later proof-theoretic studies, systems with restricted induction have played a considerable role, and in particular the question whether full induction is admitted may be relevant to the question whether a formal theory is predicatively *reducible*. See Simpson, "Predicativity: The Outer Limits." Systems with restricted induction play an important role in the work on "reverse mathematics"; see Simpson, *Subsystems of Second-Order Arithmetic*. It is still true that the systems that characterize predicative provability have unrestricted induction.

have not ruled out the possibility that methods in mathematics that are predicative in the traditional sense might still be justified from a strictly predicative point of view.

The thesis of the impredicativity of ordinary induction has been challenged in two papers by Feferman and Geoffrey Hellman.[49] I don't take them to reject the conceptual distinction between predicativity given the natural numbers and strict predicativity. They do, however, seem to undertake to show that first-order arithmetic can be developed in a strictly predicative way. That would be so if the formal theories of finite sets that they work with are strictly predicative. I'm not certain that they claim this; the difficulty is that they state that "predicativism must presuppose the concept of 'finite' in some form or other," and what they insist on is that one can do this "in a natural way without taking the natural numbers as given."[50] Whatever their view actually is, the question whether their theories are strictly predicative is of interest in its own right.

What Feferman and Hellman do is to give a formal theory of finite sets and first-order definable classes in which they can interpret arithmetic, effectively by defining Dedekind's notion of a simply infinite system, what they call an N-structure.[51] In their theory EFSC it is possible to prove the existence of an N-structure and its uniqueness up to isomorphism, that is, Dedekind's categoricity theorem.[52]

EFSC is an extension by class variables of a theory EFS of individuals and finite sets of them, whose language could be taken to be that of the two-sorted version of the theories of finite sets of §33. There are axioms for the empty set and for adding an element to a finite set, corresponding to our (H2) and (H3). However, instead of Zermelo's principle of induction on finite sets, EFS has the schema of separation

(AS) $(\exists b)(\forall x)[x \, \varepsilon \, b \leftrightarrow x \, \varepsilon \, a \wedge A]$.

In the system so far, even with the addition of classes, it would not be possible to prove that there are infinitely many individuals. The case is

---

[49] "Predicative Foundations of Arithmetic" and "Challenges to Predicative Foundations of Arithmetic."
[50] "Predicative Foundations," p. 15.
[51] The domain of such a system is obviously not a finite set; the theories with which they work contain variables for classes, and an N-structure is a class. The extension by class variables is conservative.
[52] Originally, in "Predicative Foundations," the existence of an N-structure was proved only in the system EFSC* with the additional axiom (Card) that every finite set is Dedekind finite (p. 11). Peter Aczel showed how to modify the example so that the existence of an N-structure could be proved in EFSC; see "Challenges" p. 320.

§50. Induction and impredicativity

like that of the two-sorted theory of §33. However, EFS also has a pairing function (x, y) with the axioms

(P1) $(x_1, x_2) = (y_1, y_2) \leftrightarrow x_1 = y_1 \wedge x_2 = y_2$

(P2) $(\exists u)(\forall x)(\forall y)[(x, y) \neq u]$.

The elementary Peano axioms are satisfied by taking as 0 a u given by (P2) and $Sx = (x, 0)$. This fact is used to describe an N-structure.

One's first thought about the idea of showing arithmetic to be strictly predicative by basing it in one way or another on the notion of finite set is that the most it could show is that the arithmetic that is derived or interpreted is predicative *relative to the notion of finite set*. Then the question will arise how we explain the notion of finite set itself. One obvious explanation, that a finite set is a set whose cardinality is a natural number (say, such that there is an n such that the set can be mapped 1–1 onto the numbers from 1 to n), will simply throw the question of impredicativity back to the corresponding question concerning the natural numbers. As Feferman and Hellman are well aware, a basis of arithmetic in a theory of finite sets has to understand the latter notion in some way that does not presuppose the notion of natural number.

Our own procedure in Chapter 6 was to rely on an explanation of the notion of finite set using induction that would be closely analogous to the explanation of the notion of natural number by rules that was presented in §31 and whose relevance to questions concerning induction was discussed in §47 and the present section. Given a domain of individuals, the "introduction rules" for sets of individuals would state that the empty set is a (finite) set and, if a is a set and y an individual, then $a \oplus y$ is a set whose elements are those of a and also y. The "elimination rule" would just be Zermelo's induction principle ((H4) in §33): If a predicate A holds of the empty set and, for any finite set a and individual y such that A holds of a, A also holds of $a \oplus y$, then a holds of all finite sets. This could be motivated in much the same way that induction on natural numbers is motivated, as we discussed in §47.

But then there is no more reason to believe that this explanation is strictly predicative than there is to believe that the explanation of the notion of natural number by rules is strictly predicative. So far, however, this observation is no criticism of Feferman and Hellman, because they do not offer an explanation of this form; in particular, neither of the theories EFS and EFSC has an axiom of induction on finite sets. If we compare EFS with the theory of §33, apart from the presence of pairing in the former, the important difference between them is that instead of induction EFS has the axiom schema of separation. It is even obvious that Zermelo's induction principle in its bald form will not be provable in EFS: We cannot prove that if a predicate holds for $\varnothing$ and, for any set a, whenever it holds for a it holds for $a \oplus y$ for any y, then it will hold for all sets. The reason is, however, that although the axioms of EFS contain principles only for generating finite sets (or adding finitely many elements to a given set), there is nothing more in them that expresses the intention that the variables range only over finite sets.[53]

Given, however, that it is possible to define an N-structure in EFSC, it is possible to define an inner model in which Zermelo's induction principle holds. Given an N-structure, one can define a set a to be really finite ($RF(a)$) if there is a one-one mapping of a onto $\{y: y \leq n\}$ for some $n \, \varepsilon \, N$. The mapping can be assumed to be a set, finite on Feferman and Hellman's reading, but not necessarily really finite. One can then prove in EFSC that the really finite sets satisfy Zermelo's induction principle.[54]

The conclusion we should draw is that if Feferman and Hellman's construction really offers a strictly predicative development of PA, it also offers a strictly predicative development of the theory of finite sets based on Zermelo's induction principle. But perhaps this result should make us doubtful about the claim of the development to be strictly predicative. Apart from that, I think it might avoid potential misunderstanding if we refer to the objects in the range of the set variables of EFS and EFSC simply as *sets*, rather than as finite sets as Feferman and Hellman do, since although the axioms do not force the existence of any infinite sets, they

---

[53] Feferman and Hellman point out that EFSC is essentially a subtheory of von Neumann-Bernays-Gödel set theory NBG ("Predicative foundations," p. 4) and thus is consistent with the axiom of infinity. It follows immediately that the characteristic axiom (Card) of EFSC* (see note 52) is not provable in EFSC, since it is refutable in NBG that every set is Dedekind finite. It is surprising that Feferman and Hellman do not state this, since they can hardly have failed to be aware of it.

[54] This is carried out in detail in Ferreira, "A Note on Finiteness." It should be possible also to use one of the definitions of finiteness proposed in the first quarter of the twentieth century that do not involve reference to the natural numbers or a structure isomorphic to them. Ferreira effectively carries this out for the definition of a set as finite if it has a linear ordering and every linear ordering of it has a first and last element (provided it is nonempty). This definition was proposed by Heinrich Weber in 1906. (See my "Developing Arithmetic in Set Theory," p. 205.)

Evidently, the really finite sets satisfy the principle of boundedness alluded to in "Predicative Foundations," p. 5. As Ferreira notes, the usual inductive proof of the pigeonhole principle shows that they satisfy (Card) (see note 52). Obviously, however, EFSC will have nonstandard models in which some sets satisfy $RF(a)$ and yet are infinite.

§50. Induction and impredicativity

are true in the standard set theory that allows infinite sets.[55] Then the term 'finite set' could be reserved for the really finite sets.

Then we can view Feferman and Hellman's strategy as being to develop arithmetic in a predicative theory of sets, without the presupposition that the sets involved are finite, even though it is intuitions concerning finite sets that motivate the axioms. The axioms have the character that, if we think of the notion of set involved as explained by the formal system, they do not give rise to the specific problem posed by those who have found explanations involving induction to be impredicative.

However, there is a more traditional problem raised by the theories EFS and EFSC, which Feferman and Hellman do not explicitly address.[56] The axiom of separation has the same apparently impredicative feature that it has in the setting of standard set theories: Given a set a, we have a set $\{x \; \varepsilon \; a : A\}$ for a formula A containing quantifiers over all sets in the domain.[57] In EFSC, induction for an arbitrary N-structure is given by the definition; its application to a formula A is obtained by the axiom of class comprehension, provided that A contains no bound class variables. This step does not raise an issue of impredicativity. However, Aczel's proof that the specific structure he defines is an N-structure indicates that we cannot rest so easily. We need to look at this proof in a little detail.

We define:

$$Clos^-(a) \leftrightarrow (\forall x)(Sx \, \varepsilon \, a \to x \, \varepsilon \, a)$$
$$y \leq x \leftrightarrow (\forall a)[x \, \varepsilon \, a \wedge Clos^-(a) \to y \, \varepsilon \, a].$$

The point is now somewhat clearer if we confine ourselves to EFS. We define

$$N(x) \leftrightarrow (\exists a)[(\forall y)(y \, \varepsilon \, a \leftrightarrow y \leq x) \wedge x \, \varepsilon \, a].$$

---

[55] Apart from the fact that on their intended interpretation the set variables of EFS and EFSC range over finite sets, it is likely that Feferman and Hellman choose their terminology because thought of as general set theories, these theories have some awkward weaknesses; for example, it does not seem possible to prove the existence of $a \cup b$ for arbitrary sets a, b. (It is provable if b is restricted to be really finite.) In fact, this is awkward already if the theories are thought of as theories of finite sets. However, it seems that adding it as an axiom would not compromise the claim of the theories to be strictly predicative.

[56] This is somewhat surprising in view of the fact that the relevance of this issue to that of the impredicativity of induction was already noted in George, "The Imprecision of Impredicativity."

[57] In EFSC, A is not allowed to contain bound *class* variables. But that does not affect the point in the text.

We can now prove $N(0)$ and $N(x) \to N(Sx)$. To prove that induction holds for $A(x)$, assume $A(0)$ and $(\forall x)[A(x) \to A(Sx)]$. Suppose also $N(x)$ and $\neg A(x)$. Then $\{y: y \leq x \land \neg A(y)\}$ is a set b, since it is contained in the a that witnesses $N(x)$. Evidently $x \, \varepsilon \, b$, and b is closed under predecessor. Therefore $0 \, \varepsilon \, b$, which contradicts the assumption $A(0)$.

In particular, if A contains quantifiers over N, then the set b is specified in a way that involves quantification over all sets in the domain. The matter may look different in EFSC, since in that case one proves induction for a class variable and then instantiates for it, as noted above. But the difference is only apparent unless one contents oneself with reasoning about arbitrary N-structures without troubling oneself with the question of existence.[58] Clearly, EFSC admits the existence of sets that are specified by quantification over all sets, and this assumption is used in proving the existence of an N-structure.

For this reason, I don't think that EFS and EFSC can pass muster as strictly predicative. The question arises why Feferman and Hellman seem not to be troubled by the above point. My conjecture is that because the intended domain is that of finite sets of individuals, there is no conflict with what they call "definitionism," the view that sets are the extensions of predicates that can be understood independently.[59] A finite set of individuals is of course $\{x: x = a_1 \lor \cdots \lor x = a_n\}$, where $a_1, \cdots, a_n$ are its elements. Feferman and Hellman remark, "[t]he finite sets correspond to finite lists of designators [i.e., names of the elements], and it is reasonable for the definitionist to take *this* notion – 'finite list of quasi-concrete objects' – as understood."[60] We should point out, however, that any proof that the sets definable in this way are closed under separation is nonconstructive. Suppose a is $\{x: x = a_1 \lor \cdots \lor x = a_n\}$. Then, for a formula $A(x)$, one could only actually give such a list for $\{x \, \varepsilon \, a: A(x)\}$ by deciding of each a_i whether $A(a_i)$ holds. This is in fact a reason for preferring Zermelo's induction principle over separation as an axiom schema for a theory of finite sets. Then it is quite evident that every finite set is given by such a definition. It can be proved by induction that the separation schema holds, but the nonconstructivity comes out clearly in that at the induction step one must divide cases according to whether the new

---

[58] That would suggest that Dedekind's categoricity theorem *is* strictly predicative. Nothing argued in the text implies the contrary. However, further reflection is needed on this point.
[59] Feferman used this term in "Weyl Vindicated" and possibly earlier.
[60] "Challenges," p. 323. Evidently they mean quasi-concrete in the sense of §7; such a list is close to a string of the kind that plays a central role in chapter 5.

§50. Induction and impredicativity

element satisfies A or not.[61] Thus, what requires a nonconstructive proof is the existence of the set given by separation, rather than the principle that finite sets are definable by a list.

Another possible objection to the claim of EFS and EFSC to be strictly predicative arises by asking how we understand the domains of the individual and set variables. One might raise the question what reason there is to accept a domain of individuals closed under pairing, or a domain of sets even satisfying the axioms (H2) and (H3) and unquestionably predicative instances of separation, and then going on to use classical logic. Hermann Weyl in *Das Kontinuum* was willing to apply classical logic in a language involving quantifiers over a domain and then to introduce first-order definable classes, if the domain was what he called "extensionally definite." He argued that that is the case for the natural numbers but not for the sets of natural numbers or the real numbers. But wouldn't the claim of the extensional definiteness of the individuals or sets of EFS raise questions of impredicativity again?

I don't want to press this objection, because the setting in which questions of strict predicativity have been pursued allows one to apply classical logic in a setting where one has certain generating operations but doesn't suppose one has the equivalent of Weyl's extensional definiteness for the domain. We should grant Feferman and Hellman this assumption, just as Edward Nelson, in the first serious logical work on strict predicativity, allows himself arithmetic Q without induction. It may be that Feferman and Hellman don't quite know what the domain of their variables is. But I don't think they are in this respect worse off than other researchers on predicative foundations of arithmetic, in particular those to be discussed presently. The problem of a predicative foundation of arithmetic is to arrive at subclasses of the domain that can serve as numbers or "really finite" sets and thus suffice for arithmetic.

Although I don't think Feferman and Hellman's argument succeeds in showing that, say, PA is strictly predicative, their work has advanced the discussion of such issues, first of all by clarifying what assumptions about finite sets are needed in order to interpret arithmetic, and thus by making clearer what is involved in different versions of the view, which in general terms was not in dispute between us, that arithmetic is predicative relative to the notion of finite set.

---

[61] See Appendix 1 to Chapter 6. A similar nonconstructivity will arise in showing in EFS or EFSC that the really finite sets satisfy separation.

I now turn to the logical work that in my view does show that certain fragments of arithmetic are strictly predicative. Although there is relevant work from earlier, probably the first work pursuing the program of developing arithmetic in a strictly predicative way is Edward Nelson's *Predicative Arithmetic*. This work undertakes to develop as much arithmetic as possible in formal theories that are interpretable in Q, that is, first-order arithmetic without induction but with an axiom stating that every non-zero number has a predecessor. Q thus has no induction at all; its ability to prove even very elementary generalizations about numbers is very weak.[62] It is strong in another way, however: It is the simplest natural theory for which Gödel's first incompleteness theorem holds, and it is essentially undecidable; that is, no consistent axiomatizable extension of it is decidable.[63] We might then offer as a criterion for an arithmetical theory to be predicative that it should be interpretable in Q. Nelson then showed that what is called $I\Delta_0$, the subsystem of classical first-order arithmetic PA in which induction is restricted to formulae containing only bounded quantifiers, is locally interpretable in Q, and subsequently it was shown that even an extension of $I\Delta_0$ is globally interpretable in Q. That amounts to saying that we could write down in the language of PA a predicate N_p (to be read: x is a pseudo-number) so that all theorems of $I\Delta_0$ are provable in Q when the quantifiers are restricted to N_p.[64] (The extension consists of adding a new function with its recursion equations, a function, however, that grows more slowly than exponentiation.)

It follows that by Nelson's criterion an arithmetic whose recursive functions are by a reasonable standard feasibly computable is strictly predicative. Nelson's method did not show that exponentiation could be added so as to preserve interpretability in Q, and he expressed strong doubts as to whether that could be done. Being concerned independently with feasibility, he probably would regard it as a welcome result if considerations of strict predicativity led to a limitation of arithmetic to feasibly computable functions, so that strictly predicative arithmetic and one possible conception of strictly finitist arithmetic would be very close and maybe

---

[62] See note 75 of Chapter 5.
[63] It is for this reason that Q was first formulated and applied; see Tarski, Mostowski, and Robinson.
[64] Local interpretability amounts to the claim that one can do this for any finite subset of the axioms of $I\Delta_0$.

§50. Induction and impredicativity

coincide. Subsequently it was shown that $I\Delta_0 + exp$ is not interpretable in Q, confirming Nelson's expectation.[65]

A competing apparently more direct analysis, however, leads to the conclusion that the total character of exponentiation can be proved in a strictly predicative way. This analysis pursues the natural idea that one should see how much arithmetic one can do by starting with something like Russell's ramified type theory without the axiom of reducibility.[66] Admitting 0 and S with the elementary Peano axioms amounts to assuming a version of the type-theoretic axiom of infinity. We can then use the Frege-Dedekind definition of the natural numbers, with the second-order quantifier assigned a specific order. If we write N^{k+1} when the quantifier is of order k, then the defining formula of N^k is of order k. Then the project would be to show that for some k, if we define the numbers by N^k, then we will be able to develop some arithmetic in our ramified higher-order theory. In fact, for the work reported later, ramified second-order logic is sufficient.

It was already shown by Skolem, in *Abstract Set Theory*, that we can develop the arithmetic of addition, multiplication, and order in this way. This result was extended to bounded induction by Allen Hazen, and recent work of John Burgess and Hazen has shown that exponentiation can be added; for this method the obstacles found by Nelson can be overcome. Thus the system $I\Delta_0 + exp$, which already allows as total a function that is not feasibly computable, is predicative by the Russellian criterion.[67] It follows that the Russellian criterion is more generous than Nelson's. I am inclined to conclude that Nelson's criterion is too narrow. In fact it seems to have been motivated as much by considerations of feasibility as of predicativity.

On might seek to extend the Russellian criterion so as to allow transfinite levels, using the procedure of autonomous iteration central to the Feferman-Schütte analysis of predicativity given the natural numbers.

---

[65] Hájek and Pudlák, *Metamathematics of First-Order Arithmetic*, p. 391, prove that $I\Delta_0 + exp$ is not interpretable in $I\Delta_0$, from which the statement in the text obviously follows.

$I\Delta_0 + exp$ would most readily be formulated by adding to $I\Delta_0$ the recursion equations for exponentiation and allowing exponentiation in induction axioms. However, since the predicate $z = x^y$ is definable in $I\Delta_0$ by a Δ_0 formula (ibid., p. 299), the theory just mentioned is a conservative extension of the result of adding to $I\Delta_0$ the axiom $(\forall x)(\forall y)(\exists z)(z = x^y)$. (The uniqueness of z is provable in $I\Delta_0$.)

[66] A modern formulation of this theory can readily be extracted from Church, "Comparison."

[67] Burgess and Hazen, "Predicative Logic and Formal Arithmetic."

A further result of Burgess (in the paper cited in note 67) is that if one uses predicative logic with all finite levels, one can interpret a theory with iterated exponentiation; however, bounded induction is weakened to quantifier-free induction, in a language that does not have enough recursion to derive $I\Delta_0$. This result does, however, suggest the conjecture that the addition of iterated exponentiation to $I\Delta_0 + exp$ would prove to be predicative by this extended Russellian criterion. This theory is capable of proving the consistency of Q.[68]

It is not known at this point what the limits of predicativity in arithmetic will be on a transfinite version of the Russellian criterion. It seems to me a virtual certainty that no more than first-order arithmetic will turn out to be strictly predicative and probable that strict predicativity will not extend beyond primitive recursive arithmetic.[69] Moreover, more reflection will be needed before we can conclude that an extension of Russell's hierarchy into the transfinite still answers to the intuitive idea of strict predicativity.

Already concerning the procedure by which the weaker results are obtained, one may ask how we understand the ascent of levels. This ascent is not very high; to obtain the result concerning exponentiation the number predicate used is N^3, so that in the proof the bound second-order variables used are all of level 1 or 2. Still, we should assure ourselves that there is not some hidden impredicativity in our presumed understanding of this fragment of the ramified language. Although I do not know of any, this is a point on which further reflection is needed.

Whatever the outcome of the work on strict predicativity, however, there is certainly a point at which arithmetic reasoning becomes

---

[68] This was observed by Kripke some years ago.

[69] The first would, I think, follow from a claim of Kreisel concerning finitism, that if one begins with primitive recursive arithmetic and introduces new ordinals by the procedure of autonomous iteration familiar from Feferman's work on predicativity, the limit of what one can obtain is a system of the same proof-theoretic strength as first-order arithmetic; in particular, ε_0 is the limit of the ordinals that are obtained.

We might also mention another type of theory that might be relevant to questions about strict predicativity, subsystems of the inconsistent system of Frege's *Grundgesetze*. A Fregean logic based on ramified instead of impredicative second-order logic has been shown consistent by Richard G. Heck Jr., in "The consistency of Predicative Fragments of Frege's *Grundgesetze*." Predicative theories of a Fregean character are discussed by Burgess in *Fixing Frege*, Chapter 2. Burgess carries out the construction of "Predicative Logic and Formal Arithmetic" in a Fregean theory with just two levels of "concepts." That system certainly has as much claim to be strictly predicative as its Russellian counterpart. Heck's system is more complex, and there may be nonobvious reasons for doubting that it is strictly predicative.

impredicative. An understanding of schematic generalization of predicates is part of our understanding of arithmetical reasoning, and therefore of our understanding of the number predicate itself.

§51. Predicativity and inductive definitions

The main argument of the previous section consisted essentially in applying to the case of the natural numbers a well-known argument for the impredicativity of inductive definitions. This was applied in particular to so-called "generalized inductive definitions" in which the introduction rules for an inductively defined predicate P may contain general statements involving P. A general form of such definitions consists, for a formula $A(P, x)$ containing the predicate P to be introduced but otherwise only previously understood vocabulary, of the introduction rule:

From $A(P, a)$ infer Pa.

and the induction principle (elimination rule):

From $[A(\{x : B(x)\}, a) \to B(a)]$ and Pt infer $B(t)$,

which expresses the idea that P is minimal so as to satisfy the condition given by the introduction rule. Such a means of introducing a new predicate is not always coherent; a sufficient condition is that A should be *monotonic*, that is that

$$(\forall x)[A(P, x) \wedge (\forall y)(Py \to Qy) \to A(Q, x)]$$

should hold for all P, Q. I will assume this in what follows. Monotonicity is assured if P occurs only *positively* in A, that is, if A is constructed using only $\neg$, $\wedge$, $\vee$, and quantifiers, and P occurs only within an even number of negation signs.

If P is introduced in this way, we can think of its extension as built up in stages, indexed by ordinals: At stage 0 the extension P_0 is empty; $P_{\alpha+1}a$ holds if and only if $A(P_\alpha, a)$; for a limit ordinal λ, $P_\lambda a$ holds if and only if for some $\beta < \lambda$, $P_\beta a$ holds. Monotonicity implies that if $\alpha < \beta$, then $P_\alpha a$ implies $P_\beta a$. If an α is reached such that $P_{\alpha+1}x \leftrightarrow P_\alpha x$ for all x, then P_β remains unchanged for all further β, and we can take P_α as P. Moreover, assuming the underlying domain to be a set, there will be such a "closure ordinal," for if the closure ordinal has not been reached by stage α, then α must be of cardinality no greater than that of the domain. Thus in set theory, we can prove that monotonic inductive definitions yield well-defined predicates.[70]

A simple example of such a "definition," relevant to our argument, is that of the notion of accessibility for a relation R of natural numbers, which we will assume to be a linear ordering. Intuitively, '$Acc(a)$' means that transfinite induction holds for the relation $\{\langle x, y \rangle : xRa \wedge xRy\}$; it can therefore be given an obvious second-order explicit definition. In the inductive definition, the introduction rule takes the form

From $(\forall x)[xRa \rightarrow Acc(x)]$ infer $Acc(a)$.

whereas the elimination rule or induction principle says that if $A(x)$ satisfies the above closure condition, then it holds for all accessible numbers, that is,

From $(\forall x)[xRa \rightarrow A(x)] \rightarrow A(a)$ and $Acc(t)$ infer $A(t)$.

In proof theory, beginning with Gerhard Genzten's 1936 proof of the consistency of first-order arithmetic, transfinite inductions up to certain countable ordinal numbers have been regularly used. These ordinals can be represented arithmetically by primitive recursive orderings of the natural numbers. If R is such an ordering, in many cases the notion of accessibility can be used for formal derivations of transfinite induction; standard derivations in texts on proof theory can be recast in this form.[71]

However, one comes rather quickly to need to use instances of the induction schema where the formula $A(a)$ contains 'Acc', so that, again, impredicativity enters. For this kind of reason it has been argued that this and other important generalized inductive definitions are impredicative. For example, in the course of presenting his own analysis of predicativity, Solomon Feferman offers such an argument for the impredicativity of the

---

[70] For information concerning generalized inductive definitions and their importance for proof theory see the Introduction (by Feferman and Sieg) to Buchholz, Feferman, Pohlers, and Sieg. Cf. also Aczel, "An Introduction to Inductive Definitions." This informative article is not as introductory as its title would suggest.

[71] For example, this is true of the derivations in §21 of Kurt Schütte, *Proof Theory*, which are predicative by the Feferman-Schütte criterion discussed below, and of those in §22 of Schütte's earlier *Beweistheorie*, which are not. On the latter, cf. his "Logische Abgrenzungen des Transfiniten," p. 110. To handle the ordering dealt with in §29 of *Proof Theory*, one needs arbitrary finite iteration of inductive definitions.

All these orderings are primitive recursive, and their elementary properties can be proved in primitive recursive arithmetic. It is in the problem of deriving transfinite induction that the constraint of Gödel's second incompleteness theorem shows itself. Gentzen's proof, for example, can be carried out with a single instance of induction on an ordering of order type ε_0. Thus this instance cannot be derived in first-order arithmetic.

§51. Predicativity and inductive definitions 309

usual definitions of Kleene's set O of recursive ordinal notations.[72] In the case of *Acc*, applications of it to prove transfinite induction for shorter well-orderings (such as one of type ε_0 that might be used for Gentzen's or another proof of the consistency of arithmetic) can be replaced by other methods that will be recognized as predicative. But this will no longer be true for longer orderings.[73]

The sense of "predicative" in which these last remarks hold is captured by the beautiful and persuasive analysis of predicative provability of Feferman and Schütte, already alluded to in the last section.[74] According to this analysis, generalized inductive definitions like that of '*Acc*' and O are impredicative principles of proof. I shall not enter here into the details of the analysis, which are complex. It turns on looking at transfinite iterations of different methods of enlarging the means of expression and proof of formal systems. The key condition is that if some such method of enlargement is iterated through a transfinite sequence of stages, the stages can be antecedently recognized to be well-ordered. But at no point is any restriction placed on ordinary induction on natural numbers. That was appropriate for what it intended to capture, predicativity *given the natural numbers*, where the problem of predicativity is raised in the first instance about reasoning about *sets* of natural numbers, and then perhaps extended to further reasoning about sets and functions. I repeat the quotation from Feferman given in §50:

---

[72] "Systems of Predicative Analysis," p. 5. Cited hereafter as "Systems."
[73] If the inductive definition of accessibility is our only means of proof beyond arithmetic, we need to use the elimination rule with predicates containing '*Acc*' even in some cases that are clearly predicative given the natural numbers. Does this cast doubt on our argument for the impredicativity of ordinary induction? In the present situation, such prima facie impredicativities can be replaced by other methods of proof. For example, one can use ramified second-order logic (i.e., second-order logic with the second-order variables assigned levels, so that the variables of a given level can be interpreted to range only over second-order entities defined by means of quantifiers of lower levels), where if transfinite levels are needed, they can be antecedently shown to be well founded. No such alternatives are in sight in the case of ordinary induction. Nonetheless, one might hope for greater clarity on this point from continuation of the logical work on strict predicativity described in §50.
[74] Feferman, "Systems." This is still the best source for his motivating ideas and for the statement of the most basic technical results, although his analysis is refined and extended in later papers.

Schütte offered an analysis based on the same basic ideas and obtained independently some of the relevant technical results but did not carry the matter as far as Feferman, mainly because the only form of predicative analysis that he considered was based on the ramified hierarchy. In "Logische Abgrenzungen" he presents his ideas very clearly and describes his results informally.

According to this [the predicative conception], only the natural numbers can be regarded as "given" to us. . . . In contrast, sets are created by man to act as convenient abstractions . . . from particular conditions or definitions.[75]

Feferman and others have suggested that an analysis of predicativity should describe the concepts and principles that are in some way implicit in the conception of natural number. The manner in which reasoning about classes, sets, or functions is allowed is based on the conception of classes as extensions of predicates, and so a generalization about classes (or sets) must be cashable as a generalization about the predicates of which the classes are the extensions. Such generalizations about predicates will be semantical in character, involving satisfaction or truth. Thus semantic reflection, or what Lorenzen some years ago called 'logical reflection', is the manner in which second-order entities are understood.[76] By semantic reflection I mean passing from statements using a certain vocabulary to statements applying semantic notions (reference, truth, satisfaction) to that vocabulary, thus applying what in Chapter 1 is called the method of semantic ascent. From Tarski's indefinability theorem, it follows that for classical theories semantic reflection is an enlargement of one's means of expression.[77]

One can obtain considerable strength by iterating such reflection into the transfinite, as in the construction of the ramified hierarchy. But on the Feferman-Schütte conception, the iteration is constrained by the requirement that its stages (ordinals) be given in advance as well-founded. In contrast, a generalized inductive definition like that of O or 'Acc' allows objects to fall into the extension of the introduced predicate by iteration of the introduction rule (according to the above-described iterative process), but the stages needed, if one does not take them as given from set theory, are described only by the definition itself or another comparable one. This is exactly parallel to the situation with the natural numbers.

The Feferman-Schütte conception of predicative mathematics is constructed from elements that are very basic and deeply entrenched parts of our conceptual apparatus: either first-order logic or a second-order

---

[75] "Systems," pp. 1–2. Schütte is not so explicit on the point.
[76] Lorenzen, "Logical Reflection and Formalism," p. 244. In more recent work Feferman has explored the idea of "reflective closure" of theories, to capture the idea of "closing" a theory under semantic reflection. Predicative analysis as he has previously described it turns out to be what he calls the "strong reflective closure" of first-order arithmetic. See his "Reflecting on Incompleteness."
[77] Semantic reflection and its role in mathematics are discussed in considerable detail in *Mathematics in Philosophy*, particularly Essays 1, 3, 8, and 9.

§51. Predicativity and inductive definitions 311

logic with quite minimal comprehension assumptions, the notion of natural number, and semantic reflection. Describing the limits of what can be accomplished by these means not only was a technical achievement but served to delineate a natural conceptual boundary. Some of the discussion of constructivity and predicativity in the immediately preceding period shows the lack of a clear distinction between this apparatus and a more generous conception that would extend it at least by certain generalized inductive definitions. Feferman's criticism of Lorenzen and Wang on this point is justified.[78] However, I shall argue that the distinction can be seen as one between two senses of predicativity.

If one grants a certain impredicative character to ordinary induction, the issue between Feferman and the earlier writers appears in a somewhat different light. It is hard to see how a mode of concept formation which involves a "vicious circle" in the case of generalized inductive definitions does not involve such a circle in the case of the natural numbers themselves. Granted, then, that Feferman has correctly characterized the limits of predicativity relative to the natural numbers, the case that the traditional arguments deriving from Poincaré show that this is the limit of acceptable mathematics is weakened.

Lorenzen usually characterized his position as "constructivist" or "critical"[79] and to that extent did not depend on a particular interpretation of predicativity. However, he did specifically claim that generalized inductive definitions are predicative, though to be sure in a joint paper.[80] Clearly, my view is that this claim is mistaken. However, at this point one should make another distinction. The primary sense of impredicativity applies to sets or classes, and they are said to be impredicatively defined if they are given by abstracts involving quantification over some totality of sets to which they themselves belong. This extends readily to other cases; for example, Russell's diagnosis of the semantical paradoxes involves pointing out that sentences such as 'The proposition expressed by my present utterance is false' or 'Every proposition asserted by a Cretan is false', asserted by Epimenides the Cretan, when taken naively, purport to express propositions that are in the range of their own quantifiers. In the

---

[78] "Systems," p. 5.
[79] In various of his writings, including those cited in notes 76 and 80 as well as his book *Einführung in die operative Logik und Mathematik*.
[80] Lorenzen and John Myhill, "Constructive Definition of Certain Analytic Sets of Numbers," pp. 47–8. Hao Wang also proposed that generalized inductive definitions be included in predicative theories, for example, in §5 of "Ordinal Numbers and Predicative Set Theory." On p. 644, there is intimation of a *ceteris paribus* argument like that offered here.

present case, inductive definitions are said to be impredicative because they involve introducing a predicate by rules or axiom schemata such that expressions containing the predicate itself have to be admitted as within their scope. All these cases are of the type where the question of a vicious circle was raised by Poincaré, Russell, and Weyl.

However, if we reflect on the motives of the original critique of impredicativity, an underlying conception was that classes or sets are extensions of predicates and, therefore, that the circle lay in speaking of sets that could not be extensions of predicates antecedently understood. This conception is quite clear in the writings of Poincaré and Weyl.[81] In the case of Russell, it is perhaps not so clear because of his unclear conception of the relation of propositional functions to language. However, his original adoption of a vicious-circle principle was as a guiding principle in the construction of a "no-class" theory, in which classes were to be eliminable by contextual definition, and propositional functions were treated in a completely predicative way.

Historically, what has served to defuse the critique of impredicativity is set-theoretic realism, more particularly, the abandonment of the idea that sets are extensions of predicates in a given language, so that the domain of sets one can quantify over has to be seen as potential, expanding as one's linguistic resources expand, in particular by quantifying over totalities of sets previously arrived at. Russell took the realistic attitude in a somewhat half-hearted way in introducing his axiom of reducibility, of which he said that it accomplishes "what common sense effects by the admission of classes."[82]

To return to Lorenzen, it is quite clear that he is especially concerned to avoid this set-theoretic realism, what he calls "naive" concepts of set, relation, and function.[83] His concept of set is just the one that underlies the critique of Poincaré and Weyl, even though the issue of a "vicious circle" does not occupy the center of his attention.

---

[81] Although I find it clear enough, in Poincaré it is not quite so explicit, or so clearly disengaged from other considerations such as rejection of the actual infinite, perhaps because of his negative attitude toward symbolic logic. But he makes it quite clear that he expects sets to be definable, most explicitly in the first essay of *Dernières Pensées*; see especially the criticism of Cantor's diagonal argument in §6.

For Weyl, see the analysis of the concept of set in *Das Kontinuum*, §5, and its polemical use in "Der *circulus vitiosus* in der heutigen Begründung der Analysis."

[82] "Mathematical Logic as Based on the Theory of Types," *Logic and Knowledge*, p. 81. However, that realism was central to the defense of impredicativity in the interwar period is questioned in my "Realism and the Debate on Impredicativity."

[83] "Logical Reflection and Formalism," pp. 246–247.

§51. Predicativity and inductive definitions 313

Because it is generally agreed that there is a coherent conception of "constructive" mathematics which goes beyond the predicative as characterized by Feferman and incorporates theories of generalized inductive definitions but which still does not presuppose set-theoretic realism, it might seem that our discussion can just end with the observation that Lorenzen held such a constructive conception. However, in my view there is still a remark worth making, which justifies to some extent the use of the word 'predicative' by Lorenzen and Myhill. Constructivists have not always been very explicit about the relation of higher-order entities to language; indeed, Brouwer's own radical view that mathematics is essentially independent of language works against such clarity, particularly in his notion of species (i.e., class). The term 'impredicative' was used by Poincaré because of his view of sets or classes as essentially extensions of predicates;[84] in a terminology I have used elsewhere, the language of classes serves as a means of generalizing predicate places in a language. But then it is entirely natural that the classes in the range of a generalization should be the extensions of predicates antecedently understood. Now if we call "predicative" such a view of classes, the question arises whether it is violated by inductive definitions. If it is not, it will give a sense in which Lorenzen's view is predicative, and a divergence of two possible meanings of predicativity.

Now it is characteristic of the inductive definitions we have been considering that they are introductions of *predicates* and not in the first instance definitions or characterizations of *sets*. Because semantic reflection comes so readily to us, once we have understood such a predicate as the natural-number predicate or an accessibility predicate we will almost immediately talk of its extension. However, there is an essential conceptual order here, which places the understanding of the predicate before the apprehension of its extension as an object. There is, to be sure, a subtle difference between the situation we are envisaging, where we eschew set-theoretic realism and treat the inductive rules themselves as giving us understanding of the predicate, and a situation where one assumes set theory and where, moreover, what is being introduced is a predicate of objects in a domain that has been recognized to be a set. In the latter case, the axiom of separation implies that there *is* a set that is its extension; there is therefore a proof of its existence which does not use any

---

[84] However, in "Réflexions sur les deux notes précédentes," p. 200, he credits the introduction of the term to Russell. Alexander George pointed this out ("Imprecision," p. 514), correcting an earlier assertion of mine.

semantic concepts (as indicated earlier). At least if one is prepared to hold that particular instances of a schema like the axiom of separation can be seen to be true independently of the general principle, it follows that semantic reflection does not enter into one's insight that there is such a set as $\{x: Nx \wedge Acc(x)\}$. However, this does not change the essential point: that one's understanding of the predicate is prior to the insight that the set exists.

Thus, in my opinion, Lorenzen does not violate the limitations of his own concept in admitting generalized inductive definitions, and the divergence of two senses of predicativity does indeed exist. Although this observation is a partial defense of Lorenzen against Feferman, clearly the meaning of the term 'impredicative' underlying Feferman's analysis is so firmly entrenched that it is now the more appropriate way to use the term. Feferman's own term "definitionist" could serve to describe Lorenzen's position.

In these remarks, I have bypassed an issue about the status of higher-order entities in constructive mathematics. According to many constructivists, the notion of *function* enters essentially into the interpretations of quantifiers, even over such objects as numbers. The context in which the issue arises is explanation of the meaning of statements in intuitionistic theories in terms of what would count as a proof of them, worked out formally in theories of constructions.[85] In intuitionistic mathematics, we no longer have the mutual reducibility of the notions of set and function which obtains in classical mathematics. A theory may be compatible with the idea of sets and classes as arising by semantic reflection and still depend on a notion of function of a different nature. In my view, this issue primarily affects the claim of the sort of conception I have been discussing to be constructive, which has not been our primary concern in this section. At all events, the conception of function that on this view would be required is weaker than the set-theoretic conception in that the functions that would have to be assumed are intuitively effectively calculable.

The thesis of the present section is that, as regards some fundamental issues about predicativity, the predicates introduced by generalized inductive definitions are no worse off than the natural number predicate

---

[85] One sees this clearly in W. W. Tait's characterization in "Finitism" of the mathematics that can be obtained *without* presupposing a general concept of function. The most elaborate theory of the kind I am alluding to is Per Martin-Löf's intuitionistic theory of types; see, for example, his *Intuitionistic Type Theory*.

§51. Predicativity and inductive definitions

itself. The reader may well ask whether this thesis is affected by the doubts that might be raised about the claim of §50 that ordinary induction is impredicative on the ground that there is no fact of the matter as to whether it belongs to the meaning of 'natural number'. The claim of §50 was put in such way that, we trust, it is insulated against that objection. But even if not, a purely dialectical reply to the objection is possible in the present context: If there is no such fact of the matter in the case of the concept of natural number itself, then, on the same grounds, there should be no such fact of the matter in the case of notions introduced by generalized inductive definitions. Thus, the thesis that as far as impredicativity goes ordinary induction is in the same boat as these higher inductions, still stands.

9 Reason

§52. Reason and "rational intuition"

It will be painfully obvious to the reader that what has been said so far about the epistemology of mathematics, even arithmetic, is very incomplete. In §45, it was argued that very likely intuitive knowledge has rather narrow limits. Many would hold that that is due to the very restricted character of the conception of intuition developed in Chapter 5. Occasional reference has been made to conceptions of intuition of a quite different nature, which might promise to lead further than the one we have developed. I propose in this section and the next to approach the problem of mathematical knowledge from a different direction, not beginning with intuition at all, but leading very quickly to a place where some other conceptions of intuition, in particular that of Kurt Gödel, can be situated. I will begin with some very general considerations about Reason, not at all specific to mathematics.

The relevance of Reason to our inquiry is indicated by a long tradition of regarding mathematics as rational knowledge, echoed specifically by Kant in a remark at the beginning of one of the discussions of mathematics in the *Critique of Pure Reason*. He describes mathematics as "*rational cognition* from the *construction* of concepts" (A713/B741). Construction of concepts is construction in intuition, and it is here that intuition takes its place in Kant's account of mathematics. The notion of construction of concepts in intuition is what is most distinctive of Kant's philosophy of mathematics and has been the focus of much of the discussion of its interpretation. What interests me now, however, is Kant's calling mathematics "rational cognition," so that mathematical knowledge is rational knowledge. I don't propose to explore the question how Kant would have thought of the role of reason in mathematics, although some aspects

§52. Reason and "rational intuition" 317

of Kant's specific conception of Reason will play a role in what follows. Rather, I will take up the suggestion of exploring from my own point of view how general considerations about reason, not specific to mathematics, work out in the mathematical domain.[1]

I will say that Reason is in play when actions, including assertions and related speech acts, are supported or defended by reasons. But the same characterization clearly generalizes further to beliefs and other thoughts that have something like assertive force. However, the giving of a reason might count as an exercise of reason only if it is a good reason or at least has some degree of support by further reasons. Several features of Reason should be mentioned:

a. Reasons are themselves supported by reasons. This arises naturally from the fact that in giving reasons one seeks the agreement or support of others, since reasons for one's reasons can be given in response to their being questioned. The fact that reasons are themselves supported by reasons links reason with *argument*, since iteration of such support gives rise to chains of reasoning. The natural way in which such chains arise is furthermore an entering wedge for logic, although logic is not mentioned in the characterization, and the discussion of reasons in the theory of action and in moral philosophy does not put logic at center stage. The questioning of steps in an argument and the giving of further reasons for them has the result that some premises and inferences become explicit that were not before. The buck has to stop somewhere (for the immediate argumentative purpose, if not in any final way), and one place where it characteristically stops is with simple, elementary logical inferences. In modern mathematics, the axiomatic method makes it possible to give an idealized but quite workable and highly illuminating model of proof, in which proofs are represented as deductions from well-defined axioms with the help of some definite logic, in central cases first-order quantificational logic with identity.[2] Then one has identified a central factor in mathematical knowledge that would on almost any account be attributed to reason, namely, logic.

---

[1] One of Kant's thoughts in characterizing mathematics as rational cognition is undoubtedly that it is a priori. That is not at center stage in the discussion in this chapter. Some brief remarks on the subject are made in the Preface.

[2] Of course, the distinction between the logic and the assumptions of the particular theory in question does not always coincide with the distinction between axioms and rules, as for example the rule-based formulation of the theory of natural numbers given in §31 illustrates. In the discussion that follows, we ignore this fact for the sake of simplicity and talk as if in axiomatic mathematics all nonlogical assumptions take the form of axioms, and the only primitive rules are logical.

If we ask why the buck should stop with elementary logical inferences, one part of the answer is that they are obvious. Though this is not uncontested even for all of first-order logic, as the debates surrounding constructivism showed, I shall put obviousness aside for the moment. It is noteworthy that logic combines a high degree of obviousness with a maximal degree of generality. A characteristic way of justifying a statement or action by reasons is to put it under general rules or principles.[3] This leads us to search for general reasons, and to regard generality as a virtue in reasons. Thus, their generality is another reason for treating elementary logical inferences as basic. The same logic will be common to reasoning about a wide variety of subject matters, or reasoning from different assumptions about the same subject matter. This is, of course, important because it makes possible very general results about mathematical theories and about reasoning in other domains of knowledge. Because of their high degree of obviousness and apparently maximal generality, we do not seem to be able to give a justification of the most elementary logical principles that is not to some degree circular, in that inferences codified by logic will be used in the justification.[4]

If one considers logic more broadly, one cannot but be struck by the great variety of logics that modern logicians have developed and studied. The degree to which a logic possesses the features that give logic the role sketched above will vary and will in many cases be contested. I have not undertaken in this work to delineate the boundaries of a central sense of the word 'logic'. The plurality of logics has made it possible to reason about many matters while putting aside disputes about principles. For example, a constructivist will not disagree with a classical mathematician about what is provable in a given intuitionistic or classical system. It also

---

[3] One reason why this is so is that appeal to general principles is one way of attaining objectivity. In spite of its undoubted importance, I am not emphasizing objectivity in this discussion. This contrasts with that of Thomas Nagel in *The Last Word*, although the description of reason in his introductory chapter has points in common with my own. I have not undertaken a general defense of the objectivity of mathematics, in part because it would be generally agreed upon and in part because it would take my treatment too far afield from my main themes, for example into differences such as that between classical and constructive mathematics. But, in addition, I am not quite convinced that objectivity in quite as strong a sense as Nagel claims is a feature of Reason as such, although for the most part I am in sympathy with his defense of objectivity.

It also might be remarked, as I was reminded by Ernesto Napoli, that generality is not in every context a virtue in reasons, because sometimes one needs to get at the very specific features of the case at hand.

[4] Of course this does not mean that nothing useful can be said in justification of logical principles.

§52. Reason and "rational intuition"

can mask such disputes and make it difficult to tell whether two parties using different systems of logic are disagreeing or simply talking about something different.

However, it is only in a few cases that the plurality of logics affects views about the validity of *elementary* logical inferences, which are what is described earlier as having a buck-stopping character. And cases where there is real or apparent disagreement in this domain are not as common as the existence of a plurality of logics might suggest. In the mathematical case, which most concerns us, the law of the excluded middle is the best known case where a very elementary logical principle has been controversial. In this work, we have not dealt with issues about constructivism more than in passing. In many contexts, the law can be treated simply as an assumption of classical mathematics. Another source of elementary inferences would be modal logic. Because that is typically an extension of classical nonmodal logic, the controversy it might give rise to is different from that surrounding the law of excluded middle. It is less likely to concern the correctness of elementary inferences but would rather concern the admissibility of modal concepts and the interpretation of the modal operator, as the discussion of §15 illustrates.

b. This brings me to what I take to be one of the central features of Reason, another reflection of the fact that in the giving of reasons the buck has to stop somewhere. A wide variety of statements and inferences, not just simple logical ones, are accepted without carrying the argument further. They strike those who make them as true or as evident, prior to any sense of how to construct an argument for them, at least not one that proceeds from premises that are *more* evident by inferences that are also more evident. This is not to say that there is no argument for the statement possible at all, but the possible arguments are of a more indirect character or have premises that are no more evident.[5] They might be perceptual judgments, where a relation to an external event, namely, a perception, is what prompts their acceptance. But if there is no such relation to an external event, we will say that the statement is treated as "intrinsically plausible."[6] It is a feature of Reason that some statements, even those that have the character of principles, are treated in this way, and there is no alternative to so treating some statements or inferences. In mathematics, in addition to logical inferences, these are typically axioms.

---

[5] For some statements that we might accept as having a high degree of intrinsic plausibility, we don't rely on it because they can just as well be derived from others, which are preferred as premises for various reasons.

[6] I believe I derived this phrase from John Rawls, but I have not located it in his writings.

I want to emphasize that the fact that we do rely on the intrinsic plausibility of principles, even where it is unexplained and not something all would agree on, is a fact about *Reason*, in the sense of being a fact about what counts as a reason for what. This is certainly true in practice, and I would argue that there is no way around it. There may be vanishingly few principles that we cannot avoid assuming, but if we are not to be skeptics (in the ancient sense, suspending judgment about everything), we have to accept some. This fact comes to be a fact about Reason, of course, only because inquiries (and probably also practical decisions) where principles are involved exhibit this feature when they satisfy other criteria of rationality or reasonableness.

I do not use the term 'obviousness' or 'self-evidence' to describe what I have in mind here. Both imply a higher degree of evidence than I am in general aiming at; the conception itself is intended to be neutral as to how high a degree of evidence intrinsic plausibility entails.[7] What is indispensable is enough credibility to allow inquiry to proceed and to give the more holistic considerations mentioned under the following headings (c)–(e) something to work with.[8] 'Obvious' also covers too wide a range of cases: A statement may be obvious more or less intrinsically, or perceptually, or because of the ready availability of a convincing argument for it. I prefer to use the term 'self-evident' in a rather specific sense. Roughly, a statement is self-evident if it is seen to be true by anyone who has sufficiently clearly in mind what it means or who exercises a sufficiently clear understanding of it. If one does have such an understanding, it is then only by exercising a kind of detachment that one can doubt it, if at all.[9] The notion of self-evidence is often regarded as not a very useful notion, perhaps as having no application. And one can see why: As I have formulated it, it raises the question when an understanding is sufficiently clear, and, moreover, in actual cases, such as that of the law of excluded middle, there are disputes about what meaning of the statement is *available*. But it is useful to keep the notion of self-evidence in view at least for purposes of comparison.

None of this is meant to deny that there are cases where one has more than intrinsic *plausibility*, just the cases where other writers use

---

[7] I use the term 'evidence' in the sense of the German *Evidenz* or the French *évidence*, that is, for the property of being evident. These terms as they occur in writers such as Brentano and Husserl are often translated as 'self-evidence', in my view misleadingly, independently of the particular meaning I want to give to the latter term.

[8] In this respect the conception is akin to the conception of initial credibility in Goodman, "Sense and Certainty." I owe this observation to Sidney Morgenbesser.

[9] One can presumably apply the same criterion to inferences and their validity.

§52. Reason and "rational intuition" 321

such terms as 'rational insight' or 'rational intuition'. One such writer, Laurence BonJour, illustrates the idea by three examples, "Nothing can be both red and green all over," the transitivity of tallness, and the nonexistence of round squares.[10] The cases that are proposed in general discussions are usually rather simple. One might speak in these cases of intrinsic evidence rather than intrinsic plausibility. Cases where it amounts to certainty might be relatively rare. Probably we do have certain knowledge in some cases, such as the most elementary and simplest statements in arithmetic. But it is another matter whether this certainty rests entirely on *intrinsic* evidence. A problem arises because there is long experience, both historically and in the life of the individual, with the development and application of arithmetic. This theme will be elaborated in §54, where some issues it gives rise to will be discussed.

The question arises whether intrinsic plausibility is a property of propositions or of judgments.[11] In a particular case, treating it as intrinsically plausible that p is undoubtedly a judgment, with the caveat that it is likely to be tentative and thus to fall short of what would underlie full-blown assertion. Although I am reluctant to put much weight on a notion of proposition, even if one grants that the judgment notion is the more fundamental, there is surely a derivative notion of contents that are intrinsically plausible because a judgment to that effect would be warranted. It should be obvious that intrinsic plausibility in that sense can be lost. In the eighteenth century, it would have been reasonable to treat the Euclidean parallel axiom as such (and perhaps even as evident) in application to actual space, whereas it is surely not reasonable now. More humdrum examples can probably be cited.

A natural question to ask is what limits there are on the statements that can be treated as intrinsically plausible. As I have explained matters, in mathematics they will be in the first instance axioms, because if a part of mathematics is formulated axiomatically other statements will be derived. We need to consider inferences as well as statements, and basic logical inferences will also be treated in this way; indeed, such inferences are very frequently regarded as self-evident, and in some cases that is reasonable, for example with conjunction introduction and elimination. I will not, however, be concerned with the question to what extent simple

---

[10] *In Defense of Pure Reason*, §4.2. What BonJour calls rational insight is insight into *necessity* (p. 101). That does, I believe, fit the mathematical case. But it does raise the question how he would deal with cases of what is called the "contingent a priori."

[11] This question was pressed by Elijah Chudnoff.

logical inferences are self-evident. But it does seem unavoidable to treat some logic as intrinsically plausible, as at least a minimum of logic will be presupposed in any other form of verification of knowledge with significant systematic structure.

As I have said, perceptual judgments should not be regarded as intrinsically plausible, although they share the character of being accepted in the absence of argument. The reason is that their plausible or evident character depends on their being made in the context of an event outside the judgment, namely the perception, and so is dubiously called intrinsic. The question might then arise whether statements that are intuitively known on the basis of the Hilbertian intuition of Chapters 5 and 7 might be intrinsically plausible or evident. That seems to me a borderline case. They differ from perceptual judgments in the degree of freedom to which the external event of perception of imagination could be replaced by another, but they differ from purer cases of "rational intuition" in that one could not eliminate altogether the role of an external event. The same is doubtless true in other cases where imaginative thought experiments are involved.

c. We have set out two features of Reason that seem to me fundamental: the importance of argument, which gives rise to the central role of logic, and the fact that some statements that play the role of principles are regarded as plausible (and possibly even evident) without themselves being the conclusions of arguments (or, at least, their plausibility or evidence does not rest on the availability of such arguments). A feature of the structure of knowledge that can be viewed as a third feature of Reason is the search for systematization. Logic organizes arguments so that they are application of general rules or principles, and then these principles serve to unify the case at hand with others. In mathematics, systematization is manifested in a very particular way, through the axiomatic method. That is clearly the reason why it is axioms or potential axioms whose intrinsic plausibility is primarily of interest to us. But other factors influence the selection of axioms, more or less having to do with their systematic role. And in some areas, such as elementary arithmetic, there are statements that are universally treated as derived that are probably initially at least as evident as the axioms. Examples would be simple identities like '$2 + 2 = 4$' and the commutativity and associativity of addition. So it appears that they could be described as intrinsically evident, but in the axiomatic development of arithmetic we do not rely on this.

d. Systematization makes possible a fourth feature, what I will call the dialectical relation of principles – that is, high-level generalizations,

§52. Reason and "rational intuition" 323

generally of a theoretical character – and lower-level statements or judgments that we accept, possibly on the basis of perception but not necessarily, as the case of moral judgments illustrates.[12] Statements of either level may be plausible more or less intrinsically apart from their connections with statements of the other. A suitable systematization may derive the lower-level judgments from the principles, and this can serve to reinforce both. It may derive them with some correction, so that the original lower-level judgments come to be seen as not exactly true, but then what is wrong with them should be explained. It may be that some sets of principles lead to systematizations that square much better with the lower-level judgments than others, and this is a ground for choosing those that square better.

In 1955 Nelson Goodman suggested that one should view the evident character of logic in this dialectical way.[13] There are particular inferences and arguments that we accept. Rules of deductive inference are invalid if they conflict with these judgments about the validity of particular inferences. Particular inferences are valid if they conform to the rules, and not if they do not. Goodman says that the circularity involved is "virtuous." He indicates clearly that there can be a process of mutual correction, which then has a back-and-forth character, so that harmony of accepted principles and accepted particular inferences is a kind of equilibrium. Although the picture is attractive, it is not so easy to identify the process at work in the actual history of logic, although one possible reason for this could be that with respect to elementary inferences, the decisive steps took place too far back to be documentable.[14]

Goodman's picture of the justification of logic is at least incomplete, as one can see from the fact that harmony between principles and accepted inferences could obtain for both the classical mathematician and the intuitionist and so could not yield a criterion for choosing between them. In fact, further argument was brought into play from the beginning, generally either directly philosophical or methodological or turning on questions of mathematical fruitfulness and application. That harmony obtains

---

[12] These levels need not be the highest and the lowest. But, for simplicity, I consider only two levels at a time.

[13] *Fact, Fiction, and Forecast*, ch. III, §2 (substantially the same in all editions).

[14] Michael Resnik suggests the issues surrounding the truth-functional conditional as an example (*Mathematics as a Science of Patterns*, p. 159), although this may be a case where, at least for the purposes of formalizing mathematics, we accept less than complete harmony between principles and individual judgments. A purer case might be the question of "existential import" in the theory of the syllogism.

for either intuitionistic or classical logic when all these considerations are brought in would be disputed; it has certainly not brought about complete agreement.

John Rawls's view of the justification of moral and political theories also gives a place to dialectical interplay between particular judgments and general principles. In fact, it is his discussion of his methodology in *A Theory of Justice* that brought wide attention to this phenomenon. However, to enter into a discussion of Rawls's conception of "reflective equilibrium" would take us too far afield.[15] To avoid possible misunderstanding, however, we should recall that in later writings Rawls emphasizes that reflective equilibrium involves not just particular judgments and highly general principles but also judgments and principles of all levels of generality.[16] The mature conception of reflective equilibrium is more complex than what Goodman described.

A specific incompleteness in Goodman's discussion is that it is not sufficient to regard the relation of principles and lower level judgments that we aim at as simply a matter of the former implying and only implying judgments or inferences that we are disposed to accept. Even in domains other than natural science, in mathematics in particular, principles can possess something like explanatory power, and that can be a reason for accepting them. In one of his tantalizingly brief discussions of justification of mathematical axioms by their consequences, Gödel mentions "shedding so much light on a whole field" as one of some properties of axioms such that an axiom having all of them "would have to be accepted in at least the same sense as any well-established physical theory."[17]

e. A fifth and final feature is that Reason, as embodied in what we accept as a reason for what, is an ultimate court of appeal. The reason why "self-evidence" is never the whole story about the evident character of a

---

[15] Rawls acknowledges Goodman's discussion as a model (*A Theory*, p. 20 n.). The difficulty I have found with Goodman's discussion might be addressed by the distinction between narrow and wide reflective equilibrium; see "The Independence of Moral Theory," p. 289. I am doubtful about attributing the idea of wide reflective equilibrium already to Goodman, as Resnik does (ibid.).

[16] Ibid., also *Political Liberalism*, pp. 8, 28.

[17] "What is Cantor's Continuum Problem?" (1964 version), p. 261. My attention was called to this remark by Martin, "Mathematical Evidence," p. 227. Martin argues that determinacy hypotheses in descriptive set theory satisfy the criteria that Gödel gives. His paper is an exemplary discussion of the issues about evidence in mathematics that arise from the fact that some problems in set theory are settled by axioms that are very far from self-evident or even adequately motivated by the iterative conception of set. See also §22 above.

principle is the same as the reason why judgments based on perception are in principle fallible. In the end we have to decide, on the basis of the whole of our knowledge and the mutual connections of its parts, whether to credit a given instance of apparent self-evidence or a given case of what appears to be perception. There is no perfectly general reason why judgments of either kind cannot be overridden. When Descartes, after entertaining some serious skeptical doubts, was confronted with his inability to doubt '$2+3=5$' when he had its content clearly in mind, he then arrived at the idea of a malicious demon who might deceive him even about the matters that he perceived most clearly. We might in the end say that this is not a serious possibility, but when we do come to that conclusion it will be a case of our buttressing the evidence of '$2+3=5$' by argument, not in the sense of producing a stronger proof with that conclusion but in the sense of removing a reason not to credit the reason we already have for accepting it. Something that we ask of arguments of a nice premisses-conclusion structure is that they should be robust under certain kinds of reflection. The same is true of perceptual judgments or purportedly self-evident ones.

Thomas Nagel also regards reason as the ultimate court of appeal but emphasizes that in this role the judgments involved will make a claim to be *objective*.[18] Although I think that this is at least largely correct, in some cases one has to be quite careful about this. An example would be an appeal to perception. If I claim on the basis of perception that *p*, and that is supposed to settle some question in dispute, it will, of course, not necessarily be the case that others share my perception. I may be the only one in a position to have the relevant perception, or my senses might be sharper than those of the relevant others. And some translation to account for differences of perspective will often be needed for others, or even for myself at a later time, even to express what my perception yields. For example suppose that on the basis of perception I say, "There is blood on the floor now." To convey the information to someone at a distance, I will have to specify the location of the floor, and at a later time I will have to specify the time, which will no longer be correctly called 'now'.

§53. Rational intuition and perception

Statements that are taken to be intrinsically plausible are often called intuitions; in fact, something like this is nowadays the most widespread

---

[18] *The Last Word*, e.g. p. 5.

use of the term in philosophy and is by no means confined to philosophy. Intuition in this sense would be a species of intuition *that* in the usage of §24. But so far, although the conception of intuition yielded can in principle apply to a very wide range of propositions, it is in other respects a very weak one. First, it leaves largely open the question, already broached in §24, what epistemic weight is to be given to intuitions. Already that would suggest that some tightening of the conception is necessary. Philosophers who have given great epistemic weight to intuition, such as Descartes, Husserl, and Gödel, have held that one only has intuition if one exercises considerable care to eliminate sources of illusion and error. Connected with this is a second consideration: If we think of intuition as a fundamental source of knowledge, then in theoretical matters intuitions should be stable and intersubjective, but in many inquiries what is regarded as intrinsically plausible may depend on that particular context of inquiry, and moreover disagreements in "intuitions" are very common in most fields. Third, no connection has been made between this notion of intuition and that of intuition of *objects*, so prominent in the philosophy of Kant and in writings influenced by him.[19] Such a connection was found in Chapter 5 to be a natural requirement for a propositional attitude to be sufficiently analogous to perception to be appropriately called 'intuition'. In general, intrinsic plausibility is not strongly analogous to perception, but perhaps it comes closer to being so where greater epistemic weight is accorded to it. Nonetheless, I will sometimes in what follows defer to contemporary usage and speak of intuitions, sometimes using the term 'rational intuition'.

To pursue the first two questions in a general way would carry us well beyond the limits of this work and would in my view not be very fruitful. As we go along it will become clear what I have to say about them in relation to some basic mathematical cases: arithmetic and set theory. It is also with respect to these cases that we will say something about the third question.

The most obvious disanalogy between perception and a proposition's being taken as intrinsically plausible, even to a very high degree, is that the latter occurs in the absence of an external event. But there is an analogy, stressed in the writings of Gödel on this subject, in that a proposition accepted in this way is not the conclusion of an argument. Because in interesting cases such propositions are general, one also has to make

---

[19] Among such writings should be included much that has been written on the philosophy of mathematics in the nineteenth and twentieth centuries; even writers hostile to Kant have used the term 'intuition' in a way that owes at least distant inspiration from him.

§53. Rational intuition and perception

special assumptions to discern a perception-like relation to *objects*. Gödel took rational intuition to involve perception of concepts. Someone to whom it is evident that every number has a successor but is skeptical of the whole idea of concepts is, on Gödel's view, misconstruing what is before his rational faculty. The view of predication underlying the discussion of Chapter 1 and earlier writings of mine is incompatible with that.[20] An interesting analogy was proposed by George Bealer: That reason is subject to illusions that, like perceptual illusions, persist even after they have been exposed and one no longer accepts the propositions that the illusion would lead one to.[21] In fact Kant made the same analogy with perceptual illusions in explaining his own conception of "transcendental illusion."[22] In my view the convincing examples of illusions of reason are of the general type that Kant attempted to characterize, in that they have to do with some kind or other of "absolute" or "unconditioned." An example might be the intuitions about truth and related notions that generate semantic paradoxes, where the illusion would concern either an absolute totality of all propositions or a total semantics for the language one is using or might come to use.[23]

Before we pursue these cases it is appropriate to remark on a general feature of principles in mathematics, which I will call their epistemic stratification. It bears on the question whether much significant mathematics can be done on the basis of principles that are indisputably self-evident. Principles in mathematics ascend in a rough order of abstractness and logical power, with the effect that as one ascends there is a decrease of clarity and evidence, although there will not necessarily be agreement about how much of a decrease there is with any particular step. Some traditional distinctions such as finitist/nonfinitist, constructive/nonconstructive, and predicative/impredicative distinguish dimensions on which it has been argued that there is such a decrease. Proof theory has developed a very refined analysis that enables one to answer questions as to whether a theory embodying certain principles is or is not reducible to another, and precise senses can be defined in which one theory is "stronger" than another. That enables one to place particular mathematical theories more precisely in an order significant for

---

[20] I discuss Gödel's views further in "Platonism and Mathematical Intuition" and in "Reason and Intuition," pp. 309, 312–313.
[21] Bealer, "Intuition," p. 732 of reprint, and "On the Possibility of Philosophical Knowledge," pp. 5–6.
[22] *Critique of Pure Reason*, A 297/B 353–354.
[23] See further "Reason and Intuition," pp. 310–311.

questions of clarity and evidence, although the more precise and refined it becomes the more possibility there is for dispute and uncertainty about its epistemological significance.

The epistemological significance of this stratification was first urged by Paul Bernays in "Sur le platonisme." But it has come to impose itself through the development of mathematical logic since 1930. It constitutes a major qualification on any holistic epistemology for mathematics. In many standard cases of stronger and weaker theories, the stronger theory incorporates the weaker theory, in the sense that the weaker theory can be translated directly into the stronger, and a model of the stronger theory that meets some standardness condition will contain a model of the weaker theory that also meets standardness conditions. For example, first-order arithmetic, PA, is translated directly into ZF by interpreting numbers as finite von Neumann ordinals, and the model of arithmetic that thus arises within any model of ZF is standard if the model of ZF is well founded.[24] That rules out the possibility that ZF or some stronger set theory would lead to results that contradict arithmetic without being itself inconsistent, or at least ω-inconsistent.

§54. Arithmetic

The special case that it is most natural to consider first in the light of the discussion in §§52–53 is that of arithmetic. In Chapter 5, we developed a Kantian conception of intuition whose relevance is primarily to arithmetic, but the reader may well feel that much remains to be said about what this concept accomplishes for the epistemology of arithmetic. We will attempt to fill that gap in this section. However, the best place to begin is by looking at arithmetic in the light of the discussion of Reason in §52.

The natural numbers are often claimed to be especially transparent. One version of such a claim would be the claim that the axioms of arithmetic or other elementary arithmetic statements are self-evident. Let us consider first the elementary Peano axioms: Every natural number has a successor; the successor is never 0; and if $Sx = Sy$, then $x = y$. We might agree that if one doesn't accept these statements, then one doesn't understand the concept of natural number, at least not as we do. But

---

[24] Some cases can arise in which a stronger theory seems to "refute" a weaker theory. A well-known case is that of set theory with a measurable cardinal, which implies $V \neq L$. Here, holistic considerations do have a purchase. But it should already be clear that we have not intended to deny their relevance altogether. More will be said in the section on set theory.

§54. Arithmetic

consider someone skeptical about arithmetic, for example someone who holds an extreme version of strict finitism that rejects the infinity of the natural numbers. To the claim of self-evidence, such a person will reply that the understanding of the notion of natural number that we presume is illusory, that, in the sense in which we mean it, natural numbers do not exist; we do not have the concept of natural number we claim to have. So, at least in relation to such skeptical views, the claim that these axioms are self-evident is not very helpful.

Another observation about them is that although the elementary axioms are accepted as axioms of arithmetic, there are some contexts in which one defines the arithmetic terms and proves the axioms. This is true of the standard development of arithmetic in axiomatic set theory, and it is also true of logicist constructions of arithmetic. Frege is credited with proving them in second-order logic from his criterion of identity for cardinal numbers, now popularly known as "Hume's Principle." Particularly given the power of second-order logic, it would be difficult to maintain that this derivation obtains the elementary Peano axioms from something more evident. In §47 it was argued that this is not true even if one adds induction to what is derived. Both of these cases are cases of putting the arithmetic axioms into a more general setting, in the one case of course set theory and in the other a theory of cardinality applying to whatever the variables in second-order logic range over.

These somewhat banal observations illustrate a general fact (which applies also to further arithmetic axioms): In mathematical thought and practice, the axioms of arithmetic are embedded in a rather dense network, in which their primitives may mostly play the role of primitives but do not always do so, and, likewise, the arithmetical axioms do not always play the role of premises. Moreover, the process by which they have come to be as evident as they are includes also the deduction of consequences from them, consequences, which, in some cases, were at the time when the axioms were formulated quite as evident as the axioms themselves. I have in mind such propositions as the earliest theorems proved in a development of formalized arithmetic, such as the associativity and commutativity of addition, and basic identities such as $7 + 5 = 12$ and $7 \times 8 = 56$. We might add some other kinds of considerations, such as the fact that the Dedekind-Peano axioms in the second-order setting characterize a unique structure.

A whole network of relations thus serves to buttress the evident character of the axioms of arithmetic. Some of these relations lead us into applications, and it has been argued that this exposes arithmetic to the possibility of empirical refutation. This is a complex and contested matter, which I have not gone into thus far in this work and will not pursue here. No one has offered a scenario that is at all plausible in which that would come about.[25]

This fact raises the question to what extent the evidence of the axioms of arithmetic should be described as intrinsic, that is, as obtaining without appeal to arguments in their favor. Can the *intrinsic* plausibility of a principle be increased by the development and application of a theory of which it is a part? It is not altogether easy to answer this question. In some of his later essays, Paul Bernays uses the term 'acquired evidence' to describe the evident character a proposition can have in the context of a developed conceptual scheme.[26] He maintains that evidence in mathematics is nearly all acquired in this sense. The example he develops most fully is that of Euclidean geometry.[27] There he maintains that there is something like intuitive evidence, but it is dependent on having acquired, and presumably worked with, a certain conceptual apparatus. Bernays's own view about whether such evidence can be intrinsic in our sense is not very clear. One remark hints at the negative; he writes that "evidence, originating in an intellectual situation, is relative to the implicit suppositions that such a situation includes."[28] However, the very fact that the suppositions are implicit makes the implication not so clear. The matter may turn on what is involved in a statement's being evident. In what we might call the strong sense, a proposition's being evident implies that it is known, and that implies that it is true.[29] Then it seems that the remark just quoted implies that acquired evidence is not intrinsic. But I think it more likely that Bernays' conception of evidence is weaker, and that evidence could be lost if the "implicit suppositions" are brought to light

---

[25] For some further, still brief remarks, see the Preface. Richard Tieszen has raised in correspondence the question whether the dialectical relation of principles and lower-level judgments exists only within mathematics, or whether the lower-level judgments can be empirical ones arising in application. That is a large question, which I do not attempt to answer here. I don't think it is necessarily equivalent to the question whether empirical falsification of a mathematical theory is possible, although the issues are certainly closely related.

[26] See "Quelques points de vue" and the remark about the obviousness of arithmetic in "Die Mathematik als ein zugleich Vertrautes und Unbekanntes," p. 111.

[27] "Quelques points de vue," pp. 322–323.

[28] Ibid., p. 321.

[29] These implications would hold for what I consider canonical uses of *Evidenz*, by Brentano and Husserl. Otherwise, for example, Brentano would not be able to define truth in terms of evidence in the way he does in late fragments. See for example *Wahrheit und Evidenz*, p. 139.

§54. Arithmetic

and turn out to be doubtful. One way in which it would be intrinsic is where the evidence of an axiom is connected with the understanding of the concept of a kind of mathematical object or of a structure such as that of the natural numbers. It can hardly be doubted that such understanding as we now have is the product of a long history of reasoning about numbers. But that it is increased by that history, so that the basic principles come close to evidence that is still intrinsic, is suggested by the phenomenology of arithmetical reasoning. But what we have may fall short of evidence in the strong sense, as it is still possible to doubt the coherence of the conception of number, or the consistency of theories of number.

However that may be, it may be questioned how well the dialectical picture sketched by Goodman for logic and developed by Rawls and others for moral and political theories fits the case of arithmetic. Axiom systems for arithmetic before the work of Dedekind were fragmentary and incomplete, but they did not contain axioms that had to be revised in the light of their consequences. The Dedekind-Peano axioms may not have been recognized in their role as axioms in much earlier times, but surely inferences corresponding to them were made, in particular inferences that would now be rendered as applications of induction or equivalent principles. Some trial and error was no doubt involved in arriving at the right formulations of the induction axiom, but it consisted mostly in coming to recognize induction as a central principle and giving the formulation that characterized the natural numbers. Where there might be uncertainty or controversy about induction is about whether it really needs to be a primitive principle (as in the Poincaré-Russell controversy at the beginning of the twentieth century) or what to regard as a well defined predicate and thus as falling within the scope of the principle. In the latter case, however, we do not find a modification of principles because consequences of them have been found unacceptable but, rather, questions of a more or less philosophical character, so that where there is some such dialectic is between attitudes toward mathematical axioms and rules and methodological or philosophical principles having to do with constructivity, predicativity, feasibility, and the like.[30]

---

[30] One might also find a dialectical relation of principles and lower-level mathematical judgments in the earlier history of the analysis of the concept of number, before the time of Dedekind and Frege. I have not gone into this matter and don't feel qualified to have a view of it. In the case of Kant, which I have investigated, much of his *philosophy* is certainly made dated by later developments in the foundations of arithmetic, although some basic ideas have value today. But there are not properly arithmetical claims in his

Arithmetic does already illustrate the epistemic stratification mentioned at the end of the last section. Most of us find it quite evident that exponentiation is a total function, but is it fully as evident as it is that addition and multiplication are total? As we saw in Chapter 8, philosophical reasons have been given for casting doubt on exponentiation, and a consistent position that rejects it is possible. If someone maintained that, say, arithmetic with bounded induction but with exponentiation is inconsistent, it is not clear that we could give a definite refutation that would be in no way question-begging.

Undoubtedly the axioms of arithmetic are intrinsically plausible. But the received idea that they are evident by themselves, as the application to them of the idea of rational intuition would suggest, is in an important way misleading. The evident character of the axioms is buttressed by the network of connections in which they stand, so that in that respect their evident character does not come just from their intrinsic plausibility. Moreover, such a network already existed at the time the axioms were formulated clearly in the second half of the nineteenth century. Still, the received idea has an element of truth because once formulated clearly, the axioms have not had to be revised because of consequences that mathematicians were not prepared to accept, although when philosophical considerations are brought in we can discern degrees in the evident character of arithmetical principles.

These remarks raise two questions: How does the view sketched here differ from the holistic epistemology of Quine, which has been developed in the case of mathematics in a number of writings of Michael Resnik?[31] And what is the epistemological role that this picture gives to Hilbertian intuition as described in Chapter 5? In §27, we objected to the Quinean view that it leaves unaccounted for the obviousness of elementary mathematics and logic. Resnik has challenged me to explain how my conception of intuition yields a better result on this question.[32] With regard to logic, we should not expect it to do so at all directly, because logical truth and inference do not (*pace* Frege) involve cognition of distinctively logical objects. But the question about arithmetic calls for an answer.

The formulation of the problem needs refinement; for reasons noted in §52, the term 'obvious' is not quite the right one to characterize what

---

writings that have had to be revised, unless one so considers the "metamathematical" claim that arithmetic has no axioms (*Critique of Pure Reason*, A164/B205).
[31] In particular in *Mathematics as a Science of Patterns* and "Quine and the Web of Belief."
[32] "Parsons on Mathematical Intuition and Obviousness."

§54. Arithmetic

is to be accounted for.[33] One could put the matter better by saying that the most elementary parts of arithmetic are evident to an exceptionally high degree. In the stratification of mathematical conceptions in terms of abstractness and logical power discussed in §53, the most elementary arithmetic is at the bottom.[34] We might single out the part of arithmetic that, on the Hilbertian interpretation where the domain of objects consists of strings, is intuitively evident on the view of Chapter 7. Thus it would include the elementary Peano axioms, addition and multiplication and probably some other recursive operations (but none as powerful as exponentiation), and quantifier-free induction. I leave aside whether Δ_0-induction can be added. For this domain of arithmetic, a preliminary answer to Resnik's question is offered by the argument of Chapter 7 that this much arithmetic is intuitively evident on the interpretation considered there.

Already within the Hilbertian interpretation, one takes a step that also arises in the different genetic routes to the concept of number considered in Chapter 6. That is to regard the numbers, however construed, as a domain of *objects*. I do not mean that we take the objects to constitute a set, only that we understand a predicate applying to them and can reason with it. In intuitively evident arithmetic, the extent is limited because one does not introduce full quantificational logic, even first-order, for reasons already spelled out in Hilbert's writings and indicated in §42. It follows from the view about induction advanced in §47 that the piecemeal introduction of instances of these schemata as the finitist conception of arithmetic suggests already represents gradual specification of the domain. Although I argued that, applied to the model of strings and rather sharply restricted, induction preserves intuitive knowledge, there is clearly something more than intuition involved. This is the more true when one affirms induction as a general principle, even if one has a very restrictive idea of what predicates are to be allowed.[35] In the usage of §52, we would describe individual instances as intrinsically plausible to a very high degree. We can describe this evidence as conceptual, but as we said earlier that is only informative to a limited degree.

---

[33] The formulation derives from "Mathematical Intuition," p. 151. Resnik himself expresses cautions concerning the notion of obviousness, in many respects parallel to our own, op. cit., pp. 227–228.
[34] It might be argued that some basic geometry is as well. I do not address this question.
[35] Such an idea is of course absent in §47, and the unrestricted character of induction is exploited in the discussion of uniqueness in §§48–49. But we were not concerned there with intuitive evidence or with finitism.

Applied to the model of strings, what most distinguishes the view advocated here from the holistic empiricist view is that the smaller, accessible objects of the domain are given in intuition, there is intuitive knowledge of certain ground-level propositions, and the conceptual or rational element is limited by the very low place the mathematics involved occupies on the hierarchy of levels of mathematical theories. This would continue to hold even if we admit primitive recursion in one of the ways canvassed in §44, so that we obtain a theory equivalent to primitive recursive arithmetic PRA.

It might seem that allowing free use of quantificational logic in induction, so as to have an arithmetic at least as strong as PA, has little point so long as one sticks with the model of strings, since we would have gone well beyond any plausible limit of intuitive knowledge. However, it is relevant to the structuralist view of §18, which in my view is the most plausible final view of numbers. Strings offer a domain that witnesses the possibility of the structure of numbers. But we will have taken the step of regarding the domain as definite in the sense that quantification over it yields statements that are true or false.[36] The clarity we have about what this domain is derives from the principle of induction. It might seem that this is a very weak basis on which to rely, particularly if one holds the open-ended interpretation advocated in §47. It is an explanation of a concept, where we rely on the formal notion of predicate but don't treat predicates themselves as a definite domain of objects. But it provides an elimination rule for the number predicate and so allows the derivation of general statements about numbers. And we do have quite firm ideas to the effect that certain predicates *are* well defined, which enable us to build up a body of arithmetical proofs.

It might still seem that the step that allows classical logic is questionable. Michael Dummett suggests that we should regard the concept of natural number as "indefinitely extensible" and for that reason prefer intuitionistic logic.[37] Although it was he who pointed out many years ago the vagueness introduced by characterizing the numbers in the way

---

[36] I am of course leaving intuitionism out of account and proceeding directly from a very limited use of quantification to classical logic. An intuitionistic view would proceed differently, presumably relying on the idea of construction or proof. At the level we are considering now, roughly that of first-order arithmetic, it is hard to argue that it yields greater evidence. But it would take us too far afield to argue this here.

[37] "What is Mathematics About?" pp. 442–443. However, Dummett does not here offer an argument based on the open-ended character of induction. The reference to Frege's proof that every finite cardinal has a successor suggests that he thinks that the infinity of the natural numbers itself encourages regarding the concept as indefinitely

§54. Arithmetic

we have adopted, that vagueness does not encourage the idea that the concept of natural number is the sort of concept Dummett has in mind in talking of indefinite extensibility. The difficulty is that if we come to recognize more predicates as well defined, that does not lead to the recognition of more natural numbers. If one adopts a model-theoretic perspective, the effect is the opposite, as has already been suggested in §48, as the availability of more instances of induction leads to the elimination of more nonstandard models. Of course, that perspective presupposes a lot of mathematics, but I don't think that is an objection to using it to argue for the inappropriateness of applying the conception of indefinite extensibility to the natural numbers themselves, although it enters into the conception via the idea of well-defined predicate. But an intuitionistic conception of arithmetic has to stand on its own feet, presumably relying on the conception of construction embodied in the BHK interpretation of the logical connectives, and on arguments such as Dummett's in "The Philosophical Basis of Intuitionistic Logic" and elsewhere.

Of course, we need to cut loose from the model of strings. Underlying the substitutional approach taken in §32 was a rather simple idea: We learn rules for computing and reasoning with numerical expressions, among them rules of construction of numerals. The ability to make generalizations, first, perhaps, in a language in which the possibilities of quantification are limited roughly as Hilbert proposed for finitism, is one of the things that is learned. In using these expressions in counting, establishing cardinalities and places in an order, and in making standalone statements "about numbers," we do not do anything to connect the expressions that are syntactically like singular terms with objects given otherwise. Why should one think that making any such connection is necessary? Both the substitutional interpretation of such quantification (on one way of taking it) and the related constructive idea of objects as given by canonical expressions for them are ways of making out a negative answer to this question, while still saying that natural numbers are in every essential respect objects. In this setting what the model of strings does is to make out the fact that the expressions involved exist, or can be constructed. Actually, what typically does this is a more complex syntactic construction. The objects involved are still the sort of object that the account of intuition in Chapter 5 has in view, although where

extensible. Perhaps it does, but once one has classical mathematics what reason is there for embracing the conclusion?

recursion and quantifiers are involved, knowledge of them may go beyond the limits of intuitive knowledge.

We have said that on such a view natural numbers are "in every essential respect objects." However, one of the steps described in Chapter 6 is ceasing to describe them as a syntactically segregated type but as objects that belong to a larger domain, which may contain further mathematical objects or may contain other objects, such as objects counted or ordered. There seems to be a new conceptual element in the use of the formal notion of object. However, that is not as substantial an addition as it might seem, since the notion of object is correlative to that of predicate, which was already appealed to in the general formulation of the induction principle.[38]

But one can see that what can broadly be called logic is central to the more abstract talk of numbers, as it was already to the introduction of quantifiers with reference to the model of strings. Arithmetic does not say what objects there are beyond numbers. Even if it presupposes objects to be numbered, it has no particular presupposition as to what kind of objects they are and how many of them there are. But logic plays the role, first, of providing a framework for speaking of numbers as objects among others, and, second, for speaking of the relations between numbers and whatever other objects stand in relevant relations to them.

I can now summarize the reply our work offers to the question posed by Resnik. Intuition does play a role in making arithmetic evident to the degree that it is, in that there is a ground level of arithmetic, not extending very far, that is intuitively evident. Furthermore, the objects that play the role of numbers in this low-level arithmetic can continue to do so in a more full-blooded arithmetic theory. Logical ideas, both the general idea of predicate in the formulation of the induction rule, and taking the strings to be a domain of objects that can be quantified over and reasoned about classically, are essential to this further development. Thus the conceptual or rational element in arithmetical knowledge becomes much more prominent at this point. But the role of intuition does not disappear, because it is central to our conception of a domain of objects satisfying the principles of arithmetic. When we turn to abstract talk of numbers, the new element involved is again broadly speaking logical. But an intuitive domain witnesses the possibility of the structure of numbers.

---

[38] Logic plays a similar role in the alternative, structuralist route presented in §18, where the natural numbers can be arrived at by "Dedekind abstraction" beginning with an instance of the structure such as the model of strings.

§54. Arithmetic

Apart from the role of intuition just described, and such intrinsic evidence as the more conceptual appeals possess, there is another respect in which the point of view advocated here differs from holism, namely, the fact that arithmetic as so far described is low in the hierarchy of mathematical theories.

Still, what we have said above makes clear that our view does not differ *toto coelo* from holism, because the evident character of the principles is reinforced by mathematical experience and application. The empiricist will ask whether it is possible that mathematical conclusions in this domain should be falsified by experience, presumably in application. As remarked earlier, it seems to me that no one has presented a convincing scenario according to which that occurs. But more would be involved in arguing that such falsification is impossible, and I am not attempting to offer such an argument.

It will be argued that the role we attribute to intuition can itself be replaced by purely rational evidence. Thus, in a brief discussion of the obviousness of arithmetic, Paul Bernays writes: "First, we are conscious of the freedom we have to advance from one position arrived at in the process of counting to the next one."[39] It would take some argument to show that there is no appeal here to the temporal character of experience, such as we find in Brouwer. Another apparently purely rational procedure would be to appeal to the axiom of pairing in set theory, which together with extensionality is enough to generate an infinite domain (in a slightly simpler way than in the theories of finite sets presented in §33). Just the fact that the axiom says that given *any* two objects, there is a set of which they are the elements, makes it, as we noted earlier, not presuppose intuitability of the objects. It is natural to find this axiom more evident than such axioms as Replacement and Power Set. One's first thought as to why is that, even with other axioms, it will yield only finite sets. A purely rational evidence of the axiom would be approximately as plausible as the intuition about successor appealed to in the remark of Bernays quoted above.

However, the work of §§33–34 implies that a version of the axiom sufficient for the arithmetical application has a kind of evidence that presupposes less, which can be described as linguistic. This emerges from the discussion of a relative substitutional interpretation of the theory of hereditarily finite sets. If these sets have objects of an arbitrary given domain as urelements, then such an interpretation still presupposes finite

---

[39] "Die Mathematik als ein zugleich Vertrautes und Unbekanntes," p. 111.

sequences of these objects, as noted in §35 and at the end of §38. But for the construction of numbers, only pure sets are needed. We can see these sets as objects given by canonical expressions for them, which might be constructed either in the language of the set theory of §33 or as "bracket terms" in the sense of §35 and §39.[40] Or we can interpret the theory by a term model, which is only moderately more complex than the model of strings.

§55. Set theory

The question of the justification of set-theoretic axioms, even those of ZFC before one considers large cardinals or other additions to the axioms, is a complicated one to which I cannot even attempt to do justice here. Some of the issues have already been discussed in Chapter 4. The present section will largely confine itself to the question of how the picture of rational justification developed in §52 applies to them. For someone experienced with set theory, all the axioms of ZFC have what I have called intrinsic plausibility.[41] The axiom of extensionality seems just to mark the theory as being about *sets* as opposed to attributes or other intensional objects; it is as deserving as any of the honorific title "analytic." The axiom of pairing and the existence of a null set (which in some formulations is an axiom) seem as evident as the axioms of arithmetic, and in much the same way. They were taken up at the end of the last section. It would be nice if the axiom of union could be viewed in the same light, but I do not know of an argument. It does have a natural and relatively unproblematic justification from the iterative conception of set: Given a set a, suppose it is "formed" at a stage S. All its elements were formed at stages earlier than S, and the elements of its elements also at earlier stages. Hence they are available at stage S, and $\cup\, a$ can be formed at S.[42] However, the union axiom is not quite so innocent as it seems because of the way it combines with the axiom of replacement to yield large sets.[43] But the cases

---

[40] Compare the discussion of pairing in constructive logic in my "Intuition in Constructive Mathematics," pp. 219–222.

[41] This is generally thought to extend naturally to "small" large cardinals, inaccessible and Mahlo cardinals, and probably beyond. However, how far it extends is uncertain. The matter surely depends on how great a degree of intrinsic plausibility is asked for.

[42] Shoenfield, "Axioms of Set Theory," p. 325.

[43] The availability of a simple argument like Shoenfield's is a reason why someone skeptical of the large sets yielded already in ZF is more likely to question the axiom of replacement than the axiom of union; an example is Boolos, "Must we Believe in Set Theory?" In his discussion of the axioms of set theory, Hao Wang formulates the axiom of replacement

§55. Set theory

that are really distinctive of set theory are Infinity, Replacement, Power Set, and Choice. There is a considerable literature on the justification of these axioms, in which a number of considerations play a role: conceptions of what a set is, more global considerations about the universe of sets (the "iterative conception" and ideas about the inexhaustibility of the universe of sets, which allows for extension of any explicitly described proposed universe), analogies of various kinds, methodological maxims, and considerations concerning consequences, that is, the ability of the axioms to yield the established corpus of set theory and its applications in other parts of mathematics as well as the ability to avoid untoward consequences, in particular paradoxes. If consequences have considerable weight, as I argued with respect to Replacement and Power Set in §23, then the intrinsic plausibility of these axioms falls short of intrinsic *evidence*.

A question that has to be considered is whether there is a dialectical relation of axioms and their consequences such as our general discussion of Reason would suggest. A negative answer seems to be suggested by the remarkable fact that when a set of axioms for set theory was first proposed by Zermelo in 1908, they were questioned in many ways and remained controversial for thirty years or so but emerged from the controversy hardly revised at all. The need for clearer formulations of the axiom of separation arose early, and not long afterward the axioms were found incomplete in ways that were remedied by introducing the axiom of replacement. The system ZFC as we now know it is essentially in place in Skolem's address of 1922.[44]

Nonetheless, I think we do find such a dialectical relation. We can see such a relation at work in the familiar history of the reception of the axiom of choice, even though its outcome was that the axiom as originally stated by Zermelo became a quite established part of set theory. The fact that it implies that every set can be well ordered and other existential claims that could not be cashed in by defining the set claimed to exist was regarded by many mathematicians as a reason for rejecting the axiom or at least regarding it with reserve. It did indeed clash with an antecedently reasonable idea of what a set is, roughly, the extension of an antecedently meaningful predicate. Perhaps one reason why it was that understanding

---

so that it incorporates the union axiom (*From Mathematics to Philosophy*, p. 186). He does sketch the same argument for the union axiom as Shoenfield's and says that it is "conceptually a consequence of the intended process of iteration" (ibid., p. 220 n. 4).

[44] "Einige Bemerkungen."

rather than the axiom of choice that gave way in the end was that the former already did not harmonize with the idea of a set of all sets of natural numbers, which was necessary to the most straightforward set-theoretic construction of analysis.

We also may see ZFC as having won out over various competing proposed frameworks for mathematics, many of them codifiable as subtheories of ZFC. An important apparent rival of Zermelo's axiomatization was the simple theory of types, as it emerged in the 1920s from critical discussion of the first edition of *Principia Mathematica*. But it became clear that for most mathematical purposes the syntactic type distinctions were not essential and that simple type theory was essentially a subtheory of Zermelo set theory. But other rivals that were not embeddable into ZFC in such a straightforward way, such as ideas for a type-free theory, were never made really workable. However, the jury is probably still out on whether there is a genuinely alternative framework based on category theory. Many people have observed that for most of mathematical practice a much weaker theory than ZFC is sufficient, even without the work that has been done by Harvey Friedman, Stephen Simpson, and others to identify the weakest axiomatic framework sufficient for given classical theorems.[45] But there is no convincing ground on which the part left out should be ruled out of court as genuine mathematics. But this is a case where "intuitions" concerning axioms, what is known about what follows from them (and what does not), and philosophical and methodological "intuitions" would enter into the justification of any view one could reasonably hold.

The phenomenon of decreasing clarity and evidence for more abstract and powerful principles is of course in evidence in set theory; indeed, it is the rise of set theory that first brought it to the consciousness of researchers in the foundations of mathematics, and where one finds the basis for the most convincing case for its being unavoidable. This is most evident in the case of large cardinals, and especially "large" large cardinals like measurables and above. But already the axioms of Power Set and Replacement are found insufficiently evident by many.[46] In §23, I

---

[45] See Simpson, *Subsystems*.

[46] On Replacement, see the paper by Boolos cited in note 39. Predicativism includes skepticism about the axiom of Power Set. The distinction drawn on the basis of remarks of Gödel between intrinsic and extrinsic justifications of axioms of set theory parallels roughly that between considerations of intrinsic plausibility and possibly evidence, on the one hand, and a posteriori considerations, on the other. As regards Gödel, however, he seems to have had in mind by "intrinsic necessity" being "implied by the concept of

§55. Set theory

argued that a posteriori considerations, that is, their having the right consequences, are an essential part of their justification. If one grants some force to the more "intuitive" or direct arguments for them, then we have a case where the plausibility of principles is strengthened by the consequences at lower levels that they yield. Historically, the axiom of Replacement was introduced as an axiom when it was found that certain natural constructions, such as that leading to the cardinal $\aleph_\omega$, could not be carried out on the basis of Zermelo's original axioms. Replacement was formulated as the principle underlying these constructions.

In the higher reaches of set theory, involving large large cardinals, whose existence is incompatible with $V = L$, the role of intrinsic plausibility is much diminished, and some of the sophisticated defenses of the assumptions made rely entirely on a posteriori considerations. This has been laid out at some length by a number of writers, so that there is no need to go into it in any detail here. As noted in §22, a case along these lines for the determinacy hypotheses employed in descriptive set theory and thereby for the strong axioms of infinity that have been shown to imply them has been found convincing by set theorists and by those philosophers who are best informed about set theory.[47] However, it was established some time ago that strong axioms of infinity of the sort

---

set." Gödel's own conceptual realism implies that such a status only yields truth if the concept involved exists (in the further sense of being coherent).

In a discussion of arguments for various axioms of set theory in *Naturalism in Mathematics*, Penelope Maddy also distinguishes intrinsic and extrinsic justifications (p. 37). Her use of 'extrinsic' seems very close to ours of 'a posteriori', but I do not find this entirely evident.

In the discussion of the status of Replacement in §23, we granted "a certain conceptual character" to reflection principles, where the first-order reflection principle stated there implies Replacement (given the other axioms, even without Infinity). That is, it could be seen as a working out of the idea that the universe of sets in inexhaustible in the particular sense of being undefinable, or perhaps not characterizable from below. When we consider what is involved in the formulation of stronger, higher-order principles, questions arise about how we understand the higher-order quantification. The second-order principle of Bernays, "Zur Frage der Unendlichkeitsschemata," is less problematic than the higher-order principles introduced by Tait, "Constructing Cardinals from Below." In any event, the limitative result of Koellner (see note 38 of Chapter 4) implies that even Tait's principles do not yield cardinals incompatible with $V = L$.

In his discussion in §2 of "On the Question of Absolute Undecidability," Koellner proposes to reconstruct Gödel's views by assuming that even higher-order reflection principles are intrinsically justified. He has made clear to me that that is not his personal view; furthermore, he states (p. 165) that what he is offering is a "rational reconstruction" of Gödel's view.

[47] For example Martin, "Mathematical Evidence." Further discussion, with citation of more recent mathematical results, is in Koellner, op. cit., §3.

considered up to now are unable to decide the Continuum Hypothesis. That problem has been advanced, although not solved, by recent work, especially that of W. Hugh Woodin, so that many set theorists are hopeful that a generally accepted solution of that problem will be found.[48] But apart from the purely mathematical difficulties, many problems of methodology and interpretation remain in this area.

[48] See Woodin, "The Continuum Hypothesis, I and II," "The Continuum Hypothesis," and "Set Theory after Russell." See also Kai Hauser, "Is Cantor's Continuum Problem Inherently Vague?" and Koellner, "On the Question," §§4–5.

Bibliography

References and this bibliography follow the same conventions as in *Mathematics in Philosophy*. Works are cited by author and title only, the latter often abbreviated and occasionally (especially in cross-reference within the bibliography) omitted. The following principles govern citations:

1. Works of an author are listed alphabetically by title, rather than chronologically, to facilitate locating the entry.

2. Books are cited in the latest edition listed in this bibliography, unless the contrary is explicitly stated.

3. For articles, reprintings in collections of the author's own papers are listed, but reprintings in anthologies are not, with the exception of a few essential ones such as Benacerraf and Putnam's *Philosophy of Mathematics: Selected Readings*. Where there is a collection of the author's papers, reference is to the reprint in that collection.

4. Special conventions are observed for a few classic authors:

Frege. Writings published in his lifetime are cited in the original pagination. This is given in the German editions that constitute a *de facto Collected Works*: Thiel's edition of *Grundlagen*, the Olms reprint of *Grundgesetze*, and the two collections edited by Angelelli, *Begriffsschrift und andere Aufsätze* and *Kleine Schriften*. It is also given in the canonical English editions, Austin's translation of *Grundlagen* and the *Collected Papers*, as well as in Beaney's *Frege Reader* (Oxford: Blackwell, 1997) and in the older Geach and Black collection.

Gödel is cited according to the *Collected Works*.

Kant. The *Critique of Pure Reason* is cited in the standard way according to the original editions A (first edition, 1781) and B (second edition, 1787). Other writings are cited according to the Academy edition *Gesammelte Schriften*. These paginations are given in the better other editions in English and German, including the translations of Kemp Smith and Guyer and Wood of the first *Critique*. Quotations from that work are in the Guyer and Wood translation.

Ackermann, Wilhelm. "Die Widerspruchsfreiheit der allgemeinen Mengenlehre."
    Mathematische Annalen 114 (1937), 305–315.

Aczel, Peter. "An Introduction to Inductive Definitions." In Barwise, *Handbook*, pp. 739–782.

Barwise, Jon. *Admissible Sets and Structures*. Berlin: Springer, 1975.

———, ed. *Handbook of Mathematical Logic*. Amsterdam: North-Holland, 1977.

Bealer, George. "Intuition." In Donald M. Borchert (editor in chief), *The Encyclopedia of Philosophy Supplement*, pp. 268–269. New York: Macmillan Reference USA, 1996. Reprinted in Borchert, *Encyclopedia* 2nd ed., IV, 732–733.

———. "On the Possibility of Philosophical Knowledge." **Philosophical Perspectives** 10 (1996), 1–34.

Beeson, Michael. *Foundations of Constructive Mathematics*. Berlin: Springer, 1985.

Benacerraf, Paul. "Mathematical Truth." **The Journal of Philosophy** 70 (1973), 661–679. Reprinted in Benacerraf and Putnam, pp. 403–420. (Not in 1st ed.)

———. "What Numbers Could Not Be." **Philosophical Review** 74 (1965), 47–73. Reprinted in Benacerraf and Putnam, pp. 272–294. (Not in 1st ed.) Cited according to reprint.

Benacerraf, Paul, and Hilary Putnam, eds. *Philosophy of Mathematics: Selected Readings*. Englewood Cliffs, NJ: Prentice Hall, 1964. 2nd ed. Cambridge University Press, 1983 (published 1984).

Bernays, Paul. *Abhandlungen zur Philosophie der Mathematik*. Darmstadt: Wissenschaftliche Buchgesellschaft, 1976.

———. "Bemerkungen zum Paradoxon von Thoralf Skolem." **Avhandlingen uitgitt av Det Norske Videnskaps-Akademi i Oslo**, I. Mat.-Naturv. Klasse, 1957, No. 5, 3–9. Reprinted in *Abhandlungen*, pp. 113–118.

———. "Die Mathematik als ein zugleich Vertrautes und Unbekanntes." **Synthese** 9 (1955), 465–471. Reprinted in *Abhandlungen*, pp. 107–112.

———. "Mathematische Existenz und Widerspruchsfreiheit." In *Études de philosophie des sciences, en hommage à F. Gonseth à l'occasion de son soixantième anniversaire*, pp. 11–25. Neuchâtel: Griffon, 1950. Reprinted in *Abhandlungen*, pp. 92–106.

———. "Die Philosophie der Mathematik und die Hilbertsche Beweistheorie." *Blätter für deutsche Philosophie* 4 (1930), 326–367. Reprinted with postscript in *Abhandlungen*, pp. 17–61.

———. "Quelques points essentiels de la métamathématique." **L'enseignement mathématique** 52 (1935), 70–95.

———. "Quelques points de vue concernant le problème de l'évidence," **Synthese** 5 (1946), 321–326. (The German translation in *Abhandlungen* is by Bernays but was evidently done in 1974.)

———. "Sur le platonisme dans les mathématiques." **L'enseignement mathématique** 52 (1935), 52–69. Translation in Benacerraf and Putnam, pp. 258–271 of 2nd ed. (The German version in *Abhandlungen* is a translation from the French.)

———. "A System of Axiomatic Set Theory, Part II." **The Journal of Symbolic Logic** 6 (1941), 1–17. Reprinted in Müller, *Sets and Classes*, pp. 14–30.

———. "A System of Axiomatic Set Theory, Part VII." **The Journal of Symbolic Logic** 19 (1954), 81–96. Reprinted in Müller, *Sets and Classes*, pp. 102–119.

———. "Über Hilberts Gedanken zur Grundlegung der Arithmetik," **Jahresbericht der Deutschen Mathematiker-Vereinigung** 31 (1922), 10–19.

———. "Zur Frage der Unendlichkeitsschemata in der axiomatischen Mengenlehre." In *Essays on the Foundations of Mathematics*, dedicated to Professor A. A. Fraenkel on his 70th birthday, pp. 3–49. Jerusalem: Magnes Press, 1961. Translation, "partially revised," in Müller, *Sets and Classes*, pp. 121–172.

———. See also Hilbert and Bernays.

Black, Max. "The elusiveness of sets." **Review of Metaphysics** 24 (1970–1971), 614–636.

Bolzano, Bernard. *Paradoxien des Unendlichen*. Ed. by F. Příhonský. Leipzig 1851.

———. *Wissenschaftslehre*. 4 vols. Sulzach 1837.

BonJour, Laurence. In *Defense of Pure Reason. A Rationalist Account of a priori Justification*. Cambridge University Press, 1998.

Boolos, George. "Iteration Again." **Philosophical Topics** 17 (1989), 5–21. Reprinted in LLL.

———. "The Iterative Conception of Set." **The Journal of Philosophy** 68 (1971), 215–231. Reprinted in LLL. Also reprinted in Benacerraf and Putnam, 2nd ed.

———. *Logic, Logic, and Logic*. Edited by Richard Jeffrey. With Introductions and Afterword by John P. Burgess. Cambridge, MA: Harvard University Press, 1998. Hereafter LLL.

———. *The Logic of Provability*. Cambridge University Press, 1993.

———. "Must we believe in set theory?" In LLL, pp. 120–132. Also in Sher and Tieszen, pp. 257–268.

———. "Nominalist Platonism." **Philosophical Review** 94 (1985), 327–344. Reprinted in LLL.

———. "Reply to Charles Parsons, 'Sets and Classes'." In LLL, pp. 30–36. (Written and presented in 1974.)

———. "To Be is to be the Value of a Variable (or to be Some Values of Some Variables)." **The Journal of Philosophy** 81 (1984), 430–449. Reprinted in LLL.

———, and Richard C. Jeffrey. *Computability and Logic*. Cambridge University Press, 1974. 2nd ed., 1981. 3rd ed., 1990. 4th ed., with John P. Burgess, 2002.

Borchert, Donald M. (Editor in chief). *The Encyclopedia of Philosophy*, Second Edition. 10 vols. Detroit: Macmillan Reference USA, 2006. (For first edition, see Edwards.)

Brentano, Franz. *Wahrheit und Evidenz. Erkenntnistheoretische Abhandlungen und Briefe*. Edited by Oskar Kraus. Leipzig: Meiner, 1930. Reprinted Hamburg: Meiner, 1962, 1974.

Bromberger, Sylvain. *On What We Know We Don't Know. Explanation, Theory, Linguistics, and How Questions Shape Them*. Chicago: University of Chicago Press, and Stanford: Center for the Study of Language and Information, 1992.

———. "Types and Tokens in Linguistics." In Alexander George (ed.), *Reflections on Chomsky*, pp. 58–89. Cambridge, Mass., MIT Press, 1989. Reprinted in *On What We Know We Don't Know*, pp. 170–208.

———, and Morris Halle. "The Ontology of Phonology." In Bromberger, *On What We Know We Don't Know*, pp. 209–228.

Brouwer, L. E. J. *Collected Works, volume 1: Philosophy and Intuitionistic Mathematics*. Edited by Arend Heyting. Amsterdam: North-Holland, 1975.

———. "Mathematik, Wissenschaft, und Sprache." **Monatshefte für Mathematik und Physik** 36 (1929), 153–164. Reprinted, *Collected Works*, vol. 1.

———. *Over de grondslagen der wiskunde*. Amsterdam and Leipzig: Maass and van Suchtelen, 1907. Critical edition in Dirk van Dalen (ed.), *L. E. J. Brouwer en de grondslagen van de wiskunde* (Utrecht: Epsilon, 2001), pp. 39–147. English translation (of the main text) in *Collected Works*, vol. 1.

Buchholz, Wilfried, Solomon Feferman, Wolfram Pohlers, and Wilfried Sieg, *Iterated Inductive Definitions and Subsystems of Analysis*. Lecture Notes in Mathematics 897. Berlin etc.: Springer, 1981.

Burgess, John P. "*E pluribus unum*: Plural Logic and Set Theory." **Philosophia Mathematica (III)** 12 (2004), 193–221.

———. *Fixing Frege*. Princeton University Press, 2005.

———. "Mathematics and *Bleak House*." **Philosophia Mathematica (III)** 12 (2004), 18–36.

———. Review of Shapiro, *Philosophy of Mathematics*. **Notre Dame Journal of Formal Logic** 40 (1999), 283–291.

———, and Allen Hazen. "Predicative Logic and Formal Arithmetic." **Notre Dame Journal of Formal Logic** 39 (1998), 1–17.

Burgess, John P., and Gideon Rosen. *A Subject with no Object. Strategies for Nominalistic Interpretation of Mathematics*. Oxford: Clarendon Press, 1997.

Cantor, Georg. *Gesammelte Abhandlungen mathematischen und philosophischen Inhalts*. Edited by Ernst Zermelo. Berlin: Springer, 1932. Reprinted Hildesheim: Olms, 1962.

Carnap, Rudolf. *Logische Syntax der Sprache*. Vienna: Springer, 1934.

———. *Logical Syntax of Language*. Translated by Amethe Smeaton. London: Kegan Paul, 1937. (Contains two sections not in the German edition.)

Carson, Emily. "On realism in set theory." **Philosophia Mathematica (III)** 4 (1996), 3–17.

Cartwright, Richard L. "Speaking of Everything." **Noûs** 28 (1994), 1–20.

Church, Alonzo. "Comparison of Russell's Solution of the Semantical Paradoxes with that of Tarski." **The Journal of Symbolic Logic** 41 (1976), 747–760. Reprinted in Martin, *Recent Essays*.

———. "A Formulation of the Logic of Sense and Denotation." In *Structure, Method, and Meaning: Essays in Honor of Henry M. Sheffer*, pp. 3–24. New York: Liberal Arts Press, 1951.

———. "A Formulation of the Simple Theory of Types." **The Journal of Symbolic Logic** 5 (1940), 56–68.

Cohn-Vossen, S. See Hilbert and Cohn-Vossen.

Collins, George E., and J. D. Halpern. "On the Interpretability of Arithmetic in Set Theory." **Notre Dame Journal of Formal Logic** 11 (1970), 477–483.

Dales, H. G., and G. Oliveri (eds.). *Truth in Mathematics*. Oxford: Clarendon Press, 1998.

Davidson, Donald. *Inquiries into Truth and Interpretation*. Oxford: Clarendon Press, 1984.

———. "On Saying That." **Synthese** 19 (1968), 130–146. Reprinted in *Inquiries*.

———. "Radical Interpretation." **Dialectica** 27 (1973), 313–328. Reprinted in *Inquiries*.

Dedekind, Richard. *Gesammelte mathematische Werke*. 3 vols. Edited by Robert Fricke, Emmy Noether, and Öystein Ore. Braunschweig: Vieweg, 1930–32.

———. *Was sind und was sollen die Zahlen?* Braunschweig: Vieweg, 1888, 3rd ed. 1911. Translation in Ewald, II, 787–833.

Demopoulos, William (ed.). *Frege's Philosophy of Mathematics*. Cambridge, MA: Harvard University Press, 1995.

Descartes, René. *The Philosophical Writings of Descartes*. Translated by John Cottingham, Robert Stoothoff, and Dugald Murdoch. 2 vols. Cambridge University Press, 1985.

Donnellan, Keith. "Reference and Definite Descriptions." **Philosophical Review** 75 (1966), 281–304.

Drake, F. R. *Set Theory. An Introduction to Large Cardinals*. Amsterdam: North-Holland, 1974.

Dummett, Michael. *Frege: Philosophy of Language*. London: Duckworth, 1973. 2nd ed., 1981.

———. *Frege: Philosophy of Mathematics*. Cambridge, Mass.: Harvard University Press, 1991.

———. "The Philosophical Basis of Intuitionistic Logic." In H. E. Rose and J. C. Shepherdson (eds.), *Logic Colloquium '73*, pp. 5–40. Amsterdam: North-Holland, 1975. Reprinted in *Truth and Other Enigmas*, pp. 215–247, and in Benacerraf and Putnam (not in 1st ed.).

———. "The Philosophical Significance of Gödel's Theorem." **Ratio** 5 (1963), 140–155. Reprinted in *Truth and Other Enigmas*, pp. 186–201.

———. "Platonism." In *Truth and Other Enigmas*, pp. 202–214. (Text of an address given in 1967.)

———. *The Seas of Language*. Oxford: Clarendon Press, 1993.

———. *Truth and Other Enigmas*. London: Duckworth, 1978.

———. "Wang's Paradox." **Synthese** 30 (1975), 301–324. Reprinted in *Truth and Other Enigmas*, pp. 248–268.

———. "What is Mathematics About?" In George, *Mathematics and Mind*, pp. 11–26. Also in *The Seas of Language*, pp. 429–445.

Edwards, Paul (ed.). *The Encyclopedia of Philosophy*. 7 vols. New York: Free Press/Macmillan, 1967. (For second edition, see Borchert.)

Evans, Gareth. "Can there be Vague Objects?" **Analysis** 38 (1978), 208. Reprinted in *Collected Papers* (Oxford: Clarendon Press, 1985), pp. 176–177.

Ewald, William (ed.). *From Kant to Hilbert. A Source Book in the Foundations of Mathematics*. 2 vols. Oxford: Clarendon Press, 1996.

Feferman, Solomon. "Definedness." *Erkenntnis* 43 (1995), 295–320.

———. *In the Light of Logic*. Oxford University Press, 1998.

———. "Reflecting on Incompleteness." **The Journal of Symbolic Logic** 56 (1991), 1–49.

———. "Systems of Predicative Analysis." **The Journal of Symbolic Logic** 29 (1964), 1–30.

———. "Weyl vindicated. Das Kontinuum 70 years later." In Carlo Cellucci and Giovanni Sambin (eds.), *Temi e prospettive della logica e della scienza contemporanee*, vol. 1, pp. 59–93. Bologna: CLUEB, 1988. Reprinted with corrrections and additions in *In the Light of Logic*.

See also Buchholz, Feferman, Pohlers, and Sieg.

See also Gödel.

Feferman, Solomon, and Geoffrey Hellman . "Predicative Foundations of Arithmetic." **Journal of Philosophical Logic** 22 (1995), 1–17.

———. "Challenges to Predicative Foundations of Arithmetic." In Sher and Tieszen, pp. 317–335.

Ferreira, Fernando. "A Note on Finiteness in the Predicative Foundations of Arithmetic." **Journal of Philosophical Logic** 28 (1999), 165–174.

Field, Hartry. "Are Our Logical and Mathematical Concepts Highly Indeterminate?" In Peter A. French, Theodore E. Uehling, Jr., and Howard K. Wettstein (eds.), *Midwest Studies in Philosophy, volume 19: Philosophical Naturalism*, pp. 391–429. Minneapolis: University of Minnesota Press, 1994.

———. "Is Mathematical Knowledge Just Logical Knowledge?" **Philosophical Review** 93 (1984), 509–552. Reprinted, with revisions and postscript, in *Realism, Mathematics, and Modality*, pp. 79–124.

———. "On Conservativeness and Incompleteness." **The Journal of Philosophy** 82 (1985), 239–260. Reprinted in *Realism, Mathematics, andModality*, pp. 125–146.

———. "Realism, Mathematics, and Modality." **Philosophical Topics** 19 (1988), 57–107. Reprinted in *Realism, Mathematics, and Modality*, pp. 227–281.

———. *Realism, Mathematics, and Modality*. Oxford: Blackwell, 1989.

———. *Science without Numbers: A Defense of Nominalism*. Princeton University Press, 1980.

Bibliography

———. *Truth and the Absence of Fact.* Oxford: Clarendon Press, 2001.
———. "Which Undecidable Mathematical Sentences have Determinate Truth Values?" In Dales and Oliveri, pp. 291–310. Reprinted with Postscript in *Truth and the Absence of Fact*, pp. 332–360.
Fraenkel, (Abraham) Adolf. *Einleitung in die Mengenlehre.* Berlin: Springer, 1919. 2nd ed., revised and enlarged, 1923. 3d ed., further revised and enlarged, 1928.
Frege, Gottlob. *Begriffschrift. Eine der arithmetischen nachgebildete Formelsprache des reinen Denkens.* Halle: Nebert, 1879. Reprinted in *Begriffschrift und andere Aufsätze.* Translation in van Heijenoort.
———. *Begriffschrift und andere Aufsätze.* Edited by Ignacio Angelelli. Hildesheim: Olms, 1964.
———. *Collected Papers on Mathematics, Logic, and Philosophy.* Edited by Brian McGuinness. Translated by Max Black and others. Oxford: Blackwell, 1984.
———. *Frege on the Foundations of Geometry and Formal Theories in Arithmetic.* Translated and edited by Eike-Henner W. Kluge. New Haven: Yale University Press, 1971.
———. "Der Gedanke: Eine logische Untersuchung." **Beiträge zur Philosophie des deutschen Idealismus** 1 (1918–19), 58–77. Reprinted in *Kleine Schriften.* Translation in *Collected Papers.*
———. *The Foundations of Arithmetic.* Translated by J. L. Austin. Contains the German text. Oxford: Blackwell, 1950. 2nd ed. 1953. (The pagination of the German text is exactly the same as that of the original, and thus that of the translation is very close.)
———. *Funktion und Begriff.* Jena: Pohle, 1891. Reprinted in *Kleine Schriften.* Translation in *Collected Papers.*
———. *Grundgesetze der Arithmetik.* Jena: Pohle. Vol. I, 1893. Vol. II, 1903. Reprinted Hildesheim: Olms, 1962, 1998.
———. *Die Grundlagen der Arithmetik.* Breslau: Koebner, 1884. Centenary edition, with amplifying texts, critically edited by Christian Thiel. Hamburg: Meiner, 1986. English translation *Foundations.*
———. *Kleine Schriften.* Edited by Ignacio Angelelli. Hildesheim: Olms, 1967. Translated as *Collected Papers.*
———. *Logical Investigations.* Edited by Peter Geach. Translated by Geach and R. H. Stoothoff. Oxford: Blackwell, 1977.
———. *Nachgelassene Schriften.* Edited by Hans Hermes, Friedrich Kambartel, and Friedrich Kaulbach. Hamburg: Meiner, 1969. 2nd. ed., revised, with an appendix (lecture notes) ed. with an introduction by Lothar Kreiser. Hamburg: Meiner, 1983.
———. Review of Husserl, *Philosophie der Arithmetik.* **Zeitschrift für Philosophie und philosophische Kritik** 103 (1894), 313–332. Reprinted in *Kleine Schriften.* Translation in *Collected Papers.*

―――. *Wissenschaftlicher Briefwechsel*. Edited by Gottfried Gabriel, Hans Hermes, Friedrich Kambartel, Christian Thiel, and Albert Veraart. Hamburg: Meiner, 1976.

Friedman, Michael. *Kant and the Exact Sciences*. Cambridge, MA: Harvard University Press, 1992.

Gallin, Daniel. *Intensional and Higher-Order Modal Logic*. Amsterdam: North-Holland, 1975.

Gandy, Robin O. "Limitations to Mathematical Knowledge." In D. van Dalen, D. Lascar, and T. J. Smiley (eds.), *Logic Colloquium '80*, pp. 129–145. Amsterdam: North-Holland, 1982.

George, Alexander. "The Imprecision of Impredicativity." **Mind** N. S. 96 (1987), 514–518.

―――,(ed.). *Mathematics and Mind*. Oxford University Press, 1994.

―――, and Daniel J. Velleman. "Two Conceptions of Natural Number." In Dales and Oliveri, pp. 311–327.

Givant, Steven, and Alfred Tarski. "Peano Arithmetic and the Zermelo-like Theory of Sets with Finite Ranks." Abstract 77T-E51. **Notices of the Americal Mathematical Society** 24 (1977), A-437.

See also Tarski and Givant.

Gödel, Kurt. *Collected Works*. Volume I, *Publications 1929–1936*. Edited by Solomon Feferman, John W. Dawson, Jr., Stephen C. Kleene, Gregory H. Moore, Robert M. Solovay, and Jean van Heijenoort. New York and Oxford: Oxford University Press, 1986.

―――. *Collected Works*. Volume II, *Publications 1938–1974*. Edited by Solomon Feferman, John W. Dawson, Jr., Stephen C. Kleene, Gregory H. Moore, Robert M. Solovay, and Jean van Heijenoort. New York and Oxford: Oxford University Press, 1990.

―――. *Collected Works*. Volume III, *Unpublished Essays and Lectures*. Edited by Solomon Feferman, John W. Dawson, Jr., Warren Goldfarb, Charles Parsons, and Robert M. Solovay. New York and Oxford: Oxford University Press, 1995.

―――. *Collected Works*. Volume IV, *Correspondence, A-G*. Edited by Solomon Feferman, John W. Dawson, Jr., Warren Goldfarb, Charles Parsons, and Wilfried Sieg. Oxford: Clarendon Press, 2003.

―――. *Collected Works*. Volume V, *Correspondence, H-Z*. Edited by Solomon Feferman, John W. Dawson, Jr., Warren Goldfarb, Charles Parsons, and Wilfried Sieg. Oxford: Clarendon Press, 2003.

―――. *The Consistency of the Continuum Hypothesis*. Princeton University Press, 1940. Reprinted in *Collected Works*, II, 33–101.

―――. "Is Mathematics Syntax of Language?" Unfinished paper written between 1953 and 1959. Versions III and V in *Collected Works*, III, 334–362.

―――. "On an Extension of Finitary Mathematics which has not yet been Used." Revised English version, with additional notes, of "Über eine noch nicht benützte Erweiterung." In *Collected Works*, II, 271–280.

———. "The Present Situation in the Foundations of Mathematics." Lecture to the Mathematical Association of America, 1933. In *Collected Works*, III, 45–53.

———. "Russell's Mathematical Logic." In P. A. Schilpp (ed.), *The Philosophy of Bertrand Russell* (Evanston: Northwestern University, 1944), pp. 125–153. Reprinted in *Collected Works*, II, 119–141. Also reprinted in Benacerraf and Putnam, pp. 442–470.

———. "Some Basic Theorems on the Foundations of Mathematics and their Philosophical Implications." Josiah Willard Gibbs Lecture, American Mathematical Society, 1951. In *Collected Works*, III, 304–323. (The word 'philosophical' was omitted from the title of the published version. See *Collected Works*, IV, 636.)

———. "Über eine bisher noch nicht benützte Erweiterung des finiten Standpunktes." *Dialectica* 12 (1958), 280–287. Reprinted, with English translation, in *Collected Works*, II, 240–251.

———. "Vortrag bei Zilsel." Lecture to an informal circle, Vienna, 1938. In *Collected Works*, III, 86–113. With English translation.

———. "What is Cantor's Continuum Problem?" **American Mathematical Monthly** 54 (1947), 515–525. Revised and expanded in Benacerraf and Putnam, 1st ed., pp. 258–271, also in 2nd ed., pp. 470–485. Both versions reprinted in *Collected Works*, II, 176–187 and 254–270 respectively.

———. "Zur intuitionistischen Arithmetik und Zahlentheorie." **Ergebnisse eines mathematischen Kolloquiums** 4 (1933), 34–38. Reprinted, with English translation, in *Collected Works*, I, 286–295.

Goldfarb, Warren. See Gödel.

Goldfarb, Warren, and Thomas Ricketts. "Carnap and the Philosophy of Mathematics." In David Bell and Wilhelm Vossenkuhl (eds.), *Wissenschaft und Subjektivität*, pp. 61–78. Berlin: Akademie-Verlag, 1992.

Goodman, Nelson. *Fact, Fiction, and Forecast*. Cambridge, MA: Harvard University Press, 1955. 4th ed., 1983.

———. *Problems and Projects*. Indianapolis: Bobbs-Merrill, 1972.

———. "Sense and Certainty." **The Philosophical Review** 61 (1952), 160–167. Reprinted in *Problems and Projects*, pp. 60–68.

———. *The Structure of Appearance*. Cambridge, MA: Harvard University Press, 1951. 3rd ed. Dordrecht: Reidel, 1977.

Goodman, Nelson, and W. V. Quine. "Steps toward a Constructive Nominalism." **The Journal of Symbolic Logic** 12 (1947), 105–122. Reprinted in Goodman, *Problems and Projects*, pp. 173–198.

Goodman, Nicolas D. "The Knowing Mathematician." **Synthese** 60 (1984), 21–38.

Goodstein, R. L. *Recursive Number Theory*. Amsterdam: North-Holland, 1957.

Gottlieb, Dale. *Ontological Economy. Substitutional Quantification and Mathematics*. Oxford: Clarendon Press, 1980.

Grossmann, Reinhardt. "Meinong's Doctrine of the *Aussersein* of the Pure Object." **Noûs** 8 (1974), 67–82.

Hájek, Petr, and Pavel Pudlák. *Metamathematics of First-Order Arithmetic*. Berlin, Heidelberg, New York: Springer-Verlag, 1993.

Hale, Bob. "Structuralism's Unpaid Epistemological Debts." **Philosophia Mathematica (III)** 4 (1996), 124–147.

_____, and Crispin Wright. *The Reason's Proper Study. Essays towards a Neo-Fregean Philosophy of Mathematics*. Oxford: Clarendon Press, 2001.

_____. "Benacerraf's Dilemma Revisited." **European Journal of Philosophy** 10 (2002), 101–129.

Halle, Morris. See Bromberger and Halle.

Hart, W. D. "Benacerraf's Dilemma." **Crítica** 23 (1991), 87–103.

_____. Review of Steiner, *Mathematical Knowledge*. **The Journal of Philosophy** 74 (1977), 118–129.

_____. "Skolem's Promises and Paradoxes." **The Journal of Philosophy** 67 (1970), 98–109.

Hauser, Kai. "Is Cantor's Continuum Problem Inherently Vague?" **Philosophia Mathematica (III)** 10 (2002), 257–285.

Heck, Richard G., Jr. "The consistency of predicative fragments of Frege's *Grundgesetze der Arithmetik*." **History and Philosophy of Logic** 17 (1996), 209–220.

Hellman, Geoffrey. *Mathematics without Numbers: Toward a Modal-Structural Interpretation*. Oxford: Clarendon Press, 1989.

See also Feferman and Hellman.

Higginbotham, James. "Linguistic Theory and Davidson's Program in Semantics." In Ernest Lepore (ed.), *Truth and Interpretation*. Oxford: Blackwell, 1986.

_____. "On Higher-Order Logic and Natural Language." **Proceedings of the British Academy** 95 (1998), 1–27. Somewhat abridged version in Sher and Tieszen, pp. 75–99.

Hilbert, David. *Grundlagen der Geometrie*. Leipzig: Teubner, 1899. 11th ed., ed. by Paul Bernays, Stuttgart: Teubner, 1972. English translation of 10th ed. by Leo Unger, La Salle, Ill.: Open Court, 1971, 2nd ed. 1992.

_____. "Neubegründung der Mathematik (Erste Mitteilung)." *Abhandlungen aus dem mathematischen Seminar der HamburgischenUniversität* 1 (1922), 157–177. Translation in Mancosu.

_____. "Über das Unendliche." **Mathematische Annalen** 95 (1926), 161–190. Translation in van Heijenoort. Partial translation in Benacerraf and Putnam. (Citations of a translation are of the one in van Heijenoort.)

_____, and Paul Bernays. *Grundlagen der Mathematik*. Berlin: Springer, vol. I, 1934; vol. II, 1939. 2nd ed., vol. I, 1968; vol. II, 1970.

_____, and S. Cohn-Vossen. *Anschauliche Geometrie*. Berlin: Springer, 1932.

Hodes, Harold T. "Logicism and the Ontological Commitments of Arithmetic." **The Journal of Philosophy** 81 (1984), 123–149.

Husserl, Edmund. *Erfahrung und Urteil*. Edited by Ludwig Landgrebe. Hamburg: Claassen, 1948.

———. *Formale und transzendentale Logik*. Halle: Niemeyer, 1929; reprinted Tübingen: Niemeyer, 1981. Critical edition edited by Paul Janssen. Husserliana, vol. 17. The Hague: Nijhoff, 1974.

———. *Logical Investigations*. Translated by J. N. Findlay. London: Routledge and Kegan Paul, 1970. Translation of second edition of *Logische Untersuchungen*.

———. *Logische Untersuchungen*. Halle: Niemeyer, 1900–01, 2nd ed. 1913–20. Critical edition of volume 1 edited by Elmar Holenstein. Husserliana, vol. 18. The Hague: Nijhoff, 1975. Critical edition of volume 2 edited by Ursula Panzer. Husserliana, vol. 19. The Hague: Nijhoff, 1984. Page references prefixed A are to the 1st ed., B to the 2nd ed., with the subscript indicating the part of volume 2 in the 2nd ed.

———. *Philosophie der Arithmetik, mit ergänzenden Texten (1890–1901)*. Edited by Lothar Eley. Hussserliana, vol. 12. The Hague: Nijhoff, 1970. (*Philosophie der Arithmetik* first published Halle: Pfeffer, 1891.)

Isles, David. "What Evidence Is There that 2^{65536} Is a Natural Number?" **Notre Dame Journal of Formal Logic** 33 (1992), 465–480.

Jeffrey, Richard C. See Boolos and Jeffrey.

Kant, Immanuel. *Critique of Pure Reason*. Translated by Norman Kemp Smith. London: Macmillan, 1929. 2nd ed., 1933.

———. *Critique of Pure Reason*. Translated by Paul Guyer and Allen W. Wood. Cambridge University Press, 1997.

———. *Gesammelte Schriften*. Edited by the Preussische Akademie der Wissenschaften and its successors. Berlin: Reimer, later de Gruyter, 1902–. Cited as *Ak*.

———. *Kritik der reinen Vernunft*. Edited by Jens Timmermann. Hamburg: Meiner, 1998.

———. *Prolegomena to any Future Metaphysics*. Translated by Lewis White Beck (revising earlier translations). New York: Liberal Arts Press, 1956.

Kaufman, E. L., M. W. Lord, T. W. Reese, and J. Volkmann. "The Discrimination of Visual Number." **American Journal of Psychology** 62 (1949), 498–525.

Keränen, Jukka. "The Identity Problem for Realist Structuralism." **Philosophia Mathematica (III)** 9 (2001), 308–330.

Kim, Jaegwon. "The Role of Perception in A Priori Knowledge." **Philosophical Studies** 40 (1981), 339–354.

Kirby, Laurence. "Finitary Set Theory." **Notre Dame Journal of Formal Logic**, forthcoming.

Kitcher, Philip. *The Nature of Mathematical Knowledge*. New York and Oxford: Oxford University Press, 1983.

———. "The Plight of the Platonist." **Noûs** 12 (1978), 119–136.

Kleene, Stephen Cole. *Introduction to Metamathematics*. New York: Van Nostrand, 1952.

See also Gödel.

Kluge, Eike-Henner W. See Frege.

Koellner, Peter. "On the Question of Absolute Undecidability." **Philosophia Mathematica (III)** 14 (2006), 153–188.

Kreisel, G. "Informal Rigour and Completeness Proofs." In Imre Lakatos (ed.), *Problems in the Philosophy of Mathematics*, pp. 138–186. Amsterdam: North-Holland, 1967.

———. "Mathematical Logic." In T. L. Saaty (ed.), *Lectures on Modern Mathematics*, vol. III, pp. 95–195. New York: John Wiley and Sons, 1965.

———. "Ordinal logics and the characterization of informal concepts of proof." In *Proceedings of the International Congress of Mathematicians, Edinburgh 1958*, pp. 289–299. Cambridge University Press, 1960.

Kripke, Saul A. "Is There a Problem about Substitutional Quantification?" In Gareth Evans and and John McDowell (eds.). *Truth and Meaning*, pp. 325–419. Oxford: Clarendon Press, 1976.

———. "Naming and Necessity." In Donald Davidson and Gilbert Harman, eds., *Semantics of Natural Language*, pp. 253–355, 763–769. Dordrecht: Reidel, 1972. 2nd ed.: *Naming and Necessity*. Cambridge, MA: Harvard University Press, 1980.

Krivine, Jean-Louis. *Introduction to Axiomatic Set Theory*. Translated by David Miller. Dordrecht: Reidel, 1971.

Ladyman, James. See Leitgeb and Ladyman.

Lambek, Joachim, and P. J. Scott. *Introduction to Higher-Order Categorical Logic*. Cambridge University Press, 1986.

Lambert, Karel. *Meinong and the Principle of Independence: Its Place in Meinong's Theory of Objects and its Significance in Contemporary Philosophical Logic*. Cambridge University Press, 1983.

Lavine, Shaughan. "Something about Everything: Universal Quantification in the Universal Sense of Universal Quantification." In Rayo and Uzquiano, pp. 98–148.

———. *Understanding the Infinite*. Cambridge, MA: Harvard University Press, 1994.

Leibniz, Gottfried Wilhelm. *Philosophical Essays*. Translated and edited by Roger Ariew and Daniel Garber. Indianapolis: Hackett, 1989.

Leitgeb, Hannes, and James Ladyman. "Criteria of Identity and Structuralist Ontology." **Philosophia Mathematica (III)** 16 (2008), 388–396.

Lévy, Azriel. *Basic Set Theory*. Berlin: Springer, 1979.

Lewis, David. "Mathematics is Megethology." **Philosophia Mathematica (III)** 1 (1993), 3–23. Reprinted in *Papers in Philosophical Logic* (Cambridge University Press, 1998), pp. 203–229.

———. *On the Plurality of Worlds*. Oxford: Blackwell, 1986.

———. *Parts of Classes*. Oxford: Blackwell, 1991.

Linnebo, Øystein. "Epistemological Challenges to Mathematical Platonism." **Philosophical Studies** 129 (2006), 545–574.

———. "Plural Quantification." In *Stanford Encyclopedia of Philosophy*, 2004. Online at http://plato.stanford.edu.
Lorenzen, Paul. *Einführung in die operative Logik und Mathematik*. Berlin: Springer, 1955.
———. "Logical Reflection and Formalism." **The Journal of Symbolic Logic** 23 (1958), 241–249.
Lorenzen, Paul, and John Myhill. "Constructive Definition of Certain Analytic Sets of Numbers." *Ibid.* 24 (1959), 37–49.
MacBride, Fraser. "Structuralism Reconsidered." In Shapiro, *Oxford Handbook*, pp. 563–589.
Maddy, Penelope. "Believing the Axioms." **The Journal of Symbolic Logic** 53 (1988), 481–511, 736–764.
———. "Indispensability and Practice." **The Journal of Philosophy** 89 (1992), 275–289.
———. *Naturalism in Mathematics*. Oxford: Clarendon Press, 1997.
———. "Perception and Mathematical Intuition." **Philosophical Review** 89 (1980), 163–196.
———. *Realism in Mathematics*. Oxford: Clarendon Press, 1990.
Majer, Ulrich. "Geometrie und Erfahrung: Ein Vergleich der Auffassungen von Hilbert und Kant." Report no. 17/93 of the Research Group on Semantical Aspects of Space-Time Theories. Bielefeld: Zentrum für interdisziplinäre Forschung, 1993.
———. "Geometry, intuition, and experience: From Kant to Husserl." **Erkenntnis** 42 (1995), 261–285. Revised English version of "Geometrie und Erfahrung."
Martin, Donald A. "Descriptive Set Theory: Projective Sets." In Barwise, *Handbook*, pp. 783–815.
———. "Mathematical Evidence." In Dales and Oliveri, pp. 214–231.
Martin, Robert L. (ed.). *Recent Essays on Truth and the Liar Paradox*. New York and Oxford: Oxford University Press.
Martin-Löf, Per. *Intuitionistic Type Theory*. Naples: Bibliopilis, 1984.
McCarthy, Timothy. "Platonism and Possibility." **The Journal of Philosophy** 83 (1986), 275–290.
McGee, Vann. "How we Learn Mathematical Language." **Philosophical Review** 107 (1997), 35–68.
———. "Truth by Default." **Philosophia Mathematica (III)** 9 (2001), 5–20.
Meinong, Alexius. *Gesamtausgabe*. Edited by Roderick M. Chisholm, Rudolf Haller, and Rudolf Kindlinger. Graz: Akademische Druck- und Verlagsanstalt, 1968–78.
———. "Über Gegenstandstheorie." In Meinong (ed.), *Untersuchungen zur Gegenstandstheorie und Psychologie*, pp. 1–50. Leipzig: Barth, 1904. Reprinted in *Gesamtausgabe*, II.
———. *Über die Stellung der Gegenstandstheorie im System der Wissenschaften*. Leipzig: Voigtländer, 1907. Reprinted in *Gesamtausgabe*, V.

Meyer, Robert K. "In Memoriam: Richard (Routley) Sylvan, 1935–1996." **The Bulletin of Symbolic Logic** 4 (1998), 338–340.

Montagna, Franco, and Antonella Mancini. "A Minimal Predicative Set Theory." **Notre Dame Journal of Formal Logic** 35 (1994), 186–203.

Moore, Gregory H. "Beyond First-Order Logic: The Historical Interplay between Mathematical Logic and Axiomatic Set Theory." **History and Philosophy of Logic** 1 (1980), 95–137.

See also Gödel.

Moschovakis, Yiannis N. *Descriptive Set Theory*. Amsterdam: North-Holland, 1980.

Mostowski, Andrzej. See Tarski, Mostowski, and Robinson.

Müller, Gert H. "Framing Mathematics." **Epistemologia** 4 (1981), 253–286.

―――― (ed.). *Sets and Classes: On the Work by Paul Bernays*. Amsterdam: North-Holland, 1976.

Myhill, John. "The Undefinability of the Set of Natural Numbers in the Ramified *Principia*." In George Nakhnikian (ed.), *Bertrand Russell's Philosophy*, pp. 19–27. New York: Barnes and Noble, 1974.

See also Lorenzen and Myhill.

Nagel, Thomas. *The Last Word*. New York and Oxford: Oxford University Press, 1997.

Nelson, Edward. *Predicative Arithmetic*. Mathematical Notes, 32. Princeton University Press, 1986.

Page, James. "Parsons on Mathematical Intuition." **Mind** N. S. 102 (1993), 223–232.

Parsons, Charles. "Arithmetic and the Categories." **Topoi** 3 (1984), 109–121.

――――. "Developing Arithmetic in Set Theory without the Axiom of Infinity: Some Historical Remarks." **History and Philosophy of Logic** 8 (1987), 201–213.

――――. "Finitism and intuitive knowledge." In Schirn, *The Philosophy of Mathematics Today*, pp. 249–270.

――――. "Genetic Explanation in *The Roots of Reference*." In Robert B. Barrett and Roger F. Gibson (eds.), *Perspectives on Quine*, pp. 273–290. Oxford: Blackwell, 1990.

――――. "The Impredicativity of Induction." In Leigh S. Cauman, Isaac Levi, Charles Parsons, and Robert Schwartz (eds.), *How Many Questions? Essays in Honor of Sidney Morgenbesser*, pp. 132–153. Indianapolis: Hackett, 1983. Revised and expanded version in Michael Detlefsen (ed.), *Proof, Logic, and Formalization*, pp. 139–161. London: Routledge, 1992.

――――. "Intuition and Number." In George, *Mathematics and Mind*, pp. 141–157.

――――. "Intuition in Constructive Mathematics." In Jeremy Butterfield (ed.). *Language, Mind, and Logic*, pp. 211–229. Cambridge University Press, 1986.

――――. "Mathematical Intuition." **Proceedings of the Aristotelian Society** N. S. 80 (1979–1980), 145–168.

――――. *Mathematics in Philosophy: Selected Essays*. Ithaca, NY: Cornell University Press, 1983. Paperback edition with corrections and short preface, 2005.

_____. "On Some Difficulties concerning Intuition and Intuitive Knowledge." **Mind** N. S. 102 (1993), 233–245.

_____. "Paul Bernays' Later Philosophy of Mathematics." In Costas Dimitracopoulos, Ludomir Newelski, Dag Normann, and John R. Steel (eds.), *Logic Colloquium 2005*, pp. 129–150. Lecture Notes in Logic, 28. Urbana, Association for Symbolic Logic, and Cambridge University Press, 2007.

_____. "Platonism and Mathematical Intuition in Kurt Gödel's Thought." **The Bulletin of Symbolic Logic** 1 (1995), 44–74.

_____. "Realism and the Debate on Impredicativity, 1917–1944." In Sieg, Sommer, and Talcott, pp. 372–389.

_____. "Reason and Intuition." **Synthese** 125 (2000), 299–315.

_____. Review of Kitcher, *The Nature of Mathematical Knowledge*. **Philosophical Review** 95 (1986), 129–137.

_____. "Structuralism and Metaphysics." **Philosophical Quarterly** 54 (2004), 56–77.

_____. "The Structuralist View of Mathematical Objects." **Synthese** 84 (1990), 303–346.

_____. "Substitutional Quantification and Mathematics. Review of Gottlieb, *Ontological Economy*." **British Journal for the Philosophy of Science** 33 (1982), 409–421.

_____. "The Transcendental Aesthetic." In Paul Guyer (ed.), *The Cambridge Companion to Kant*, pp. 62–100. Cambridge University Press, 1992.

_____. "The Uniqueness of the Natural Numbers." **Iyyun** 39 (1990), 13–44.

_____. "What Can We Do 'In Principle'?" In Maria Luisa Dalla Chiara, Kees Doets, Daniele Mundici, and Johan van Benthem (eds.), *Logic and Scientific Methods*, pp. 335–354. Volume One of the Tenth International Congress of Logic, Methodology, and Philosophy of Science, Florence, August 1995. Dordrecht: Kluwer, 1997.

See also Gödel.

Parsons, Terence. "The Methodology of Nonexistence." **The Journal of Philosophy** 76 (1979), 639–662.

_____. *Nonexistent Objects*. New Haven: Yale University Press, 1980.

Plantinga, Alvin. *The Nature of Necessity*. Oxford: Clarendon Press, 1974.

Pohlers, Wolfram. See Buchholz, Feferman, Pohlers, and Sieg.

Poincaré, Henri. *Dernières Pensées*. Paris: Flammarion, 1913. 2nd ed., 1926.

_____. "Réflexions sur les deux notes précédentes," **Acta Mathematica** 32 (1909), 195–200.

Previale, Flavio. "Induction and Foundation in the Theory of Hereditarily Finite Sets." **Archive for Mathematical Logic** 33 (1994), 213–241.

Pudlák, Pavel. See Hájek and Pudlák.

Putnam, Hilary. *Mathematics, Matter, and Method. Collected Philosophical Papers, Volume 1*. Cambridge University Press, 1975. 2nd ed. 1980.

_____. "Mathematics without Foundations." **The Journal of Philosophy** 64 (1967), 5–22. Reprinted in *Mathematics, Matter, and Method*, pp. 43–59, and in Benacerraf and Putnam, pp. 295–311 (not in 1st ed.). Cited according to *Mathematics, Matter, and Method*.

———. "Models and Reality." **The Journal of Symbolic Logic** 45 (1980), 464–482. Reprinted in *Realism and Reason*, pp. 1–25, and in Benacerraf and Putnam, pp. 421–444 (not in 1st ed.). Cited according to *Realism and Reason*.

———. *Realism and Reason: Collected Philosophical Papers, Volume 3*. Cambridge University Press, 1983.

———. "The Thesis that Mathematics is Logic." In Ralph Schoenman ed., *Bertrand Russell: Philosopher of the Century*, pp. 273–303. London: Allen & Unwin, 1967. Reprinted in *Mathematics, Matter, and Method*, pp. 12–42.

See also Benacerraf and Putnam.

Quine, W. V. *Methods of Logic*. New York: Holt, 1950. 3rd ed., 1972. 4th ed. Cambridge, MA: Harvard University Press, 1982.

———. "Ontological Relativity." **The Journal of Philosophy** 65 (1968), 185–212. Reprinted in *Ontological Relativity and Other Essays*, pp. 26–68.

———. *Ontological Relativity and Other Essays*. New York: Columbia University Press, 1969.

———. *Philosophy of Logic*. Englewood Cliffs, N. J.: Prentice-Hall, 1970. 2nd ed., Cambridge, MA.: Harvard University Press, 1986.

———. *The Roots of Reference*. La Salle, IL: Open Court, 1974.

———. *Set Theory and its Logic*. Cambridge, MA: The Belknap Press, Harvard University Press, 1963. Revised edition, 1969.

See also Goodman and Quine.

Rawls, John. *Collected Papers*. Edited by Samuel Freeman. Cambridge, MA: Harvard University Press, 1999.

———. "The Independence of Moral Theory." **Proceedings and Addresses of the American Philosophical Association** 48 (1974–75), 5–22. Reprinted in *Collected Papers*, pp. 286–302.

———. *Political Liberalism*. New York: Columbia University Press, 1993.

———. *A Theory of Justice*. Cambridge, MA: The Belknap Press of Harvard University Press, 1971.

Rayo, Agustín. "Word and Objects." **Noûs** 36 (2002), 436–464.

Rayo, Agustín, and Gabriel Uzquiano (eds.). *Absolute Generality*. Oxford: Clarendon Press, 2006.

Reinhardt, W. N. "Epistemic Theories and the Interpretation of Gödel's Incompleteness Theorems." **Journal of Philosophical Logic** 15 (1986), 427–484.

Resnik, Michael D. *Frege and the Philosophy of Mathematics*. Ithaca, NY: Cornell University Press, 1980.

———. "Mathematics as a Science of Patterns: Ontology and Reference." **Noûs** 15 (1981), 529–550.

———. *Mathematics as a Science of Patterns*. Oxford: Clarendon Press, 1997.

———. "Parsons on Mathematical Intuition and Obviousness." In Sher and Tieszen, pp. 219–231.

———. "Quine and the Web of Belief." In Shapiro, *Oxford Handbook*, pp. 412–436.
———. "Second-order Logic Still Wild." **The Journal of Philosophy** 85 (1988), 75–87.
Rheinwald, Rosemarie. *Der Formalismus und seine Grenzen. Untersuchungen zur neueren Philosophie der Mathematik*. Königstein/Ts.: Hain, 1984.
Ricketts, Thomas. See Goldfarb and Ricketts.
Robinson, Raphael M. See Tarski, Mostowski, and Robinson.
Rödding, Dieter. "Primitiv-rekursive Funktionen über einem Bereich endlicher Mengen." **Archiv für mathematische Logik und Grundlagenforschung** 10 (1967), 13–29.
Rosen, Gideon. See Burgess and Rosen.
Rorty, Richard. "Intuition." In Edwards, *Encyclopedia*, III, 204–212, and with addenda by George Bealer and Bruce Russell in Borchert, *Encyclopedia* 2nd ed., IV, 722–735.
Routley, Richard. *Exploring Meinong's Jungle and Beyond*. Departmental Monograph #3, Philosophy Department, Research School of Social Sciences, Australian National University, Canberra, 1980.
———, and Valerie Routley. "Rehabilitating Meinong's Theory of Objects." **Revue internationale de philosophie** 27 (1973), 224–254.
Russell, Bertrand. *Essays in Analysis*. Edited by Douglas Lackey. New York: Braziller, 1973.
———. *Logic and Knowledge*. Edited by Robert C. Marsh. London: Allen and Unwin, 1956.
———. "Mathematical Logic as Based on the Theory of Types." **American Journal of Mathematics** 30 (1908), 222–262. Reprinted in *Logic and Knowledge*, pp. 59–102, and in van Heijenoort, pp. 150–182 (with introductory note by W. V. Quine).
———. "The Philosophy of Logical Atomism." **The Monist** 28 (1918), 495–527; 29 (1919), 32–63, 190–222, 345–380. Reprinted in *Logic and Knowledge*, pp. 175–281, and in *The Philosophy of Logical Atomism and Other Essays, 1914–19*, pp. 157–244. Cited according to *Logic and Knowledge*.
———. *The Philosophy of Logical Atomism and Other Essays, 1914–19*. Edited by John G. Slater. *The Collected Papers of Bertrand Russell*, Vol. 8. London: Allen and Unwin, 1986.
———. *The Principles of Mathematics*. Cambridge University Press, 1903. 2nd ed. London: Allen and Unwin, 1937.
———. *The Problems of Philosophy*. London: Williams and Norgate, and New York: Holt, 1912. Later reprints Oxford University Press. (That of 1997 adds an introduction by John Perry.)
See also Whitehead and Russell.
Saunders, Simon. "Are Quantum Particles Objects?" **Analysis** 66 (2006), 52–63.
Schirn, Matthias (ed.). *The Philosophy of Mathematics Today*. Oxford: Clarendon Press, 1998.
Schütte, Kurt. *Beweistheorie*. Berlin: Springer, 1960.

———. "Logische Abgrenzungen des Transfiniten." In Max Käsbauer and Franz von Kutschera (eds.), *Logik und Logikkalkül*, pp. 105–114. Freiburg: Alber, 1962.

———. *Proof Theory*. Revised and expanded translation of *Beweistheorie*. Berlin: Springer, 1977.

Scott, Dana. "Axiomatizing Set Theory." In Thomas J. Jech (ed.), *Axiomatic Set Theory, part 2*, pp. 207–214. Proceedings of Symposia in Pure Mathematics, vol. 13, part 2. Providence, RI: American Mathematical Society, 1974.

Shapiro, Stewart (ed.). *Intensional Mathematics*. Amsterdam: North-Holland, 1985.

———, (ed.). *The Oxford Handbook of Philosophy of Mathematics and Logic*. New York and Oxford: Oxford University Press, 2005.

———. *Philosophy of Mathematics. Structure and Ontology*. New York and Oxford: Oxford University Press, 1997.

———. "Space, Number, and Structure: A Tale of Two Debates." **Philosophia Mathematica (III)** 4 (1996), 148–173.

Sharvy, Richard. "Why a Class Can't Change its Members." Noûs 2 (1968), 303–314.

Sher, Gila, and Richard Tieszen (eds.). *Between Logic and Intuition. Essays in honor of Charles Parsons*. Cambridge University Press, 2000.

Shoenfield, J. R. "Axioms of Set Theory." In Barwise, *Handbook*, pp. 321–344.

Sieg, Wilfried. "Fragments of Arithmetic." **Annals of Pure and Applied Logic** 28 (1985), 33–71.

Sieg, Wilfried, Richard Sommer, and Carolyn Talcott (eds.). *Reflections on the Foundations of Mathematics. Essays in honor of Solomon Feferman*. Lecture Notes in Logic, 15. Urbana, IL: Association for Symbolic Logic, and Natick, MA: A. K. Peters, 2002.

See also Buchholz, Feferman, Pohlers, and Sieg.

Simons, Peter. *Parts. A Study in Ontology*. Oxford: Clarendon Press, 1987.

Simpson, Stephen G. "Predicativity: The Outer Limits." In Sieg, Sommer, and Talcott, pp. 130–136.

———. "Reverse Mathematics." In Anil Nerode and Richard A. Shore (eds.), *Recursion Theory*, pp. 261–271. Proceedings of Symposia in Pure Mathematics, 42. Providence, RI: American Mathematical Society, 1985.

———. *Subsystems of Second-Order Arithmetic*. Springer-Verlag, 1999.

Skolem, Thoralf. *Abstract Set Theory*. Notre Dame Mathematical Lectures, 8. University of Notre Dame, 1962.

———. "Begründung der elementaren Zahlentheorie durch die rekurrierende Denkweise ohne Anwendung scheinbarer Veränderlichen mit unendlichem Ausdehnungsbereich." *Skrifter, Videnskabsakadmiet i Kristiania* I, no. 6 (1923). Reprinted in *Selected Works*, pp. 153–188. Translation in van Heijenoort, pp. 303–333.

———. "Einige Bermerkungen zur axiomatischen Begründung der Mengenlehre." In *Matematikerkongressen i Helsingfors den 4–7 Juli 1922, Den femte skandinaviska matematikerkongressen, Redogörelse*, pp. 217–232. Helsinki:

Akademiska Bokhandeln, 1923. Reprinted in *Selected Works*, pp. 137–152. Translation in van Heijenoort, pp. 290–301.

———. *Selected Works in Logic.* Edited by Jens Erik Fenstad. Oslo: Universitetsforlaget, 1970.

Sommer, Richard. See Sieg, Sommer, and Talcott.

Steiner, Mark. *Mathematical Knowledge.* Ithaca, NY: Cornell University Press, 1975.

Sylvan, Richard. See Routley.

Szmielew, Wanda, and Alfred Tarski. "Mutual Interpretability of some Essentially Undecidable Theories" (abstract). In *Proceedings of the International Congress of Mathematicians, Cambridge, Massachusetts, U.S.A., August 30–September 6, 1950*, vol. 1, p. 734. Providence, RI: American Mathematical Society, 1952.

Tait, William (W. W.). "Against Intuitionism: Constructive Mathematics is Part of Classical Mathematics." **Journal of Philosophical Logic** 12 (1983), 173–195.

———. "Constructing Cardinals from Below." In *The Provenance*, pp. 133–154.

———. "Constructive Reasoning." In B. van Rootselaar and J. F. Staal (eds.), *Logic, Methodology, and Philosophy of Science III*, pp. 185–198. Amsterdam: North-Holland, 1968.

———. "Critical Notice: Charles Parsons' *Mathematics in Philosophy*." **Philosophy of Science** 53 (1986), 588–607.

———. "Finitism." **The Journal of Philosophy** 78 (1981), 524–546. Reprinted in *The Provenance*, pp. 21–42.

———. "Frege versus Cantor and Dedekind: On the Concept of Number." In W. W. Tait (ed.), *Early Analytic Philosophy, Frege, Russell, Wittgenstein: Essays in honor of Leonard Linsky*, pp. 213–248. Chicago and La Salle, IL: Open Court, 1997. Reprinted in *The Provenance*, pp. 212–251.

———. *The Provenance of Pure Reason. Essays in the Philosophy of Mathematics and its History.* Oxford University Press, 2005.

———. "Truth and Proof: The Platonism of Mathematics." **Synthese** 69 (1986), 341–370. Reprinted with some additional notes in *The Provenance*, pp. 61–88.

Takeuti, Gaisi. "Construction of the Set Theory from the Theory of Ordinal Numbers." **Journal of the Mathematical Society of Japan** 6 (1954), 196–220.

Talcott, Carolyn. See Sieg, Sommer, and Talcott.

Tarski, Alfred. "Sur les ensembles finis." **Fundamenta Mathematicae** 6 (1924), 45–95.

———. "What is Elementary Geometry?" In Leon Henkin, Patrick Suppes, and Alfred Tarski (eds.), *The Axiomatic Method, with Special Reference to Geometry and Physics*, pp. 16–29. Proceedings of an International Symposium held at the University of California, Berkeley, December 26, 1957–January 4, 1958. Amsterdam: North-Holland, 1959.

Tarski, Alfred, Andrzej Mostowski, and Raphael M. Robinson. *Undecidable Theories.* Amsterdam: North-Holland, 1953.

Tarski, Alfred, and Steven Givant. *A Formalization of Set Theory without Variables.* American Mathematical Society Colloquium Publications, 41. Providence, RI: American Mathematical Society, 1987.
See also Givant and Tarski and Szmielew and Tarski.

Taylor, R. Gregory. "Mathematical Definability and the Paradoxes." Dissertation, Columbia University, 1983.

———. "Zermelo, Reductionism, and the Philosophy of Mathematics." **Notre Dame Journal of Formal Logic** 34 (1993), 539–563.

Thomason, Richmond H. "Modal Logic and Metaphysics." In Karel Lambert (ed.), *The Logical Way of Doing Things*, pp. 119–146. New Haven: Yale University Press, 1969.

Tieszen, Richard L. *Mathematical Intuition: Phenomenology and Mathematical Knowledge.* Dordrecht: Kluwer, 1989.
See also Sher and Tieszen.

Troelstra, A. S., and Dirk van Dalen. *Constructivism in Mathematics: An Introduction.* 2 vols. Amsterdam: North-Holland, 1988.

van Aken, James. "Axioms for the Set-Theoretic Hierarchy." **The Journal of Symbolic Logic** 51 (1986), 992–1004.

van Dalen, Dirk. See Troelstra and van Dalen and also Brouwer.

van Dantzig, D[avid]. "Is $10^{10^{10}}$ a Finite Number?" **Dialectica** 9 (1955–56), 272–277.

van Heijenoort, Jean, ed. *From Frege to Gödel: A Source Book in Mathematical Logic, 1879–1931.* Cambridge, MA: Harvard University Press, 1967.
See also Gödel.

Velleman, Daniel J. See George and Velleman.

von Neumann, Johann. "Eine Axiomatisierung der Mengenlehre." **Journal für die reine und angewandte Mathematik** 154 (1925), 219–240. Berichtigung, ibid. 155, 128. Translation in van Heijenoort, pp. 393–413.

Wang, Hao. "Between Number Theory and Set Theory." **Mathematische Annalen** 126 (1953), 385–409. Reprinted as chapter XIX of *A Survey*.

———. *A Logical Journey: From Gödel to Philosophy.* Cambridge, MA: MIT Press, 1996.

———. *From Mathematics to Philosophy.* London: Routledge and Kegan Paul, 1974.

———. "Ordinal Numbers and Predicative Set Theory." **Zeitschrift für mathematische Logik und Grundlagen der Mathematik** 5 (1959), 216–239. Reprinted as Chapter XXV of *Survey*.

———. *A Survey of Mathematical Logic.* Peking: Science Press, and Amsterdam: North-Holland, 1962. Reprinted as *Logic, Computers, and Sets.* New York: Chelsea, 1970.

———. "What is an Individual?" **Philosophical Review** 62 (1953), 413–420. Reprinted as Appendix 5 to *From Mathematics to Philosophy*.

Wetzel, Linda. "Expressions vs. Numbers." **Philosophical Topics** 17 (1989), 173–196.

Weyl, Hermann. *Das Kontinuum*. Leipzig: Veit, 1918.

———. "Der *circulus vitiosus* in der heutigen Begründung der Analysis." **Jahresbericht der deutschen Mathematiker-Vereinigung** 28 (1919), 85–92.

Whitehead, A. N., and Bertrand Russell. *Principia Mathematica*. 3 vols. Cambridge University Press, 1910–13. 2nd ed., 1925–27.

Williamson, Timothy. "Everything." In John Hawthorne and Dean Zimmerman (eds.), *Philosophical Perspectives 17: Language and Philosophical Linguistics*, pp. 415–465. Oxford: Blackwell, 2003.

Wollheim, Richard. *Art and its Objects*. New York: Harper and Row, 1968. 2nd ed. with six suppelementary essays. Cambridge University Press, 1980.

Woodin, W. Hugh. "The Continuum Hypothesis, part I." **Notices of the American Mathematical Society** 48 (2001), 567–576.

———. "The Continuum Hypothesis, part II." *Ibid.*, 681–690.

———. "The Continuum Hypothesis." In René Cori, Alexander Razborov, Stevo Todorcevic, and Carol Wood (eds.), *Logic Colloquium 2000*, pp. 143–197. Lecture Notes in Logic, 19. Urbana, IL: Association for Symbolic Logic, and Wellesley, MA: A. K. Peters, 2005.

———. "Set Theory after Russell: The Journey Back to Eden." In Godehard Link (ed.), *One Hundred Years of Russell's Paradox: Mathematics, Logic, Philosophy*, pp. 29–47. Berlin: de Gruyter, 2004.

Wright, Crispin. *Frege's Conception of Numbers as Objects*. Aberdeen University Press, 1983.

See also Hale and Wright.

Yi, Byeong-Uk. "Is Two a Property?" **The Journal of Philosophy** 96 (1999), 163–190.

———. "The Logic and Meaning of Plurals. Part I." **Journal of Philosophical Logic** 34 (2005), 459–506.

———. "The Logic and Meaning of Plurals. Part II." **Journal of Philosophical Logic** 35 (2006), 239–288.

Zach, Richard. "The Practice of Finitism: Epsilon Calculus and Consistency Proofs in Hilbert's Program." **Synthese** 137 (2003), 211–259.

Zermelo, Ernst. "Sur les ensembles finis et le principe de l'induction complète," **Acta Mathematica** 32 (1909), 183–193.

———. "Über den Begriff der Definitheit in der Axiomatik." **Fundamenta Mathematicae** 14 (1929), 339–344.

———. "Über Grenzzahlen und Mengenbereiche." *Ibid.* 16 (1930), 29–47. Translation in Ewald, II, 1219–1233.

———. "Untersuchungen über die Grundlagen der Mengenlehre I." **Mathematische Annalen** 65 (1908), 261–281. Translation in van Heijenoort, pp. 199–215.

See also Cantor.

Index

a priori knowledge, xiii–xiv, 176.
  See also intrinsic plausibility;
  Reason.
abstract objects, 1–2, 5, 33, 107
  and causality, 1n1, 5, 179
  and perception, 155
  as located, 1n1, 35n57
  intuition of, 179. *See also*
    mathematical objects; natural
    numbers
  pure, 35–36, 100, 117, 151, 222
abstract particulars, 157n37
Ackermann, Wilhelm, 200, 200n20
actuality, 5, 24, 33
  and intuition, 10
apparatus of reference, 4, 4n2, 11.
  See also ontological
  commitment; quantification;
  Quine; reference; singular terms
applications, problem of, 48, 73–79,
  102. *See also* external relations;
  structuralism
arithmetic
  and epistemic stratification, 332
  elementary axioms, 244
  knowledge of, 328–329. *See also*
    intuitive knowledge
  predicative, 298, 303–306, 306n69.
    See also impredicativity;
    predicativity
  substitutional interpretation of,
    194–197, 335. *See also*
    substitutional quantification
arithmetic truth, relativity of, 60.
  See also if-thenism
automorphisms, 107–109

Bealer, George, 327
being, 23–33, 24. *See also* Meinong;
  Parsons; Routley
Benacerraf, Paul, 51n20, 106–107,
  149n25
  Benacerraf's dilemma, xiv, 99n33
  eliminative structuralism, 51
Bernays, Paul, xii, 134n38, 227n57,
  280n20, 337. *See also* Hilbert
  epistemic stratification, 328
  finitism, 185, 235, 236n2, 238,
    239n8, 240, 243n20
  intrinsic plausibility, 330
  on intuition, 163n48, 171–172, 250
  on recursion, 254–260
  on structuralism, 42, 101, 109
  sets as quasi-combinatorial, 136

BonJour, Laurence, 321, 321n10
Boolos, George
   plural quantification, 62, 65–73, 113, 120, 126–127. *See also* comprehension principles; ontological commitment; pluralities; second-order logic
Brouwer, L.E.J., xv. *See also* intuitionism
   and iteration, 174–175
   language and mathematics, 313
   on intuition, 9, 112, 139–140, 174, 176, 337
Burgess, John P., 1n1, 305–306
   on fictionalism, 10n12
   on structuralism, 107
   plural quantification, 72–73
   predicative arithmetic, 306n69

Cantor, Georg, 128, 136n42, 167
   absolute infinity, 133
   cardinal numbers, 198
   cardinality, 75
   collections, 98, 120n5
   multiplicities, 49n18, 66, 71n73, 120, 126
cardinality, 190–199
   and counting, 190, 190n4
Carnap, Rudolf, 10, 153n31
categoricity theorem, 75, 188, 273, 279–293, 298. *See also* Dedekind
   and first-order logic, 281–283
   and second-order logic, 281
categories. *See* Kant
category theory, 109, 340
causal theory of knowledge, xiv, 99n33, 180n77
choice, axiom of, 126n20, 277, 339
Church, Alonzo, 14n16
classes, 65–66. *See also* generalization of predicate places; method of nominalization; method of semantic ascent; plural quantification
   and predicativity, 313. *See also* predicativity; substitutional quantification
   and semantic reflection, 314
   as extensions, 313. *See also* sets
collections, 98–99, 98n29, 119–120, 121n10, 132. *See also* iterative conception of set; multiplicities; pluralities
colors, perception of, 155–157
comprehension principles, 59, 61n49, 63–64, 120n5, 127, 264
   impredicative, 61, 70
   restricted, 72
concept of an object in general. *See* objects; logical concept of object; Kant
concrete objects, 2, 5
   sets of, 166–167. *See also* finite sets
constructivism, xv, 94, 196, 249, 314, 319. *See also* intuitionism
conventionalism, 153n31, 182n79

Davidson, Donald, 180n77, 216, 283
Dedekind abstraction, 47, 47n14, 104–105, 112
Dedekind, Richard, 49n18, 51, 57n38, 74, 189. *See also* Dedekind abstraction
   as eliminative structuralist, 46, 47n13, 73
   categoricity theorem, xii, 46, 75, 188, 273, 281
   on the natural numbers, 45–50, 45n10, 95
Dedekind-Peano axioms, 187, 236, 328–329, 331
deductivism. *See* if-thenism

Descartes, René, 325
    on intuition, 140, 140n5, 141, 144, 172, 247, 326
dialectical relation of principles. *See* Reason
Dummett, Michael
    impredicativity, 293–296, 294n44
    on indefinitely extensible concepts, 134, 334–335, 334n37
    on structuralism, 51, 73–79
    uniqueness of the natural numbers, 273, 279–281, 279n20, 280n21, 281n22, 287–288

eliminative structuralism, 46, 49, 50–56, 78, 95, 100, 107, 117, 126, 130n26, 280. *See also* logicism; nominalism
    and second-order logic, 61–73, 87, 97
    modal, 55–56, 91–93
Euclidean geometry, xiv, 53, 59, 330
Evans, Gareth, 166n54
existence. *See also* being; ontological commitment; quantification; reference
    and reference, 30–31
    as relative to a structure, 115
exponentiation. *See* intuitive knowledge
expression-types. *See* types; strings; strokes
external relations, 75, 77–79, 115, 222. *See also* applications; structuralism

Feferman, Solomon, 290
    predicativity, 297–303, 297n48, 308–315, 309n74, 310n76
    reflection, 310n76. *See also* semantic reflection

fictional objects, 23, 106
Field, Hartry, xivn4, 57n39, 89n18
    instrumentalism about mathematics, 10n12
    mereology, 62–65, 65n58
    nominalism, 57, 59, 95n25
    second-order logic, 62
    uniqueness of the natural numbers, 284n26, 285n27, 289–290. *See also* uniqueness of the natural numbers
figure-ground, 173, 177
finite sequences, 202–205. *See also* linguistic objects; quasi-concrete objects; strings; strokes; types
    and induction, 203
    and intuitive knowledge, 217
    length of, 204
finite sets, xii, 199–202, 298–299. *See also* quasi-concrete objects,
    and induction, 201
    and iteration, 212
    and natural numbers, 199
    arbitrary finite sets, 212
    intuition of, 166–167, 191n5, 205–218
    substitutional interpretation of, 215–218, 231–234, 337. *See also* substitutional quantification
finitism, 235–243, 236n3, 271. *See also* Bernays; Hilbert; induction; recursion
    and Σ_1-induction, 257
    and recursion, 257
first-order logic, 59–60, 87, 275
    and cardinality statements, 192–193
Fraenkel, A.A., 131
Frege arithmetic, 221
Frege, Gottlob, 53, 104, 329

criterion of identity for numbers, 192n6, 197–198, 197n16
on actuality, 6, 6n6
on collections, 121n10
on concepts, 13–22, 13n14, 17n20, 61–62, 71
on induction, 270–271
on number as applying to objects in general, 37–38
on number as property of concepts, 36–37, 210
on numbers as cardinals, 73–79
on objects, 10–11
quantification, 267
Friedman, Harvey, 340

generalization of predicate places, 13–22, 70, 132, 269, 287, 313. *See also* Boolos; method of nominalization; method of semantic ascent; ontological commitment; plural quantification; quantification; Quine; second-order logic
generalization over languages, 291
genetic method, xii, 120. *See also* Quine
Genzten, Gerhard, 308
Gödel, Kurt, 124, 133, 154, 154n32
  a posteriori justification, 124–125, 324, 324n17
  on finitism, 237–241, 237n6, 240n14
  on intuition, 141, 144, 145n17, 146–147, 147n24, 149, 149n25, 155n36, 223, 316, 326–327
Goodman, Nelson, 62, 121n10, 323–324, 331. *See also* nominalism

grammar, 105, 107, 180

Hale, Bob, 174n68
  on intuition, 139n4
Hart, W.D., xiv, xivn4
Hazen, Allen, 305
Heck, Richard G., Jr., 306n69
Hellman, Geoffrey, 91
  predicativity, 298–303
Higginbotham, James, 69–70, 70n71
higher-order logic, 22, 154n32
Hilbert, David, xi, 9, 54, 112. *See also* Bernays
  finitism, 185, 235–241, 236n2, 241n16, 272, 335
  on intuition, 9, 139–140, 159, 163n48, 171, 247, 250
  on recursion, 254–260
  quantification, 251–252
Hodes, Harold, 53, 61, 67
holistic empiricism, 334, 337. *See also* Quine
Hume's Principle, 221, 329. *See also* cardinality; Frege; Frege arithmetic
Husserl, Edmund, 41, 98n29, 157n37, 161
  intuition of finite sets, 166, 209–210
  on intuition, 143n10, 145–146, 149, 155, 173, 196, 223, 326

if-thenism, 53–56, 55n35, 58–61, 73. *See also* eliminative structuralism; logicism
  and consistency, 56, 58
  and first-order logic, 59–60
  modal, 92
  problem of nonvacuity, 54
impredicativity, 64, 113, 295–296, 309n73, 312. *See also*

comprehension principles;
plural quantification;
predicativity; second-order
logic
inaccessible cardinals, 133. *See also*
large cardinal axioms
indefinitely extensible concepts, 12,
134, 334–335
indispensability argument, 208,
208n34, 210–211
induction, xii, 102, 112, 176–177, 178,
201, 264–272, 331, 336. *See also*
finitism; impredicativity;
open-endedness; predicativity;
recursion; second-order logic;
second-order principles
and quantification, 267
and second-order logic, xiii, 20,
236, 276, 293n43
and vagueness, 20, 261, 270, 270n7,
272
as a logical principle, 253
impredicativity of, 270, 293–307,
311
indefinite iteration, 175
open-endedness of, 20–21,
269–270, 290–293
schematic character of, 269–272
inductive definitions, 265, 265n3,
291–292
predicativity of, 307–315
inexhaustibility, 133–134, 339, 341n47
intrinsic plausibility, 152, 319–322
and induction, 333
and perception, 325–328
and set theory, 338–342, 341n47
of arithmetic, 330–331
relativity of, 326
intuitability, 8–10
in principle, 261n47

intuition, 138. *See also* linguistic
objects, quasi-concrete objects;
strings; strokes; types
and arbitrary strings, 171, 173–178
and causality, 149n25
and cognitive limitations, 207
and deduction, 142, 142n9, 146n20,
247–252
and imagination, 173, 181
and induction, 175
and possibility, 261–262, 261n47
and structuralism, 151–152
and the conceptual, 142–143
and vagueness, 166–171
as analogous to perception, 8,
143–148, 152, 206
as factive, 140–143
as opposed to perception, 164–166,
169, 181
Hilbertian, 159–171
immediacy of, 8, 144
intuition *of*, xii, 143–152, 143n10,
154–159
intuition *of* vs intuition *that*, 8, 139
mathematical. *See* mathematical
intuition
of abstract objects, 179. *See also*
abstract objects; finite sets;
mathematical objects; natural
numbers
of finite sets, 166–167. *See also*
abstract objects; finite sets;
mathematical objects; natural
numbers
of mathematical objects. *See also*
abstract objects; finite sets;
mathematical objects; natural
numbers
of mathematical objects, 8–9
rational. *See* rational intuition

representation in, 9
intuitionism, xv, 104, 153, 314, 334n36
　and finitism, 251n31
　and intuitive knowledge, 251
intuitive knowledge, 162, 186, 217, 316
　and addition, 255
　and deduction, 172–173, 172n66, 224, 247–252
　and exponentiation, xii, 256–262, 256n41, 271, 332
　and finitism, 235–243
　and induction, 223, 252–260, 333. See also finitism; induction; recursion
　and intuitionism, 251
　and iteration, 177, 254
　and multiplication, 255
　and quantification, 251–252
　and recursion, 252, 254–260
　and the successor function, 244–247
　of arithmetic, 171–178, 235–262, 333–338
　of strings, 235
iterative conception of set, 38, 118, 122–124, 123n13, 128, 132, 338
　and the power set axiom, 135–137
　principle of plenitude, 135
　weak conception, 135

Julius Caesar problem, 105
justification of set theory, 122–137, 338–339
　a posteriori, 124, 131, 136, 341, 341n47
　holistic, 125, 339–340
　intuitive, 124, 126–137

Kant, Immanuel, xii, 11, 182n79, 316, 331n30
　intuition a priori, 149–150, 182
　on actuality, 5–6
　on intuition, 8–10, 138, 139, 142, 143, 143n11, 146, 146n20, 147, 149, 155, 166, 176, 181–183
　on Reason, 317n1, 327
　on the cateogories, 4–6
　on the concept of an object in general, 4–8
Keränen, Jukka, 107
Kim, Jaegwon, 1n1, 180n77
Koellner, Peter, 134n38, 340n46, 341n47
Kreisel, Georg, 21
　on finitism, 239, 241
Kripke, Saul, 80, 81n1, 84
Kronecker, Leopold, 297n48

Ladyman, James, 108
large cardinal axioms, 88, 124–125, 340–342
　and intrinsic plausibility, 90, 341–342
Lavine, Shaughan, 136n42
　uniqueness of the natural numbers, 290–293
Lawvere, F.W., 282n25
Leibniz, G.W., 145, 147
Leitgeb, Hannes, 108
Lewis, David
　on abstract-concrete distinction, 1n1
　on set theory, 129–130, 130n26
limitation of size, 130n26, 132–135, 132n34. See also replacement
linguistic objects. See also quasi-concrete objects; strings; strokes; types
　and nominalism, 159, 159n41, 161–164

Index 371

as abstract, 160
as quasi-concrete, 160
as types, 158–164, 214
incompleteness of, 183–185
perception of, 157–164, 179–181
logic, 88, 317
and intuition, 247–252
and ontological commitment, 88
as general, 250
first-order. *See* first-order logic
generality of, 248, 250, 318
intuitionistic, 250, 251
second-order. *See* second-order logic
topic neutrality, 153
logical concept of object, 3–7, 10–12, 24, 29–33, 99–100, 336
and intuition, 218
logicism, 53–56, 88
Lorenzen, Paul, 310–315
Löwenheim-Skolem theorem, 277–279

Maddy, Penelope, 110n47, 166n53, 208n34, 340n46
on collections, 121n8
perception of sets, 167n56, 181n78, 207–211
Martin, Donald A., 324n17, 341n47
Martin-Löf, Per, 104, 196n11, 282n25, 314n85
mathematical induction. *See* induction
mathematical intuition, 139, 148–152
and causality, 149
mathematical knowledge. *See also* intrinsic plausibility; intuitive knowledge; Reason;
and proof, 82
conventionalism, 152–153
empiricism, 153

mathematical modality. *See* mathematical possibility; modality
mathematical objects, 1
and causality, 1n1, 149, 180n77
as positions in structures, 41
as potential, 91
elimination of, 2, 96, 99n33, 215
incompleteness of, 25n33, 106, 151
intuition of, 9, 148–152
nature of, 2, 40–43, 90, 100–116, 151
pure, 36, 42, 107, 117, 151
mathematical possibility, 90–92, 114, 176, 178, 261–262, 261n47
mathematics
epistemic stratification of, 327–328
generality of, 90
necessity of, 89–92
obviousness of, 153, 332
ontological commitment, 154
topic neutrality, 153
McCarthy, Timothy, 92n21, 97n27
McGee, Vann, 227n57, 289n35
meaning intentions, 145
Meinong, Alexius, 23–29, 25n33, 26n38, 28n42
on false propositions, 28n43
mereology, 62, 129
See also Lewis; nominalism; second-order logic; set theory
method of nominalization, 15–22, 62, 62n50. *See also* generalization of predicate places; method of semantic ascent; plural quantification; second-order logic
method of semantic ascent, 19–22, 62n50, 70, 310. *See also* generalization of predicate places; method of

nominalization; plural
quantification; second-order
logic
modal logic, 319
  quantificational, 127
modal nominalism, 56, 59, 61, 92–100,
  161, 241n16
modalism, 92–100, 92n22, 94n24
modality, 55, 80
  absolute, 87
  absolute *vs.* nonabsolute, 83
  de re, 92n21
  epistemic, 81n1
  epistemic vs nonepistemic, 80
  logical, 86–89, 89n18
  mathematical, 86–92, 93, 114–115.
    See also mathematical
    possibility
  metaphysical, 83–85, 90–91
  physical, 84, 91, 98
  possible worlds, 96
  provability, 81–82
multiple reductions, 42, 48–50,
  102–106. *See also* eliminative
  structuralism
multiplicities, 66, 70, 120, 126. *See also*
  collections; pluralities; sets
  rigidity of, 127, 127n23
Myhill, John, 293n43, 294n44, 313

Nagel, Thomas, 318n3, 325
natural numbers
  and impredicativity, 293–307.
    See also arithmetic;
    impredicativity; induction;
    predicativity; recursion
  and induction, 186–187, 264–272,
    272n10. *See also* schemata;
    second-order principles
  and iteration, 177, 186, 260
  and substitutional quantification,
    195
  and vagueness, 273. *See also*
    open-endedness; vagueness
  as a progression, 188–189
  as a type, 101, 104–105, 219, 269
  as cardinals, 38n61, 73–79, 101,
    189–190, 196, 222
  as objects, 106–107, 333
  as ordinals, 74, 101, 189, 198–199,
    222
  as pure abstract objects, 36, 224
  as quasi-quasi concrete, 37
  as *sui generis*, 101–103
  as types, 206n27, 211n39
  as types of finite sets, 202, 205
  as types of sequences, 205
  genetic account of, 205, 218, 224
  intuition of, 154, 186, 195, 207,
    222–224
  uniqueness of. *See* uniqueness of
    the natural numbers
naturalistic epistemology, xiv, 99n33
Nelson, Edward, 294n44, 303–305
nominalism, xi, 56–61, 72, 96, 247.
    See also eliminative
    structuralism; logicism
  about linguistic objects, 159.
    See also linguistic objects;
    strings; strokes; types
  and first-order logic, 59–61
  and second-order logic, 61–73
  and syntax. *See* nominalistic syntax
  and types, 179
  modal. *See* modal nominalism
  problem of nonvacuity, 57–59,
    93. *See also* problem of
    nonvacuity
nominalistic syntax, 56, 58, 161
  and consistency, 58, 58n43

noneliminative structuralism, 51,
    100–116, 221
  and the natural numbers, 189
nonstandard models, 102, 188,
    272–279, 275n14, 276n16, 292.
    See also uniqueness of the
    natural numbers
  and induction, 278
nuclear properties, 26n38
numbers. *See also* natural numbers;
    numerals; ordinals
  and cardinality, 190–199
  as cardinals, 196
  as objects, 194–199
  as ordinals, 198
  as types of finite sets, 202
  reference to, 197
  sameness of, 194
numerals, 191, 194

objective reality, 8
objectivity, 11, 318n3
objects
  abstract. *See* abstract objects
  actual, 10, 28, 57
  and existence, 23–33
  and logic, 3, 7, 10–12, 24, 336. *See
    also* apparatus of reference
  as opposed to entities, 3, 13–22. *See
    also* being; Meinong
  canonical expressions for, 195, 335
  classification of, 2
  concept of, 3–8, 98, 218. *See also*
    Kant; logical concept of object.
  concrete. *See* concrete objects
  identity of indiscernibles, 108,
    108n43, 108n44. *See also*
    automorphisms
  linguistic. *See* linguistic objects

mathematical. *See* mathematical
    objects
  nature of, 3, 90, 99–100
  number of, 37. *See also* cardinality
  quasi-concrete. *See* quasi-concrete
    objects
ontological commitment, 4, 51, 62,
    66–68, 70–73, 126–127, 195, 211,
    215–217. *See also* apparatus of
    reference; existence;
    generalization of predicate
    places; method of
    nominalization; method of
    semantic ascent; plural
    quantification; second-order
    logic; substitutional
    quantification
  and first-order logic, 11–12
  and reference, 29–31
open-endedness, xiii, 20–21, 63, 267.
    See also indefinitely extensible
    concepts; induction;
    nonstandard models; schemata;
    uniqueness of the natural
    numbers; vagueness
ordinals, 123, 308
  as generalized types, 217

Parsons, Terence, 25, 26n38
perception. *See* intuition
physicalism, 57. *See also* nominalism
Plantinga, Alvin, 84, 84n8
plural logic, 62, 72, 120
plural quantification, 65–73. *See also*
    generalization of predicate
    places; method of
    nominalization; method of
    semantic ascent; second-order
    logic

and ontological commitment, 62, 126
pluralities, 70, 126–127, 132
   rigidity of, 127, 127n23
Poincaré, Henri
   on induction, xii, 297
   predicativity, 296, 312, 312n81, 313
possible worlds, 83, 96. *See also* modality
potential totality. *See* indefinitely extensible concepts
power set, axiom of, 130, 134, 135–137. *See also* indefinitely extensible concepts; inexhaustibility; iterative conception of set; justification of set theory; open-endedness; predicativity; reflection principles
   a posteriori justification, 136
predicates. *See* generalization of predicate places; method of nominalization; method of semantic ascent; open-endedness; second-order logic
   and open-endedness, 63, 267
predication, 30–32. *See also* generalization of predicate places; method of nominalization; method of semantic ascent; second-order logic
   and reference, 13–22, 209, 327
predicativism, 134. *See also* imprelicativity, predicativity
predicativity, 61, 196, 260, 296–297, 311
   and exponentiation, 303–304
   as opposed to constructivity, 313

given the natural numbers, 309
primitive recursive arithmetic, 237, 243
   and logic, 249–252
   as logic-free, 249, 259
problem of nonvacuity, 48–50, 54–55, 93, 95, 97–100, 117, 219–222. *See also* eliminative structuralism; nominalism
progressions, 188–190
   second-order definable operations, 189
projective determinacy, 125, 125n17
provability logic, 81–82
Putnam, Hilary, xiv, 56n36, 97
   if-thenism, 53–54, 59–60
   mathematical modality, 89n18, 92
   nonstandard models, 288
   on second-order logic, 96–97, 97n27

quantification. *See also* apparatus of reference;
   and existence, 23–33, 25n33. *See also* being; existence
   and finitism, 251–252
   domain of, 267–269
   numerical, 193
   over numbers, 220–222
   plural. *See* plural quantification
   second-order. *See* second-order logic
   substitutional. *See* substitutional quantification
quantifiers
   generalized, 71–72, 88, 193. *See also* ontological commitment
   numerical, 193
quasi-concrete objects, 9, 33–39, 43, 113, 183, 242, 242n17, 242n18

and intuition, 155–159. *See also*
    Hilbertian intuition
  and perception, 35
  and structuralism, 115, 151–152
  and their representations, 34–37
Quine, W.V., xii, xiii, 84n8, 121
  existence, 23
  mathematical knowledge, 153, 332
  on objects, 10–12
  on structuralism, 42, 125
  on types, 35
  ontological commitment, 70–72
  semantic ascent, 20

rational intuition, 321, 325–328, 337.
    See also intrinsic plausibility;
    intuitive knowledge; Reason
  and deduction, 326–327
rational knowledge, 316. *See also*
    rational intuition
Rawls, John, 324, 324n15, 331
Reason. xiii, 147, 316–328. *See also*
    intrinsic plausibility
  and argument, 317–319
  and aritthmetic, 328–338
  and logic, 317–319
  and systematization, 322
  as court of appeal, 324–325
  dialectical relation of principles,
    322–324, 330n25, 331n30
recursion, 254–260, 266
  and impredicativity, 271, 294–295
reference, 4, 107. *See also* apparatus of
    reference; ontological
    commitment
  and context, 103–104
  of predicates, 13–22, 209, 327
  to nonexistent objects, 23–33
reflection principles, 133–134, 134n38,
    340n47. *See also*

inexhaustability; iterative
    conception of set; limitation of
    size; open-endedness;
    replacement
reflective equilibrium, 324
replacement, axiom of, 130–134, 339.
    See also iterative conception of
    set; justification of set theory;
    limitation of size
  and reflection principles, 133–134
Resnik, Michael, 184n82, 323n14,
    332–333, 336
  on structuralism, 40–41, 115n57
  set-theoretic conception of
    structure, 44
Rheinwald, Rosemarie, 55n35, 92n42
Rorty, Richard
  on intuition, 138, 138n1, 140, 142
Rosen, Gideon, 1n1
Routley, Richard, 26n36, 26n38, 29–33,
    33n53
  neutral quantification, 29–30
Russell, Bertrand, 73, 104
  class as many, 66, 71n73, 120
  on false propositions, 28n43
  on induction, 297
  on Meinong, 25, 32
  predicativity, 311–312
Russell's paradox, 17, 22, 32n51, 121,
    122, 126, 127. *See also*
    inexhaustibility;
    open-endedness

schemata, xiii, 19, 269, 290–293
Schultz, Johann, 8n9
Schütte, Kurt, 309, 309n74
second-order logic, xiii, 13, 22, 52–53,
    73, 75, 88–96, 265, 270, 329
  and generalization of predicate
    places, 70

and impredicativity, 270–293
and ontological commitment, 61
monadic, 66
ontological commitment, 66–73
plural interpretation. *See* plural quantification
second-order principles, xiii, 20, 64, 132, 276
self-evidence, 320, 320n7. *See also* intrinsic plausibility
semantic reflection, 21–22, 271, 277, 285, 288, 310, 310n76, 313. *See also* indefinitely extensible concepts; open-endedness; schemata
sense-qualities
perception of, 155–157, 179
separation, axiom of, 18–19, 21, 300–303, 313
See also iterative conception of set; justification of set theory
impredicativity, 301–303, 301n56
set theory
and epistemic stratification, 340–342
and mereology, 128–130
and structuralism, 97, 100, 111, 117–119, 123, 128–129. *See also* structuralism; structures
as a framework for mathematics, 110, 110n47
intrinsic plausibility of, xivn3, 338–339. *See also* intrinsic plausibility; justification of set theory; large cardinal axioms
justification of. *See* justification of set theory
modal, 129n23, 132n33
objectivity of, 124
sets

and vagueness, 166–167, 166n54
as collections, 98–100, 113, 119–120, 122–124, 129, 206, 218
as constituted by their elements, 118–120, 122, 126, 128–130
as extensions, 113, 119–122, 135, 312, 339
as pluralities, 113, 119–122, 126–127
as trees, 128–129
concept of, 38
extensionality of, 129, 338
finite. *See* finite sets
inexhaustibility of, 133–135, 339
intuition of, 212–216
iterative conception of. *See* iterative conception of set
ontological conception of, 117n1, 118–124, 129–130
quasi-concrete, 35
structuralist conception of, 125, 128. *See also* structuralism; structures
urelements, 118n2
Shapiro, Stewart, 51–52
Shoenfield, Joseph, 122, 122n13, 124, 126, 131
simply infinite system, 45, 52, 75, 188, 298. *See also* Dedekind
initial element of, 76–77
Simpson, Stephen G., 297n48, 340
singular terms, 3–4. *See also* apparatus of reference
Skolem, Thoralf, 131, 279n20
relativity of set theory, 278
Sosein, 23, 27. *See also* Meinong, being
Spinoza, Baruch de, 147
Steiner, Mark, 1n1, 180n77
strict finitism, 261, 282n25, 329
strings. *See also* linguistic objects; strokes; types

and induction, 175–178, 178n75, 236
and iteration, 175–178
arbitrary, 173–175
as sequences, 168
as types, 168–171, 168n58
infinity of, 176–178
intuition of, 185
intuitive knowledge of, 235
stroke-arithmetic, 159, 235. *See also* Bernays; Hilbert; intuition; strings; strokes
strokes. *See also* linguistic objects; quasi-concrete objects; strings; types
and nominalism, 160–171
and vagueness, 163–164, 163n48, 168–171
arbitrary string of, 173–175
intuition of, 165–166
perception of, 164
strings of, 159–161, 169–170
structuralism, xi, 40–143, 197, 280
and applications, 48, 73–79
and context, 103–104
and convention, 77, 77n82
and external relations, 75, 78–79, 222
and incompleteness, 106, 183. *See also* abstract objects; mathematical objects; objects
and intuition, 222–224. *See also* quasi-concrete objects
and limitation of size, 135
and modality, 83, 189
and second-order logic, 52–53, 189. *See also* eliminative structuralism; second-order logic

and set theory, 97–100, 111, 117–137. *See also* set theory; structures
and the nature of mathematical objects, 40–43, 90–92. *See also* mathematical objects
eliminative. *See* eliminative structuralism
in re and *ante rem*, 51–52
noneliminative. *See* noneliminative structuralism
problem of nonvacuity, 48–49, 93, 219–222, 334
structuralist view of mathematical objects. *See* structuralism
structures, 40–43, 43n8, 110–116
as basic mathematical objects, 41, 44, 50, 111
basic and constructed, 108, 108n43
metalinguistic conception of, 111–113, 115n57
set-theoretic conception of, 43–50, 114–116
substitutional quantification, 27, 194–196, 215–218, 335. *See also* generalization of predicate places; method of nominalization; method of semantic ascent; second-order logic
and ontological commitment, 195, 195n10, 215–217
Sylvan, Richard. *See* Routley, Richard
systematic ambiguity, 21, 115n57. *See also* inexhaustibility; open-endedness; schemata

Tait, W.W., 104, 112, 134n38, 149n25, 211n39, 258n43, 282n25

on Dedekind, 47, 47n13
on finitism, 239–241, 253, 314n85
Tieszen, Richard, 330n25
transfinite induction, 308–311, 308n71
type theory, 14n16, 105, 340
   ramified, 305–307
types, 34. *See also* linguistic objects; strings; strokes
   and causality, 179
   and nominalism, 160, 179
   and vagueness, 162–164, 167–171, 169n60
   criteria of identity, 161
   intuition of, 162, 165–166
   paradigmatic role of, 162–163
types and tokens, 158
   contrasted with elements and sets, 214–215

uniqueness of the natural numbers,. 60, 188, 272–293. *See also* nonstandard models; open-endedness; schemata; second-order principles; second-order logic
   and communication, 279–293
   and first-order logic, 275
   and induction, 275–277
   and radical interpretation, 283–285
   and second-order logic, 188

vague objects, 166n54, 169n60
vagueness
   of induction, 20–21, 261, 270, 272. *See also* Dummett; indefinitely extensible concepts; open-endedness; schemata
   of laws of logic, 20–21
Wang, Hao, 132, 311
Wetzel, Linda, 160n44
Weyl, Hermann, 296–297, 303, 312
Wirklichkeit, 5–8. *See also* actuality
Wollheim, Richard, 184
Woodin, W. Hugh, 342
Wright, Crispin, 174n68
   on intuition, 139n4

Zach, Richard, 238
Zermelo, Ernst, 18, 49n18, 122
   definite property, 18n21, 19n22

www.ingramcontent.com/pod-product-compliance
Ingram Content Group UK Ltd.
Pitfield, Milton Keynes, MK11 3LW, UK
UKHW032104190125
453752UK00004B/9